New Trends in Lipid and Lipoprotein Analyses

New Trends in Lipid and Lipoprotein Analyses

Editors

J.-L. Sebedio
Institut National de la Recherche Agronomique
Dijon, France

Edward G. Perkins
University of Illinois
Urbana, Illinois

Champaign, Illinois

AOCS Mission Statement

To be a forum for the exchange of ideas, information, and experience among those with a professional interest in the science and technology of fats, oils, and related substances in ways that promote personal excellence and provide high standards of quality.

AOCS Books and Special Publications Committee

E. Perkins, chairperson, University of Illinois, Urbana, Illinois
T. Applewhite, Austin, Texas
J. Bauer, Texas A & M University, College Station, Texas
T. Foglia, USDA—ERRC, Philadelphia, Pennsylvania
M. Mossoba, Food and Drug Administration, Washington, D.C.
Y.-S. Huang, Ross Laboratories, Columbus, Ohio
G. Maerker, Oreland, Pennsylvania
G. Nelson, Western Regional Research Center, San Francisco, California
F. Orthoefer, Riceland Foods Inc., Stuttgart, Arkansas
J. Rattray, University of Guelph, Guelph, Ontario
A. Sinclair, Deakin University, Geelong, Victoria, Australia
T. Smouse, Archer Daniels Midland Co., Decatur, Illinois
G. Szajer, Akzo Chemicals, Dobbs Ferry, New York
L. Witting, State College, Pennsylvania

Copyright © 1995 by AOCS Press. All rights reserved. No part of this book may be reproduced or transmitted in any form or by any means without written permission of the publisher.

The paper used in this book is acid-free and falls within the guidelines established to ensure permanence and durability.

LC Number: 95–13519

Printed in the United States of America with vegetable oil–based inks.

00 99 98 97 96 95   5 4 3 2 1

Preface

It has become a cliché in science that real advances can only be made through developments in methodology, but it is difficult to deny the truth of such an observation. This is as valid for lipid research as for any other branch of science. Therefore, those who were able to attend the excellent meeting "New Trends in Lipids and Lipoprotein Analysis" in the delightful surroundings of La Grande Motte in France have the advantage of a flying start in being able to take back to their own labs the results of many excellent presentations on new aspects of lipid analysis. Readers of this book now have the opportunity to follow suit. Most of the contributors are from Europe, so it is hoped that a slightly different perspective is given from those found in other AOCS Press books in related areas.

It is impossible to discuss all the chapters in this book in detail, but in the opening chapter by Frank Gunstone on the history of lipid analysis, you will find much to entertain and stimulate. Subsequent chapters discuss extraction methodology, TLC (including a chapter on the Iatroscan TLC/FID system), HPLC (detectors, lipid class separations, and silver ion and size exclusion chromatography), structural analysis of triacylglycerols (including stereospecific analysis and mass spectrometry) and GC (optimization, geometrical isomers, trans fatty acids, derivatization, mass spectrometry and FTIR spectroscopy). The newer technique of 13C NMR spectroscopy applied to lipids is balanced by an account of classical methodology for structural analysis of fatty acids. Sensory analysis of fats and oils is an emphasized topic (often neglected in other works of this kind). A role for sterol oxides in human disease states has become quite evident; therefore, sensitive analytical methods that do not generate artifacts are included. Finally, there is a group of chapters dealing with clinical aspects of lipid methodology, for example, the use of stable isotopes in studying the metabolism of lipids in humans and lipoprotein separations (size exclusion, electrophoresis and preparative ultracentrifugation methods).

In short, we have a comprehensive series of reviews on a wide range of topics related to lipid analysis. The authors are authorities in their fields, and this is reflected in the quality of the finished chapters. Please read and enjoy.

J.-L. Sebedio
Institut National de la Recherche Agronomique
Dijon, France

E.G. Perkins
Department of Food Science
University of Illinois
Urbana, Illinois

Contents

Preface v

Chapter 1   **Lipid Analysis: A Brief Historical Survey**
Frank D. Gunstone 1

Chapter 2   **Extraction of Lipids for Analytical Purposes**
Anne Castera 10

Chapter 3   **Thin-Layer Chromatography of Lipids**
Vijai K.S. Shukla 17

Chapter 4   **Utilization of Thin-Layer Chromatography-Flame Ionization Detection for Lipid Analyses**
J.-L. Sebedio 24

Chapter 5   **High-Performance Liquid Chromatography: Normal-Phase, Reverse-Phase Detection Methodology**
Vijai K.S. Shukla 38

Chapter 6   **Lipid Class Separations Using High-Performance Liquid Chromatography**
William W. Christie 45

Chapter 7   **Silver-Ion High-Performance Liquid Chromatography**
William W. Christie 59

Chapter 8   **Utilization of Silver-Ion High-Performance Liquid Chromatography for Separation of the Geometrical Isomers of α-Linolenic Acid**
Pierre Juanéda 75

Chapter 9   **High-Performance Size-Exclusion Chromatography Applied to the Analysis of Edible Fats**
M. Carmen Dobarganes and Gloria Márquez-Ruiz 81

Chapter 10   **Stereospecific Analysis of Triacyl-*sn*-Glycerols**
W.W. Christie 93

Chapter 11	**Structural Analysis of Peanut Oil Triacylglycerols** *J.A. Bézard and B.G. Semporé*	106
Chapter 12	**Gas–Liquid Chromatography: Choice and Optimization of Operating Conditions** *F.X. Mordret and J.L. Coustille*	133
Chapter 13	**Recent Applications of Capillary Gas–Liquid Chromatography to Some Difficult Separations of Positional or Geometrical Isomers of Unsaturated Fatty Acids** *Robert L. Wolff*	147
Chapter 14	**Determination of *trans* Fatty Acids in Dietary Fats** *W.M.N. Ratnayake*	181
Chapter 15	**Gas Chromatography–Mass Spectrometry and Tandem Mass Spectrometry in the Analysis of Fatty Acids** *Jean-Luc Le Quéré*	191
Chapter 16	**Mechanism for Separation of Triacylglycerols in Oils by Liquid Chromatography: Identification by Mass Spectrometry** *S. Héron, J. Bleton, and A. Tchapla*	205
Chapter 17	**Gas Chromatography-Fourier Transform Infrared Spectrometry in the Analysis of Fatty Acids** *Jean-Luc Le Quéré*	232
Chapter 18	**Contribution of Grignard Reagents in the Analysis of Short-Chain Fatty Acids** *Michel Pina, Catherine Ozenne, Gilles Lamberet, Didier Montet, and Jean Graille*	242
Chapter 19	**Information About Fatty Acids and Lipids Derived by ^{13}C Nuclear Magnetic Resonance Spectroscopy** *Frank D. Gunstone*	250
Chapter 20	**Sensory Assessment of Fats and Oils** *Renée Raoux and Odile Morin*	265
Chapter 21	**Classical Chemical Techniques for Fatty Acid Analysis** *J.-L. Sebedio*	277

Chapter 22	**Analysis of Sterol Oxides in Food and Blood** *Lars-Åke Appelqvist* 290
Chapter 23	**Utilization of Stable Isotopes to Study Lipid Metabolism in Humans** *B. Descomps* 299
Chapter 24	**Utilization of Stable Isotopes to Study the Compartmental Metabolism of Polyunsaturated Fatty Acids: An *In Vivo* Study Using ^{13}C-Docosahexaenoic Acid** *M. Croset, N. Brossard, J. Lecerf, C. Pachiaudi, S. Normand, V. Chirouze, J.P. Riou, J.L. Tayot, and M. Lagarde* 309
Chapter 25	**A Quick Method for Sterols Titration in Complex Media** *D. Pioch, P. Lozano, C. Frater, and J. Graille* 317
Chapter 26	**Some Improvements in Contaminants Analytical Methodology** *F. Lacoste, A. Castera, J.L. Perrin, and J.L. Coustille* 323
Chapter 27	**Size Exclusion Chromatography Applied to the Analysis of Lipoproteins** *Philip J. Barter* 337
Chapter 28	**Determination of Lipoprotein-Size Distribution by Polyacrylamide Gradient Gel Electrophoresis** *Laurent Lagrost* 347
Chapter 29	**Preparative Ultracentrifugation of Plasma Lipoproteins: A Critical Overview** *P. Michel Laplaud* 357
	Index 371

Chapter 1
Lipid Analysis: A Brief Historical Survey

Frank D. Gunstone

School of Chemistry, The University, St. Andrews, Fife KY16 9ST, UK.

Introduction

Natural lipids are mixtures and two questions have to be addressed—what structures? and how much of each? The analytical study of lipids therefore is both qualitative and quantitative. Such a study can also be carried out at various levels and any or all of the following questions may have to be addressed.

1. What classes of lipids are present?
2. What component acids, alcohols, or long-chain bases are present?
3. What molecular species, taken in groups or individually, are present within each class?

The number of possible triacylglycerols (and other lipid classes) rises rapidly with the number of fatty acids present in the pool(s) from which the triacylglycerols are assembled. The necessary formulae to calculate these are given along with some examples in Table 1.1. The number of fatty acids present generally increases in the order seed oils < animal fats < fish oils, but the numbers depend also on the thoroughness of the analytical investigation.

These numbers quickly become quite large and the analyst has to ask "What use can be made of these data if I present results for so many individual molecular species? Will they have to be grouped in some way to produce figures that are capable of quick and sensible interpretation and which can be compared with data for other oils?" Other articles in this monograph will indicate how detailed these studies can be and in what way further progress is expected in the near future.

The objective in this first chapter is to look back and see by what pathways lipid analysts have arrived at the present position. This will be done in three stages:

TABLE 1.1
Relation Between the Number of Fatty Acids and the Possible Number of Triacylglycerols

Number of fatty acids	n	5	10	20
All enantiomers	n^3	125	1000	8000
Optical isomers not distinguished	$\frac{(n^3+n^2)}{2}$	75	550	4200
No isomers considered	$\frac{(n^3+3n^2+2n)}{6}$	35	220	1540

1. Lipid analysis at the time of Chevreul (1811–1923),
2. Lipid analysis at the time of Hilditch (1926–1951), and
3. The development of chromatographic procedures.

Chevreul

Michael E. Chevreul is considered by many to be the father of fat chemistry. Oils and fats had been known for centuries and used as sources of food, as well as for cleaning and as cosmetics, lubricants, and illuminants, but Chevreul is credited with the first attempt to understand these substances in terms of their chemical components (1).

Chevreul (1786–1889) was a French chemist who lived for over 100 years. He was born just before the death of his compatriot Antoine Lavoisier (born in 1743 and guillotined in 1794) while Hilditch (1886–1965) was born in Chevreul's centennial year. Chevreul was a contemporary of great chemists such as John Dalton (1766–1844), Friedrich Wöhler (1800–1882), Jean Dumas (1800–1884), and Justus von Liebig (1803–1873). He had a wide range of interests and is acknowledged by art historians for his contribution to the understanding of color. He is best remembered by chemists for his careful study of the saponification process which he carried out during the years 1811–1823.

In considering Chevreul's work, it must be remembered that at that time organic compounds were designated only by empirical and molecular formulae. The concept of homologous series was developed by Kekulé, Couper, and others from the late 1850s, 30 or 40 years after Chevreul completed his investigations. The atomic weight of oxygen was still thought to be 8, so that when fatty acid structures were first derived the carboxyl group was written CO_4H. It is important not to credit Chevreul with levels of interpretation and understanding which were not open to him when he made his investigations 170–180 years ago.

Chevreul made five important conclusions about the nature of fats and fatty acids from his study of the saponification process.

1. Fats contain glycerine and acidic substances. The word "contain" does not describe anything about the mode of association of these components beyond indicating that they are not present in the free state and become apparent only after they are unbound.
2. Fats themselves are not homogeneous entities but mixtures of compounds. They were separated by crystallization into a solid component which Chevreul termed margarin and a liquid component designated elain (later olein).
3. When fats are treated with alkali, glycerine and fatty acids (after acidification) are produced.
4. Some fats give volatile fatty acids (butyric). All fats give nonvolatile acids which can be separated into a liquid member (oleic) and a solid member. The

latter varied with the fat under investigation and Chevreul described solid acids melting at 60°C (margaric) and at 70°C (stearic).

5. Water, not oxygen as was previously thought, is involved in the saponification process.

As already indicated, despite these advances Chevreul could make no structural assignments. Although glycerol had been discovered in 1779 by Scheele (1742–1786) and was known as Scheele's sweet principle, it was not identified until 74 years later in 1853 by Berthelot and Wurtz.

Chevreul based his very thorough study on a limited range of experimental techniques. These included repeated crystallization from a range of solvents and checking purity of crystalline solids by melting point and by combustion analysis (measuring percent carbon and hydrogen). His experimental equipment was confined to glassware and a balance.

Hilditch

Hilditch (1886–1965) was three years old when Chevreul died. He was trained at University College, London and at Jena and Geneva and after a period in industry (at Joseph Crosfield and Son, Warrington) was invited by the University of Liverpool in 1926 to be the first holder of the Chair of Industrial Chemistry. He continued in that post until his retirement in 1951. The study of oils and fats was considered to be an appropriate topic for the chair-holder because of the concentration of industrial interest in these materials in the Liverpool area (2–4).

By the time Hilditch took up his academic appointment, there had been further developments in understanding the chemical nature of fats and their constituent fatty acids. The chemical structure of glycerol had been established, the general concept of homologous series in aliphatic compounds was well accepted, and the saturated acids and the majority of unsaturated acids had been correctly identified—oleic acid in 1893, linoleic acid in 1906, and α-linolenic acid in 1909. The solution of iodine monochloride in acetic acid, devised by the Dutch chemist Wijs and widely employed as the basis of a successful method of measuring average unsaturation, was first described in 1898.

In the century between Chevreul's pioneer work and 1926 when Hilditch took up his appointment in Liverpool, the methods of structure determination were only slightly changed. Experimental methods had been refined and improved, but it was the general understanding of organic structure which provided the biggest change in the background for investigations of the early 20th century and which added confidence to the conclusions obtained and to the growing body of knowledge. But while the understanding of chemical structure had grown, the understanding of reaction mechanisms was still in its infancy. The stereochemistry of addition and elimination reactions—so important in the isolation of pure unsaturated acids for structure determination—had yet to be explained. This ignorance produced confusion and hindered progress in isolating and identifying unsaturated fatty acids.

TABLE 1.2
Methods of Isolation and Identification of Fatty Acids in the Second Quarter of the Twentieth Century

---
Isolation
　Crystallization of acids or their salts (Pb, Li, etc.)
　Fractional distillation of methyl esters

Identification
　Iodine value—a measure of average unsaturation
　Saponification equivalent—a measure of average molecular weight
　Hydrogenation and recognition of the perhydro acid from its melting point, mixed melting point, and equivalent
　Oxidative cleavage and identification of the oxidation products to indicate double-bond position
---

One of Hilditch's greatest contributions to lipid chemistry was his collection of quantitative analytical data for a wide range of fats and oils of both plant and animal origin. His book, *The Chemical Composition of Natural Fats,* (published by Chapman and Hall) appeared in four editions in 1940, 1947, 1956, and 1964, and in the last edition provided information on the composition of almost 1500 fats and oils (5). This book was a classic in its time and remained a valuable resource for lipid chemists for many years. It is still cited occasionally in research publications. Methods of isolating and identifying fatty acids in Hilditch's time are set out in Table 1.2. Though satisfactory for saturated, monoene, diene, and triene acids, these methods did not work well with acids having more than three unsaturated centers.

It is interesting, 45 years later, to read the lecture which Hilditch delivered in 1948 to the Chemical Society (now the Royal Society of Chemistry) in London entitled "Structural Relationships in the Natural Unsaturated Higher Fatty Acids" and published in that same year (6). Two features stand out. First is the number of unsaturated fatty acid structures available for consideration by Hilditch at that time was "nearly forty." That number is now at least five times as large. Second, the structure of the "polyethenoid acids of marine-animal fats," determined by the oxidation process then available, were largely incorrect. For example, eicosapentaenoic acid ($20{:}5n{-}3$) was believed to have pentaene unsaturation at 4,8,12,15,18 rather than 5,8,11,14,17, and docosahexaenoic acid ($22{:}6n{-}3$) was considered to have one of five possible structures. The structure now accepted for this acid was not included in that list. The correct structures of these polyene acids, the recognition that the most important of them are wholly methylene-interrupted in their patterns of unsaturation, and that they exist in families which were later shown to express biosynthetic relationships, arose from the work of Klenk and others mainly in the period from 1950 to 1955 (7).

What about the quantitative analysis of the component acids of oils and fats as performed before the days of chromatography (8)? The object was to separate the acids or esters into fractions which were simple enough to be analyzed by the measurement of iodine value (mean unsaturation) and saponification equivalent (mean chain length).

The mixed fatty acids were separated into 2–4 fractions, with one type of acid being concentrated in each. Steam distillation separated volatile from nonvolatile acids, crystallization of lead salts separated saturated from unsaturated acids, crystallization of lithium salts separated nonpolyene from polyene acids, and crystallization of the acids themselves from the appropriate solvents (acetone, petroleum ether, or diethyl ether) at temperatures down to −70°C gave concentrates of saturated, monoene, and polyene acids. Each fraction was then subjected to further examination.

The acids were converted into methyl esters and distilled under reduced pressure through a column packed with glass helices. Typically 5–20 fractions of around 3 g were collected. Separation by distillation was mainly by chain length with double bonds having only a minor effect. So C_{16} esters were separated from C_{18} esters. The C_{18} fractions declined in iodine value as the polyene esters were slightly more volatile than oleate and stearate.

The iodine values and saponification equivalents of all the ester fractions (20–80) were then determined. This involved weighing (on a free swinging balance using weights and a rider), a chemical reaction, and a titration. Finally there was a lot of arithmetic calculation done with logarithms.

A new technique which had been reported in the United States and was developed in Hilditch's laboratory by John Riley involved alkali-isomerization by which the mixed acids/esters were treated with alkali at elevated temperatures (170 or 180°C) under prescribed conditions. Double-bond migration in polyenes gave conjugated isomers—linoleate gave diene and linolenate gave a mixture of diene and triene—these could be measured quantitatively by ultraviolet spectroscopy (9). This was probably the first general use of a spectroscopic method in lipid chemistry. The first spectrometer used by the author was very antiquated and took a lot of handling, including the development of photographic plates.

These procedures were applied in a study of linseed oil carried out in Hilditch's laboratory and published in 1946 (10). This investigation started with 267 g of linseed oil, required three low-temperature crystallizations of the acids, four esterifications and fractional distillations of the methyl esters, and measurement of about 50 iodine values and saponification equivalents. It took about 3 weeks to complete. Glyceride analysis was carried out in a similar way, except that the oil/fat was first separated by crystallization into 2–5 fractions, each of which was then examined in the ways described for linseed oil. In addition, the glyceride fractions were oxidized and (neutral) fully saturated glycerides, which resist oxidation, were separated from the acidic products resulting from oxidation of unsaturated glycerides.

Shortly after Hilditch retired from his chair in Liverpool, the identification of the first natural epoxy acid (vernolic) in the seed oil of *Vernonia anthelmintica* (11) was reported. Vernolic acid was shown to be *cis*-12, 13-epoxyoleic acid. Following the normal methods at that time, the vernonia oil was hydrolyzed to mixed fatty acids, the saponification equivalent (314.1) was measured, and these acids were converted to methyl esters by reaction with methanol containing hydrogen chloride or sulfuric acid catalyst. Methylation should have raised the saponification equivalent by 14. In the first methylation procedure the saponification equivalent was unexpectedly low (255.6), and in the second it was unexpectedly high (350.7), suggesting the presence of some unusual feature in the molecule, but then there was no convenient spectroscopic way of identifying this.

The methanol-hydrogen chloride product, dissolved in alcohol, was observed to give a precipitate when ethanolic potassium hydroxide was added. This was water soluble and proved to be potassium chloride, indicating the presence of readily available chlorine in the esterification product.

The results could be explained by the presence of an epoxide group in the original acids. The chlorohydrin (produced with hydrogen chloride) reacted with a second molecule of alkali, while the methoxy hydroxy ester (produced with sulfuric acid) had a higher molecular weight than the epoxy acid (+46 mass units). Once this part of the problem was understood, the full structure was readily identified by oxidation with potassium permanganate and with potassium periodate which react with the double bond and epoxide functions. The degradation products were identified by equivalent and/or melting point. Apart from one minor use of UV spectroscopy to demonstrate double-bond migration during periodate oxidation, the entire study was carried out with the kind of equipment used by Chevreul—glassware, a burette, and a balance.

Chromatography

Chromatography was first described in 1903 and 1906 (12,13) by Mikhail Tswett (1872–1919). Tswett was a Russian who studied in Switzerland, returned to Russia, worked in Warsaw (then a part of the Russian empire) at the Institute of Plant Physiology and was interested in plant pigments such as chlorophyll. Tswett used glass columns and powdered calcium carbonate was his favorite adsorbent. He also used inulin and sucrose. The technique, though subsequently used by a few chemists, was largely forgotten until it was rediscovered in 1930–31 by a group working with carotenoids in Heidelberg. This group consisted of Richard Kuhn (1900–1967), Alfred Winterstein (1889–1960), and Edgar Lederer (1908–1988) (14). All these early studies were with colored compounds, and this is reflected in the name given to this separation technique—chromatography.

The next key step in this saga comes about 10 years later with a publication by A.J.P. Martin and R.L.M. Synge in 1941 from the Wool Industries Research Association in Leeds, England (15). This paper "151. A New Form of Chromatogram Employing Two Liquid Phases: 1. A theory of chromatography, 2. Application to the micro-determination of the higher monoamino-acids in proteins" described partition chromatography as opposed to the adsorption chromatography systems used previously, and attention was drawn to the fact that "the mobile phase need not be a liquid but may be a vapour" though this concept was not developed experimentally in the 1941 paper. Martin and Synge shared the Nobel prize in 1952 for their development of new chromatographic procedures.

The year 1941 was not the best time to disseminate new ideas, and the *Biochemical Journal* was probably not widely read by analysts or by lipid chemists. It was not until 1952 that A.T. James and A.J.P. Martin published another paper in the same journal: "Gas-Liquid Partition Chromatography: The Separation and Micro-Estimation of Volatile Fatty Acids from Formic Acid to Dodecanoic Acid" (16). In order to detect the acids, the eluate was bubbled through water in a titration cell which was kept neutral by the manual addition of measured volumes of alkali. Martin later developed improved detectors of more general application (gas-density balance and thermal-conductivity units). In the paper, the separation of a range of C_2 to C_5 acids (acetic, propionic, isobutyric, butyric, $\alpha\alpha$-dimethylpropionic, isovaleric, α-methylbutyric, and valeric) was described and illustrated. Good separations were only achieved when stearic acid was added to the stationary phase. In 1956 James and Martin published another paper "Gas-Liquid Chromatography: The Separation and Identification of the Methyl Esters of Saturated and Unsaturated Acids from Formic Acid to *n*-Octadecanoic Acid." (17). They used either "Apiezon M vacuum stopcock grease. (Shell Chemicals Ltd.)" or an "extract of an aromatic character from a heavy lubricating oil (supplied by British Petroleum Ltd)" as the stationary phase. When separation of saturated and unsaturated acids was not satisfactory, the latter were modified by hydrogenation, oxidation, or bromination followed by a second chromatographic separation.

The work of James and Martin aroused considerable interest in both academic and industrial laboratories and many homemade gas chromatographs were made during the 1950s. Progress was more marked once commercial instruments became available. One of the first early models was a Pye instrument fitted with an argon ionization detector which contained a ^{90}Sr source. The straight column was about 4 feet long and stood vertically in a heating unit. It was attached to the detector at its lower end and closed with a ground-glass stopper at the top. Samples were injected from a micropipette by removing and then replacing the stopper. At first the stationary phase was nonpolar (Apiezon L) but this was followed by a number of polar phases. The results were obtained on a recorder trace, and peak areas were assessed by one of a number of manual operations, most often triangulation. Some investigators cut up the paper trace and weighed the various parts.

Gradually these primitive instruments were converted to the sophisticated but user-friendly equipment now in use. Changes included improved stationary phases of varying polarity and greater thermal stability, improved detectors, integrated recording devices, automatic injection systems, and capillary chromatography. These are now considered to be required instrumentation in any lipid laboratory, and this general availability is assumed in the new chromatographic developments described elsewhere in this text. From about 1980 onwards it has become increasingly common to attach a gas chromatographic system to either Fourier Transform Infrared (FTIR) or Mass Spectrometric (MS) devices. These are known as hyphenated techniques.

While these developments were occurring (1960 onwards), thin-layer chromatography (TLC) was also developing as a useful technique in lipid laboratories. This procedure arose from work by Stahl in the late 1950s and from the provision of simple commercial equipment. Though simpler (and cheaper) than gas-chromatography, thin-layer chromatography did not lend itself as easily to quantitative study. Much of what was undertaken by TLC is now achieved by HPLC, but TLC remains a useful semipreparative technique for separating lipids mainly on the basis of polarity. Its value was enhanced by incorporation of silver nitrate into the adsorbent which led to separation on the basis of unsaturation (18). Silver ion chromatography is still important and is used in several modes.

The first publication on HPLC is credited to Horvath and Lipsky (19) in 1966. Commercial instruments became available from about 1980 onwards and are now widely used in lipid laboratories. The selection of a suitable detector presented some difficulty, but the mass detector now finds general acceptance.

Gas chromatography and HPLC are largely complementary: the former is better for methyl esters while HPLC is better for separation of lipid classes and for the separation of individual members within a class. These two chromatographic procedures are now the basis of virtually all lipid analytical investigations.

Obviously there will be further developments in the existing chromatographic procedures and additional exploitation of their possibilities. One instrument manufacturer recently indicated that the GC market is now stable and greater sales (and therefore development) is more likely with HPLC. A general feature of development in analytical instrumentation is that they will become more sophisticated (and more expensive) but more automated and more user-friendly. The manipulative skills which were necessary before the advent of GC are largely outmoded and are destined to become even more so.

And what is next? Some interesting developments are described in the remaining chapters of this book. Others will arise from the skillful use of hyphenated techniques.

Acknowledgments

The author thanks N.U. Olsson and B.G. Herslöf for permission to use material taken from his article in *Contemporary Lipid Analysis* (Lipid Teknik, Stockholm, 1992).

References

1. Costa, A.B. (1962) In *Michel-Eugene Chevreul: Pioneer of Organic Chemistry,* Dept. of History, University of Wisconsin.
2. Gunstone, F.D. (1965) *J. Am. Oil Chem. Soc. 42:*474A.
3. Gunstone, F.D. (1982) *Dictionary of National Biography,* Oxford University Press.
4. Morton, R.A. (1966) *Biographical Memoirs of Fellows of the Royal Society, 12:*259.
5. Hilditch, T.P., and Williams, P.N. (1964) *The Chemical Constitution of Natural Fats,* 4th edn., Chapman and Hall, London.
6. Hilditch, T.P. (1948) *J. Chem. Soc.* 243–252.
7. Klenk, E., and Lindlar, F. (1955) *Hoppe-Seyl Z. 301:*156.

8. Gunstone, F.D. (1978) *Trends in Biochemical Science* 54, Elsevier, Amsterdam.
9. Hilditch, T.P., Morton, A.R., and Riley, J.P. (1945) *Analyst, 70:*68.
10. Gunstone, F.D., and Hilditch, T.P. (1946) *J. Soc. Chem. Ind., 65:*8.
11. Gunstone, F.D. (1954) *J. Chem. Soc.,* 1611–1616.
12. Tswett, M.S. (1903) *Tr. Protok, Varshav. Obshch. Estestvoispyt Otd. Biol.,* 14.
13. Tswett, M.S. (1906) *Ber, Dtsch, Botan. Ges.,* 316–323, 324–393.
14. Kuhn, R., Winterstein, A., and Lederer, E. (1931) *Z. Physiol. Chem. 197:*141–60.
15. Martin, A.J.P., and Synge, A.L.M. (1941) *Biochem. J. 35:*1358–1368.
16. James, A.T., and Martin, A.J.P. (1952) *Biochem. J. 50:*679–990.
17. James, A.T., and Martin, A.J.P. (1956) *Biochem. J. 63:*144–152.
18. Morris, L.J. (1966) *J. Lipid Research 7:*717–732.
19. Horváth, Cs., and Lipsky, S.R. (1966) *Nature* (London) *211*:748–749.

Chapter 2

Extraction of Lipids for Analytical Purposes

Anne Castera

Institut des Corps Gras, Rue Monge, Parc Industriel, 33600 Pessac, France.

Introduction

The analysis of lipids from a food or biological product needs a preliminary step of extraction. The choice of the extraction method is very important to ensure accurate results for the further characterization of lipid material.

The extraction of fat can be performed for quantitative determination of fat content or to simply obtain lipids for further investigation by chemical or chromatographic methods. A wide range of procedures has appeared in the literature, some of which, generally the quantitative ones, are standardized by national and international organizations for one type of product. The analyst must choose the most appropriate extraction method and decide what kind of "fat" is to be determined.

First, fats and lipids must be defined. Lipids are generally classed in two groups: the simple and the complex lipids. Neutral, simple lipids or "free" fat include esters of fatty acids and glycerol (triglycerides) and unsaponifiable matter. These components are easily soluble in nonpolar organic solvents such as hexane, light petroleum, or supercritical carbon dioxide.

Complex or polar lipids include phospholipids, glycolipids, lipoproteins, oxidized or polar glycerides, and free fatty acids. Because of the polar nature of some of their functional groups, complex lipids are preferentially extracted by or are soluble in polar solvents, such as methanol. Their quantitative extraction needs a chemical hydrolysis agent (acid or base) for breaking ionic and hydrogen bonds associated with proteins, carbohydrates, or metals.

The definition of "extracted fat" depends directly on the extraction procedure used, and each method defines fat content under the operating conditions of the test. Consider that fat includes "all the fatty chemical components obtained in an analytical extract, soluble in a small volume of warm light petroleum and relatively non-volatile at 100°C" (1). Whatever method is used, the extract contains a small quantity of co-extractant nonfat material: water, lactose, and hydrophobic proteins.

Extraction with organic liquid solvent is the most useful analytical fat isolation procedure. The theoretical mechanisms are very complicated: they include complex formation between extracted compounds and solvent molecules; Van der Waals' interactions; ion-exchange in the presence of an acid or base, and cleaving covalent, hydrogen, or electrostatic bonds (2,3).

An experimental analytical extraction procedure always involves four different steps.

1. *Extraction* by mixing, grinding, dispersing, or boiling the sample in the solvent. The contact surface between liquid or solid matrix and solvent and the extraction time will be optimized to enable efficient solvent penetration.
2. *Separation* of the organic phase from the nonextractable matter by decantation, centrifugation, or filtration.
3. *Purification* of the organic phase by washing, drying, filtration, adsorption, or reextraction.
4. *Desolventization* of the extract by distillation or evaporation under a nitrogen stream.

The specificity of each method mainly depends on the choice of the solvent: its polarity; boiling temperature; water or other solvents' miscibility; solubility parameters, such as Hildebrand Constant; volatility; and toxicity.

Methods for the quantitative determination of "total" fat usually involve an acid or basic digestion step of the sample prior to solvent extraction to free bound and complex lipids (1,4). A time-consuming reflux procedure is generally proposed for extracting the "free" fat. Such operations are indicated in standard methods for the determination of fat content in meat and meat products, milk and dairy products, animal feeds, cereal products, and oilseeds. The three most common solvents specified in standard methods are hexane, light petroleum, and diethylether.

Hydrolysis or heating are not compatible with characterizing lipids because of the risk of degradation or oxidation. Formation of free fatty acids and esterification or alcoholysis reactions are mentioned by several authors (1,5–9). The quality of the solvents and reagents must also be verified (10–12). Heating or washing are not advisable in order to minimize risks of losing partially hydrosoluble or volatile components, such as short-chain free fatty acids.

To sum up, the "ideal" extraction method should be quantitative and selective without altering the structural or chemical composition of the lipids. The other parameters for choosing a procedure are the toxicity and inflammability of the solvents; the simplicity, time, and cost of the method; the repeatability and reproducibility of the results, and the automatization possibilities. Standardization and field of application should also be considered (1,13,14).

A complete list of solvent extraction methods generally known by the name of their original proposer will not be produced in this chapter. Lumley and Colwell (1) indicate more than 100 references of international standard methods. Nevertheless, they can be classified within one of the three following basic procedures. The principle, field of application, advantages, and disadvantages of each method are described in Table 2.1.

1. *Reflux methods:* A Soxhlet or similar continuous extraction apparatus is used (12,14–17)

TABLE 2.1
Analytical Methods for the Extraction and Determination of Fat

Extraction method	Principle	Field of application	Remarks
Reflux method (Soxhlet type)	- Grinding and drying the test sample - Fat extraction during several hours in a suitable reflux apparatus with hexane, light petroleum, or diethyl ether (ethanol in the case of milk powder)	- Solid or semisolid products (oilseeds, meat products, feedstuffs, milk powder)	- Standard quantitative methods - Time-consuming - No extraction of polar lipids - Unsuitable for wet products or liquids - No purification step - Thermal destruction of hydroperoxides
Acid hydrolysis–gravimetric methods (SBR type)	- HCl hydrolysis of the test sample - Fat extraction in Mojonnier or Soxhlet type apparatus with ethanol, diethyl ether, and light petroleum (SBR methods) or hexane after filtration the digest (Weibull–Berntrop method)	- Dairy products - Meat products - Feedstuffs - Cereal products and vegetables	- Standard quantitative methods - Time-consuming - Contamination of the extract with ether-extractable carbohydrates (lactose) - Hydrolysis of triglycerides - Extraction of polar lipids - Other mineral acids: formic acid, pure acetic acid
Acid hydrolysis–volumetric methods (Van Gulik type)	- HCl hydrolysis (Gerber French method) or H_2SO_4 hydrolysis (Babcock U.S. method) - Addition of isoamyl alcohol and centrifugation in a butyrometer - Lecture of the liquid fat volume	- Milk and some dairy products - Suitable for oilseeds, meat, seafood	- Rapid method - Good repeatability - Only quantitative method - Important hydrolysis of triglycerides - Formation of isoamyl esters
Basic hydrolysis–gravimetric methods (Rose-Gottlieb type)	- NH_4OH hydrolysis of the test sample in ethanol - Fat extraction with diethyl ether and light petroleum	- Only milk and dairy products	- Standard quantitative method - Important hydrolysis of triglycerides - Extraction of polar lipids - Loss of free fatty acids in the aqueous phase

(continued)

Nonheating methods Purification by washing (Folch type)	- Dispersion of the test sample in a mixture of organic solvents: *2:1 chloroform-methanol (Folch method) *1:2:0.8 chloroform-methanol-water (Bligh and Dyer method) *2:1 dichloromethane-methanol (Chen method) *3:2 hexane-isopropanol (Radin method) *hexane-methanol (Emery method) - Filtration, decantation after washing the organic phase with NaCl or KCl solution	- Biological samples - Meat, fish - All food products	- Biochemical reference method (Folch) - Extraction of polar lipids - No degradation of fat - Risk of contamination with hydrophobic proteins or carbohydrates - Formation of emulsions during washing - Chloroform is toxic and pro-oxidative
Nonheating methods purification on dry column	- Maxwell method: *Grinding the test sample with anhydrous Na_2SO_4 + Celite 545 + $CaHPO_4$ *Extraction of fat by elution in a glass column with a 9:1 dichloromethane-methanol mixture *Direct evaporation of solvent under nitrogen stream - Wolff and Castera method: *Dispersion of the test sample in 3:2 hexane-isopropanol mixture *Filtration through anhydrous Na_2SO_4 + Celite 545 dry column	- All food products	- Standard method - No degradation of fat - Extraction of polar lipids - Contamination with carbohydrates by using isopropanol - Time-consuming elution when water is present - Celite can be pro-oxidative

2. *Digestion step methods:* The Schmid-Bondzynski-Ratzlaff (SBR) or Weibull-Berntrop methods involve a hydrochloric acid digestion prior to solvent extraction and gravimetric determination. Similar volumetric methods for dairy products are based on the Van Gulik butyrometric procedure (18). The Rose-Gottlieb method is used mainly for dairy products also and involves precipitation and solubilization of proteins by ethanol and ammonia.

3. *Nonheating methods:* These methods are based on the original procedure proposed by Folch et al. (19) for the extraction of brain lipids and animal tissues. This method involves mixing the sample with a chloroform-methanol (2:1, v/v) mixture and washing the organic phase with saline solution. The lower (chloroform) phase contains lipids and the upper (aqueous and methanol) phase contains water and nonlipid material. The procedure is relatively rapid, and the extracted fat can be used for chemical or physical investigations, for instance, peroxide value determination. The polar complex lipids, such as phospholipids and glycolipids, are extracted efficiently when compared with nonpolar solvent extraction methods.

The hexane-isopropanol (3:2, v/v) mixture proposed by Hara and Radin (20) is less efficient but is also less toxic. Different experimental studies dealing with a comparison of extraction methods have shown that the efficiency increases with the polarity of the solvent system while the selectivity decreases. A wide range of other solvent mixtures have been proposed: chloroform-methanol-water, dichloromethane-methanol, and hexane-methanol (8,13,21,22).

Another nonheating extraction and purification method was proposed by Maxwell and Marmer (23,24) who developed a "dry column" procedure applied to meat tissues. The ground sample is mixed with anhydrous sodium sulfate and Celite 545, and packed into a glass column. Neutral lipids are extracted by a first elution using dichloromethane and then polar lipids are eluted with a dichloromethane-methanol (9:1, v/v) mixture. This sequential column procedure is very simple, rapid, and selective (1,3,25,26). Calcium hydrogen phosphate is added to remove proteins from the extract.

Wolff and Castera (5–7) later proposed another dry column method using a prior extraction step by dispersing the sample in a hexane-isopropanol (3:2, v/v) mixture. The extract passes through an anhydrous sodium sulfate bed to absorb residual moisture and a Celite 545 bed to remove impurities. This procedure gives acceptable results for a wide range of food and feed products (fat content from 0.5–80% and moisture from 0–90%) and is standardized by the French organization (AFNOR) as an extraction method compatible with characterization of fat.

Finally, we wish to indicate that laboratories now use some automatic solvent extraction apparatus for routine analysis. A Soxhlet-type extractor with or without a digestion unit is proposed by Büchi and Tecator companies (Soxtec System H') (4,27,28) and a nonheating Equilibrium Extractor is commercialized by Tecator.

Some physical methods have been developed for a rapid routine and nondestructive determination of the fat content (1,29). The near infrared reflectance (NIR) can give the composition of a food product by the measurement of diffuse reflectance and the difference in absorption between 1725 nm and 1650 nm has a high correlation with fat content. For instance, the Technicon Infralyser instrument can give correct results depending on the homogeneity of the sample and the particle size (30). The technique is applicable for meat products, milk, and cocoa powder with specific calibrations.

Wide line Nuclear Magnetic Resonance (NMR) has also been applied to oilseeds. Other physical methods have been used to evaluate fat content, such as X-Ray analysis, titration, and refractometry, but none are really fat extraction methods.

An extraction process using supercritical fluids has been developed recently for the purification or fractionation of biological substrates without changing their physical or chemical properties. Supercritical carbon dioxide (SC–CO_2) is the most useful supercritical fluid because it is nontoxic, inexpensive, nonexplosive, and its critical point is easy to reach (about 31°C at 73 bar). Supercritical carbon dioxide can extract apolar compounds, such as fat and oil, and more specifically triglycerides. However, proteins, carbohydrates, mineral salts, or polar lipids (phospholipids or glycolipids) are not or are only slightly soluble in SC–CO_2. Supercritical carbon dioxide is also a more versatile solvent than liquid organic solvents because the extraction conditions and selectivity can be varied over a wide range of temperatures and pressures. This is particularly relevant in the near-critical area, where small changes in pressure and temperature correspond to large density variations. Another advantage is the easy removal of the solvent from the extract by isothermal pressure reduction or isobaric heating.

Accordingly, analysts and laboratories have taken great interest in SC–CO_2 extraction for sample preparation or directly coupled with a chromatographic technique (31–34). A lot of results have been published in the case of selective extraction of apolar traces (pesticides, PAH, hydrocarbons, and volatile or aromatic compounds) from vegetables, spices, or soils, but the technique is less accurate for a quantitative determination of fat content in a food or biological product. The yield of a SC–CO_2 extraction depends greatly on the composition of the matrix particularly its water content, the surface area, and the experimental conditions (temperature, pressure, CO_2 flow-rate, extraction time, addition of cosolvent or adsorbant). Some applications have been developed for the specific extraction of sterols or short-chain free fatty acids. Commercial instrument manufacturers propose relatively expensive analytical supercritical fluid extractors (Hewlett-Packard, Fisons, Dionex, Isco). It is expected that the analytical use of this technique will be developed in the near future.

In conclusion, the overview of the standard and nonstandard methods available for the extraction and quantification of fat has proved the difficulty and complexity of this analytical step. On the other hand, demands for fat determination increase more and more as the control of specifications and nutritional value of new products is increased. No one extraction procedure is universally acceptable for all the food and feed products and for every analytical investigation. Consequently, new developments and techniques will continue to be proposed and studied.

References

1. Lumley, L.D., and Colwell, R.K. (1991) In *Analysis of Oilseeds, Fat, and Fatty Foods,* Rossel, J.B., and Pritchard, J.R.L., Elsevier Applied Science Ed., p. 227.
2. Camilleri, C. (1986) *Ind. Alim. Agric. 103:*1011.
3. Dieffenbacher, A., Durieux, B., and Jordan, C. (1988) *Rev. Fr. Corps Gras 35:*495.
4. Halvarson, H., and Alstin, F. (1981) *Intern. Laboratory* 11–12, 102.
5. Wolff, R.L., and Castera-Rossignol, A. (1987) *Rev. Fr. Corps Gras 34:*123.
6. Wolff, R.L., and Fabien, J.R. (1989) *Le Lait 69:*33.
7. Castera, A., Mordret, F., and Chazan, J.B. (1989) In *Actes of First Eurolipid Congress,* Angers, June 6–9, Etig Ed., p. 272.
8. Sheppard, A.J., Hubbart, W.D., and Prosser, A.R. (1974) *J. Am. Oil Chem. Soc. 51:*416.
9. De Jong, C., and Badings, H.T. (1990) *J. of High Resolution Chromatography 13:*94.
10. Mulry, M.C., Schmid, R.H., and Kirk, J.R. (1983) *J. Ass. Off. Anal. Chem. 66:*746.
11. Schmid, P., Hunter, E., and Calvert, J. (1973) *Physiol. Chem. and Physics* 151.
12. Tagaki, Y., and Yanagita, T. (1985) *Bull. Fac. Agr. Saga. Univ. 59:*23.
13. Chen, I.S., Shen, C.S.J., and Sheppard, A.J. (1981) *J. Am. Oil Chem. Soc. 58:*599.
14. Delpecii, P., Guezel, M., Leclerc, B., and Kahane, E. (1966) *Rev. Fr. Corps Gras 13:*615.
15. De Koning, A.J., Evans, A.A., Heydenrych, C., De Purcell, C.J., and Wessels, J.P.H. (1985) *J. Sci. Food Agric. 36:*177.
16. Abraham, G., Hron, R.J., and Koltun, S.P. (1988) *J. Am. Oil Chem. Soc. 65:*129.
17. Sukhija, P.S., and Palmquist, D.L. (1988) *J. Agric. Food Chem. 36:*1202.
18. Barbano, D.M., Clark, J.L., and Dunham, C.E. (1988) *J. Ass. Off. Anal. Chem. 71:*898.
19. Folch, J., Sloane-Stanley, G.H., and Lees, M. (1957) *J. Of Bioch. Chem. 226:*497.
20. Hara, A., and Radin, N.S. (1978) *Anal. Biochem. 90:*420.
21. Warren, M.W., Brown, H.G., and Davis, D.R. (1988) *J. Am. Oil Chem. Soc. 65:*1136.
22. Bligh, E.D., and Dyer, W.J. (1959) *Can. J. Biochem. Biophysiol. 37:*911.
23. Maxwell, R.J., Marmer, W.N., Zubillaga, M.P., and Dalickas, G.A. (1980) *J. Assoc. Off. Anal. Chem. 63:*600.
24. Maxwell, R.J. (1984) *J. Ass. Off. Anal. Chem. 67:*878.
25. Durieux, B., and Dieffenbacher, A. (1989) In *Actes of First Eurolipid Congress,* Angers, June 6–9, Etig Ed., p. 264.
26. Adnan, M., Argoudelis, C.J., Tobias, J., Marmer, W.M., and Maxwell, R.J. (1981) *J. Am. Oil Chem. Soc. 58:*550.
27. Randall, E.L. (1974) *J. Assoc. Off. Anal. Chem. 57:*1165.
28. Lerique, D., Robert, P., and Boulet, P. (1986) *Rev. Alim. Animale 394:*39.
29. Kopp, J. (1988) *Viandes et Produits Carnés 9:*163.
30. Jeunet, R., and Grappin, R. (1985) *Technique Laitiére 1003:*53.
31. Castera, A., Auge, P., and Coustille, J.L. (1989) In *Actes of First Eurolipid Congress,* Angers, June 6–9, Etig Ed., p. 1099.
32. Hawthorne, S.B., (1990) *J. Chromatographic Sc. 28:*2.
33. King, J.W. (1990) *J. Chromatographic Sc. 28:*9.
34. Majors, R.E. (1991) *LC-GC Intl. 4:*10.

Chapter 3
Thin-Layer Chromatography of Lipids

Vijai K.S. Shukla

International Food Science Centre A/S, P.O. Box 44, Sønderskowej 7, DK-8520 Lystrup, Denmark.

Introduction

Chromatography is a separation technique of great resolving power and complexity. Chromatographic separations are carried out by mechanical manipulations involving a few of the general physical properties of molecules. The major properties involved are

1. The tendency of a molecule to dissolve in a liquid (solubility);
2. The tendency for a molecule to attach itself to a finely divided solid (adsorption); and
3. The tendency for a molecule to enter the vapor state or evaporate (volatility).

In chromatography, mixtures of substances to be separated are placed in a dynamic experimental situation where they can exhibit two of these properties. This may involve using the same property twice, such as solubility in two different liquids, or it may involve two different properties entirely.

Thin-Layer Chromatography (TLC) was described and routinely used by Kirchner et al. in 1951 (1). Only 5 years later, Stahl (2) introduced standardized equipment and techniques and the method found rapid acceptance for qualitative analysis. Thin-layer chromatography of lipids gained importance in lipid analysis through the citation classic paper of H.R. Mangold entitled "Thin-layer chromatography of Lipids" which appeared in the *Journal of American Oil Chemists' Society* in 1964 (3). Since then, TLC provided an ever-improving tool to fractionate lipid classes and through combination with Gas Chromatography (GC) and High-Performance Liquid Chromatography (HPLC) providing advanced insights into molecular species of lipids.

Separations by HPLC and TLC occur by essentially the same physical method. The two techniques are frequently considered competitors when it could be more realistic to consider them as complimentary techniques (4,5). Through HPLC it is easier to generate large numbers of theoretical plates and simpler to automate the analytical procedure than is the case in TLC. On the other hand, TLC offers a much higher sample throughput due to the possibility of performing separations simultaneously, and it can handle cruder samples since the separation medium is used only once. The detection process in HPLC is dynamic, while in TLC it can be considered static with the TLC plate acting as a storage detector. The separation can be evaluated at intermediate times during the separation, as is commonly done in multiple development, or can be evaluated sequentially by different detection techniques at the completion of the separation. In TLC it is relatively simple to apply chemical reagents to enhance detection sensitivity and selectivity unrestricted by the time constraints that must apply to a dynamic detection system.

Thin-Layer Chromatography provides separation of a wide variety of compounds with different polarities on a single plate. The separations achieved by HPLC through gradient elution can possibly be produced by TLC using a single mixed solvent system. The unique feature of TLC is that a mixed solvent system behaves like a gradient elution during travel on a plate because differential evaporation from the plate may occur. This adds to the potential of TLC as a cost-effective, simple, and easy chromatographic technique.

Unlike HPLC, silica is still the stationary phase material preferred for modern TLC by a large majority of chromatographers, and reversed phase thin-layer chromatography (RPTLC) occupies a rather modest second place. A general guideline on adsorbents and solvent mixtures used for fractionating lipids is listed in Table 3.1.

Complex lipid mixtures, for example, phospho- and glycolipids extracted from natural sources, are preferentially separated by two-dimensional TLC, using a neutral/basic solvent in the first, and an acidic solvent in the second dimension. The plates are developed in a lid-covered glass chamber containing the solvent. To

TABLE 3.1
Common Adsorbents and Solvents in Thin-Layer Chromatography

Material	Application
Adsorbents	
Silica gel with calcium sulfate (G)	Neutral lipids
Silica gel (H)	All lipids
Silica gel with magnesium-activated zinc Silicate (F_{254})	Detection
Silica gel with paraffins or silicone oil	Reversed phase
Silica gel with silver nitrate	Double bonds
Silica gel with boric acid	Vicinal diols
Magnesium silicate	Waxes
Solvents	
Hexane-diethylether-acetic acid (or formic acid)	Neutral lipids
Hexane-diethylether	
Hexane-diethylether	Hydrocarbons and waxes
Pentane-diethylacetate	
Chloroform-methanol-water	
Chloroform-methanol-ammonia	Phospholipids
Chloroform-methanol-acetic acid-water	
Chloroform-methanol-water	
Chloroform-acetone-methanol-acetic acid-water	Glycolipids
Diisobutylketone-acetic acid	

TABLE 3.2
Characteristics of Thin-Layer Chromatography and High-Performance
Thin-Layer Chromatography

Parameter	TLC	HPTLC
Height of a theoretical plate	30 μm	12 μm
Sample volume applicable	1–5 1μL	0.1 μL
Diameter of spots	3–6 mm	0.1–0.5 mm
Separation distance	10–16 cm	1–3 cm
Time required	30–200 min	2–3 min
Sensitivity	5.0–50 ng	0.5–5 ng

increase the speed, these chambers may be lined with filter paper to saturate the air with solvent vapor, and occasionally improved resolution is observed. Often a sandwich configuration of plates is used to improve speed and resolution, but this technique has not found broad application in the lipid field. When the solvent front is close to the top of the plate, the plate is removed from the chamber and residual solvent is evaporated under an inert atmosphere. After visualization of spots, migration distance of the lipid can be determined and is expressed in R_f value (retention factor), derived from the ratio of migration distance of the lipid to the migration distance of the solvent. Recently, high-performance thin-layer chromatography (HPTLC) has been employed in lipid analysis, some basic characteristics are summarized in Table 3.2 (6).

Due to the low sample capacity of the thin layers used in HPTLC, overloading is imminent; however, the speed and improved accuracy in quantitative determinations outweigh this disadvantage in routine analysis. This method requires a horizontal developing chamber of special configuration, which is available from several companies.

Detection Methodology

Thin-layer chromatography offers several alternatives for qualitative and quantitative assays, from applying corrosive reagents to the use of mild fluorescent dyes or even water. Lipids after fractionation on silica gel can be quantitated in situ by densitometric or fluorimetric scanning of the plate, or the lipids can be recovered from the silica gel for preparative or analytical purposes. Quantitation by coupling TLC with flame ionization is well practiced today; however, coupling with mass spectrometry still needs further development before being applicable to routine analysis. Isotope techniques have also found special applications.

Non-Destructive Methods

Rhodamine 6G and 2',7'-dichlorofluorescein in water or aqueous alcohol solution sprayed onto the plate render all lipids visible when viewed under ultraviolet-light. With the former reagent lipids appear as pink spots, with the latter as yellow spots.

Lipids can be recovered from the silica gel by elution with an organic solvent, subsequent filtration of the solution through a bed of glass-wool removes the dyes by adsorption. Silica gels containing inorganic fluorescent dyes are also on the market. After fractionation of lipids on such adsorbents, individual spots can be easily detected as well as quantitated by fluorimetry.

Lipids can be visualized as brown spots when plates are exposed to iodine vapors for a few minutes. Though transient in staining, unsaturated lipids may not be resistant enough to iodine; caution is necessary in preparative work. Water serves as the simplest nondestructive spray when bulk amounts of lipids are separated, as fractions are easily recognized as white zones against a translucent background.

Destructive Methods

Spraying plates with 50% sulfuric acid and subsequent charring at 180°C for up to 1 h may be the most popular detection method. Lipids appear as black spots, and transient coloration from pink to purple to black indicates steroidal compounds. Charring followed by direct densitometric scanning of the TLC plate is a well established method for quantitation of lipids Several companies sell equipment for automated densitometric evaluation and fluorimetric measurements in one instrument.

A number of corrosive spray reagents are well suited for group-specific determinations, as for phospholipids in general, plasmalogens, glycolipids, fatty acids, and esters. Moreover, specific stains are known for phosphatidylethanolamine and -serine as well as for phosphatidylcholine and sphingomyelin. An exhaustive list of specific and nonspecific reagents for detection of lipids is shown in Tables 3.3 and 3.4, respectively.

Application. A typical application of TLC in resolving several lipids is shown in Figures 3.1 and 3.2, respectively. First, complex polar lipids are resolved using

TABLE 3.3
Specific Reagents for Detection of Lipids

Reagent	Specificity	Manipulation	Spots
Molybdic anhydride/H_2SO_4/powdered Mo	Phospholipids		Blue
Sodium molybdate/HCl/hydrazine	Phospholipids		Blue
Ammonium pentachlorooxymolybdate H_2SO_4	Phospholipids		Blue
Ninhydrin/BuOH/MeOH	Aminophospholipids	120°C	Purple
Basic bismuth nitrate AcOH KI	Choline-containing lipids		Orange
2,4-dinitrophenylhydrazine/H_2SO_4	Plasmalogens	110°C	Yellow-orange
Anthrone thiourea H_2SO_4	Glycolipids	110°C	Red-purple
α-naphthol MeOH H_2SO_4 EtOH	Glycolipids	60°C	Purple
Anisaldehyde EtOH H_2SO_4 AcOH	Glycolipids	90°C	Red
Resorcinol HCl cupric sulfate	Gangliosides	95°C	Blue-purple

TABLE 3.4
Nonspecific Reagents for Detection of Lipids

Reagent		Manipulation	Spots
Destructive			
50%	H_2SO_4 in MeOH	110°C	Brown-black
3%	Cupric Acetate in 8% H_3PO_4	180°C	Black
8%	Cupric Acetate in 8% H_3PO_4	160°C	Black
5%	Molybdophosphoric acid in EtOH	120°C	Blue
0.04%	Bromothymol blue in 0.1 NaOH	—	Blue-green
0.03%	Coomasie brilliant blue R in 20% NaOH	—	Blue
Nondestructive			
Iodine vapor			
0.001%	Rhodamine CG in acetone	UV	Brown
0.01%	Fluorescein in EtOH	UV	Fluorescence
0.01%	2',7',-dichlorofluorescein in EtOH water	UV	Fluorescence
	Water	Translucent background	White

chloroform and acetone (96:4, v/v) and the upper cut of the nonpolar lipids are subjected to hexane:ether:acetic acid (90:10:0.5, v/v/v) leading to the separation of individual nonpolar components (Figure 3.2).

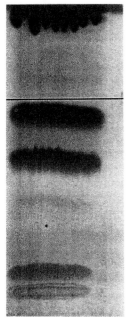

Solvent System
Chloroform: Acetone
(96:4)

1.3 Diglycerides

1.2 Diglycerides

Free Fatty Acids

Monoglycerides
Polar Components

Fig. 3.1. Plate produced with solvent system chloroform:acetone (96:4, v/v).

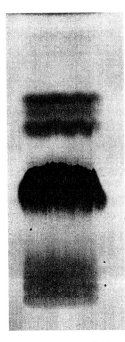

Solvent System
Hexane:Ether:Acetic Acid
(90:10:0.5)

Unsaponifiable material
from shea fat

triglycerides

Sterols
Unidentified

Fig. 3.2. Plate produced with solvent system using hexane:ether:acetic acid (90:10:0.5, v/v/v).

A commercially available instrument for quantitative measurements combines TLC with flame ionization. Thin-layer chromatography is carried out on so-called "Chromarods" (silica gel fused and sintered onto 0.9-mm diameter quartz rods), which are then passed through the detector. Multiple use of chromarods has successfully been used in lipid analysis (7).

Densitometers

The current trend in quantitative methodology for TLC is in situ scanning using linear and video scanners, although lasers have been used for fluorescence measurements. In a linear scanner, light passes through the layer and is captured by a photomultiplier, from which energy is converted to the signal that appears on the recording device or integrator. The same operations can be performed in the reflectance model by changing the position of the light or sensor. Scanners, that include monochromators, can scan low in the UV range. Such instruments offer detection limits in the low nanogram range for absorbance of colored or charred zones. Fluorescence detection produces limits in the subpicogram range, particularly when labelling with fluorescent probes.

Imaging detectors provide rapid acquisition by analyzing the whole TLC plate at once. Imaging detectors obtain information in three ways: transmittance, reflectance, and fluorescence. Few videoscanners are commercially available. Videoscanners readily perform fluorescence scanning, but at present cannot perform spectral analyses. They are well suited for fast, routine methods: the videoscanner can analyze >20 samples on a TLC layer in the same time it takes for one HPLC assay.

TLC–MS

Perhaps one of the latest advances in TLC is the coupling of TLC with mass spectrometry (MS), made possible by the development of small, high-performance mass spectrometers. These detection instruments will certainly enhance the capabilities of TLC as applied to lipids, and as these instruments become more popular, further advantages will emerge.

Conclusions

Thin-layer chromatography is a complimentary technique to HPLC providing much higher sample throughput while simultaneously achieving separations. Precision and accuracy, combined with the unique separating capability, speed, simplicity, and cost of modern TLC have made it a very powerful tool in modern lipid analysis.

References

1. Kirchner, J.G., Miller, J.M., and Keller, G.J. (1951) *Anal. Chem. 23:*420.
2. Stahl, E., Schroeter, G., Kraft, G., and Reuz, R. (1951) *Pharmazie 11:*633.
3. Mangold, H.K. (1964) *J. Am. Oil Chem. Soc. 41:*762.
4. Kaiser, R.E. (1986) *Planar Chromatography,* vol. 1, pp. 47 and 165, Hüthig, Heidelberg.
5. Brinkman, UA Th. (1988) *J. Planar Chromatography 1:*150.
6. Schwedt, G. (1979) *Chromomatographische Trennmethoden,* Thieme-Verlag, Stuttgart.
7. Fraser, A.J., and Taggart, C.T. (1988) *J. Chromatogr. 439:*404.

Chapter 4

Utilization of Thin-Layer Chromatography-Flame Ionization Detection for Lipid Analyses

J.-L. Sebedio

Institut National de la Recherche Agronomique, Station de Recherches sur la Qualité des Aliments de l'Homme, Unité de Nutrition Lipidique, 17 rue Sully, 21034 Dijon, France.

Introduction

Since the development of the thin layer chromatography-flame ionization detection (TLC-FID) system in the early 1970s, over 400 scientific papers dealing with this technique have been published. This technique combines the efficiency of thin-layer chromatographic separations using silica-gel-coated chromarods, and the sensitivity of the flame ionization detector (FID) commonly used for gas-liquid chromatography. Reviews dealing with the basic principles, techniques, and applications were published by Ackman in 1981 (1) and by Ranny (2). This technique is not only applicable to lipid analysis, but also to a great variety of compounds, such as steroids, alkaloids, surfactants, lubricants, stabilizers, drugs, pesticides, cosmetics, polymers, and heavy oils. Overviews of the applications of this technique in different fields have been published by Ranny (2) and, more recently, by Ackman et al. (3).

Since the first apparatus was introduced, several developments have taken place and the latest model (Mark V) seems to have very little in common with the earliest. Modifications (4) include the introduction of an improved detector, an automatic sample spotter, new chromarods (SIII), and data processing. This chapter will emphasize how these new developments have improved the reproducibility and enabled sample size reduction compared with that used with the older models.

For quantitative analysis, one must take into account different factors such as reproducibility, detection limit, and detector responses. The different techniques to improve these factors will be reviewed before describing the major applications of the TLC-FID system in the lipid field.

Reproducibility

When the first TLC-FID instrument was marketed, there were controversies concerning the acceptability of the quantitative analyses published (51). This was due to the high standard deviations which were obtained, especially for low sample loads (6), and to a high rod-to-rod variability (7). Since then, changes were made in order to improve the reproducibility (new rods, SIII instead of SII, a new detector, and an automatic autospotter).

In Table 4.1 are the results, compiled by Iatron Laboratories, of some reproducibility studies carried out using three standards, cholesterol ester (Chol.E), triglyceride (TG), and cholesterol (Cho.); two Iatroscan models (Mark III and Mark IV); and two types of Chromarod (SII and SIII). The combination of the rod SII and the Mark III shows some important rod-to-rod variation for all types of molecules. This variation was reduced with the utilization of SIII Chromarods and the Mark IV, probably owing to the technical improvement in the design of the instrument's FID. For that combination, the coefficients of variation for Chol.E, TG, and Cho. were 1.0, 0.6, and 1.6, respectively, which were the best values obtained simultaneously. Unfortunately, the amount of sample spotted on each rod was not reported.

Among all the factors which can influence the precision and the reproducibility of the analysis (1,3–8), great importance should be given to sample spotting. The volume of solution and the size of the sample spot must always be as small as possible. In general, solvents with a low boiling point and a low polarity should be chosen (3). The sample solution should be applied in several aliquots in order to minimize the size of the spot. One microliter disposable pipets (Microcaps, Drummond, USA) generally have been utilized. To avoid sample spreading from the point of application when spotting large volumes, the lipid material is refocused into a small band. This can be achieved by developing in acetone to just above the origin (9). The utilization of a semiautomatic sample spotter can, however, be a very good alternative as the instrument can apply the sample in successive aliquots, allowing the solvent to evaporate prior to the spotting of the next measured amount of sample; this avoids diffusion of the sample spot.

TABLE 4.1
Quantitative Analyses of a Mixture of a Cholesterol Ester (Chol.E), Triglyceride (TG), and Cholesterol (Cho.) Using Combinations of Different Types of Chromarods and Iatroscan Models.

Rod	Mark III SII			Mark III SIII			Mark IV SII			Mark IV SIII		
	Chol.E	TG	Cho.	Chol.E	TG	Cho.	Chol.E	TG	Cho.	Chol.E	TG	Cho.
1	37.1	36.2	25.5	35.2	36.5	26.7	33.9	38.5	24.6	34.4	38.2	24.7
2	37.9	31.5	26.4	34.6	36.4	28.0	33.9	38.9	24.6	34.8	38.8	24.8
3	37.0	36.1	26.1	34.9	36.3	27.6	35.0	38.8	23.7	35.1	38.7	23.9
4	37.5	35.4	26.1	34.6	36.3	27.8	35.0	38.4	24.1	34.9	38.7	24.8
5	37.6	35.9	25.6	34.3	36.6	27.9	35.1	38.0	24.5	34.2	38.6	24.8
6	37.0	35.9	26.1	34.3	36.6	27.7	34.7	38.6	24.2	34.6	38.6	24.3
7	37.1	36.5	25.3	33.7	36.5	28.6	35.0	38.9	23.7	34.4	38.2	25.0
8	35.6	36.7	26.8	34.6	36.1	28.0	34.1	38.5	24.9	34.0	38.4	24.8
9	37.4	35.7	26.0	35.1	36.5	27.0	33.9	39.1	24.4	34.8	38.7	24.3
10	37.5	36.0	25.1	34.7	36.5	27.6	33.7	39.0	24.4	35.0	38.7	24.0
x	37.2	35.9	25.9	34.6	36.4	27.7	34.4	38.7	24.3	34.6	38.6	24.5
SD	0.63	0.61	0.52	0.44	0.16	0.53	0.58	0.33	0.39	0.36	0.22	0.38
CV%	1.7	1.7	2.0	1.3	0.4	1.9	1.7	0.9	1.6	1.0	0.6	1.6

Source: From Iatron Laboratories.

TABLE 4.2
Analyses (Area) of a Mixture of Cholesterol Ester (Chol.E), Free Fatty Acid (FFA), Triglyceride (TG), Cholesterol (Cho.), and Diglyceride (DG): Interrod Variation (Chromarod SIII, Mark IV)

Rod	Chol.E	FFA	TG	Cho.	DG
1	3 829	308	39 907	2 470	411
2	4 038	370	43 997	2 843	467
3	3 863	350	43 927	2 937	475
4	3 929	373	43 320	3 129	471
5	3 618	238	40 021	2 690	546
6	3 758	318	41 360	2 741	495
7	3 997	368	42 076	2 791	547
8	3 577	—	39 952	2 733	516
9	3 800	417	40 864	2 775	463
10	4 072	411	41 820	2 847	497
Average area	3 846	350	41 724	2 796	489
CV%	4.3	15.9	3.8	6.1	8.4

Source: From Sebedio and Juaneda. (11).

Reproducibility tests carried out in our laboratory in Dijon were effected using a mixture of free fatty acid (FFA) 50 ng, diglyceride (DG) 100 ng, Cho. 300 ng, Chol.E 350 ng, and TG 4200 ng. This is a typical mixture which can be extracted from a biological sample. Good coefficients of variation (CV) were obtained ranging from 3.8% for TG to 15.9% for FFA (Table 4.2). The CV increased when the amount of sample spotted on the rod was reduced. Considering the small rod-to-rod variation, the practice of considering each rod as an analytical entity, as for chromarods SII (1,4), does not seem to be necessary for chromarods SIII because for this type of sample no statistical differences were observed for the 10 rods. However, it is a good practice when opening a new set of rods to test them with the type of sample to be analyzed.

Detection Limit

When using the Mark III and Chromarods SII, 5–20 mg of sample seems to be the usual amount spotted on the rods. The utilization of SIII rods as well as the new electrode of the Mark IV and Mark V enables a smaller quantity of sample to be spotted. The detection limits of different types of lipid are reported in Table 4.3. The different lipid classes studied were Chol., Chol.E, TG, FFA, DG, and phospholipids (PL). The phospholipid sample was a mixture of phospholipids such as those commonly found in rat liver lipids. The amount of sample spotted on the rods ranged from 20–1024 ng. The samples were spotted on the rods which were developed using a mixture of hexane-diethyl ether-formic acid (97:3:1 v/v) and scanned in a Mark IV system.

TABLE 4.3
Detection Limit of Different Lipid Classes (Chromarod SIII, Iatroscan Mark IV)

	Amount Spotted (ng)						
	20	40	80	160	2540	5120	10240
Cholesterol	X	X	X	X		X	X
Cholesterol ester	X	X	X	X		X	X
Tristearin	X	X	X	X		X	
Triolein	ND[a]	ND	X	X		X	X
Trilinolein	X	X	X	X		X	X
Stearic acid	X	X	X	X		X	X
Oleic acid	ND	X	X	X		X	X
Linoleic acid	X	X	X	X		X	X
Distearin	X	X	X	X		X	X
Diolein	ND	X	X	X		X	X
Phospholipids	X	X	X	X		X	X

[a]ND = not detected.
Source: Adapted from Sebedio and Juaneda (11).

All the components were detected and could be quantified from 80–5120 ng. One type of component, tristearin could not be used at high concentrations on the SIII rods. Tristearin spotted at 10μg was not eluted by the solvent mixture and remained at the origin. At low concentrations (40 ng) most of the samples except triolein could be detected. For the smallest amount of sample studied, 9 out of 13 components were detected and integrated. It should also be emphasized that the detection limits of some components were lower than 20 ng; the signal given by the FID was, however, not clean enough for satisfactory integration.

According to results published by the Iatron Laboratories (Table 4.4), the detection limit could be improved by using the new Mark V model. For example the detection limit of FFA, Chol.E, and PL was 7.8 ng while the detection limit for TG and Chol. were 3.9 and 1.9 ng, respectively. However, in order to obtain a good signal-to-noise ratio, it has been recommended to spot at least 10 ng.

TABLE 4.4
Detection Limit (Area) of Different Types of Lipid (Chromarod SIII, Iatroscan Mark V)

Weight (ng)	Cho.	TG	FFA	Chol.E	PL
1000	5894	4915	3935	5212	8256
500	3464	2189	2197	2443	2653
250	1773	922	1156	noise	1746
125	871	499	599	471	964
62.5	382	196	386	207	337
31.3	208	102	182	103	163
15.6	58	48	50	55	38
7.8	45	29	54	28	30
3.9	43	9			
1.9	27				

Source: From Iatron Laboratories.

Detector Response

Many studies have been performed using the Mark III model and the detector response has been shown to depend on the chemical nature of the component and not to vary linearly with the amount of sample spotted on the rods (9–12). Quantitative analysis of lipid mixtures therefore requires the utilization of FID correction factors which take into account the nature of the component and its quantity on the rod, as well as in some cases, the utilization of an internal standard (3). Generally, a linear relationship between the amount of sample spotted and the detector response would not pass through the origin. Different models have been tested: $y = a + bx + cx^2$; $y = a + bx + cx^2 + dx^3$; and $y = ax^b$ (1,13,14), where x represents the amount of sample spotted on the rods and y is the FID response. For many components the power curve, $y = ax^b$, described the FID response as a function of the amount of sample spotted on the rods quite well. Similar determination coefficients were generally obtained for the different models tested (14).

In order to determine the influence of the new collector electrode (Mark IV model) on the FID response, the calibration curves were determined for different lipid molecules: Chol., FFA, TG, Chol.E, and DG. The quantities spotted on the rods were 20, 40, 80, 160, 320, 640, 1280, 2560, 5120 and 10,240 ng except for components which could not be detected at 20, 40, or 10,240 ng, as already indicated. Similar FID responses were obtained as previously described for the "old" electrode. The new electrode was much more sensitive, but there was still no linear relationship between the amount of sample spotted on the rod and the FID response over the large concentration range studied.

It is, of course, always possible to calculate calibration curves in this sample range. The power curve gave good correlation coefficients (Table 4.5). The utilization of such curves can, however, be dangerous when calculating response factors, even if the correlation coefficients were excellent. In Table 4.6, for example, data on stearic acid, oleic acid, and trilinolein have been reported. The values represent the FID response (area) either measured, or calculated using the equations reported in Table 4.5 for the different classes. In some cases (e.g., stearic acid), the values

TABLE 4.5
Calibration Curves for Different Lipid Classes (Chromarod SIII, Iatroscan Mark IV)

Cholesterol	$y = ax^{(b)}$	Correlation Coefficient [r^2]
Cholesterol	$y = 3.313x^{1.619}$	0.9988
Cholesterol ester	$y = 183.343x^{0.7594}$	0.9896
Stearic acid	$y = 1.736x^{1.1812}$	0.9994
Oleic acid	$y = 0.667x^{1.2920}$	0.9990
Triolein	$y = 0.498x^{1.3230}$	0.9947
Trilinolein	$y = 0.171x^{1.4296}$	0.9959
Distearin	$y = 4.037x^{1.616}$	0.9933
Diolein	$y = 0.710x^{1.253}$	0.9956

Source: Adapted from Sebedio and Juaneda (11).

TABLE 4.6
Comparison Between the Experimental Area (M) and the Area (C) Calculated from the Calibration Curves for Stearic Acid, Oleic Acid, Trilinolein as a Function of the Amount of Sample Spotted

		Amount of Sample Spotted (ng)				
		20	80	320	1,280	5,120
Stearic	M	62	253	1,637	9,441	41,141
acid	C	60	307	1,580	8,124	41,777
Oleic	M	ND	182	961	6,883	41,761
acid	C	32	192	1,150	6,897	41,353
Trilinolein	M	39	122	571	4,822	35,018
	C	12	90	652	4,732	34,337

Source: Adapted from Sebedio and Juaneda (11).

calculated using the curve agree with the experimental values. In most cases, and especially for the lower amounts of sample spotted on the rods, the calculated values and the experimental ones are quite different. In this case, the quantitative analyses, especially for the compounds present in small quantities in the mixture, would be incorrect. If the FID response of the different types of components is considered (Figure 4.1), it is always possible to find in the sample range different sections where there is a linear relationship between the FID response and the amount of sample spotted on the rods. Consequently, when analyzing an unknown lipid

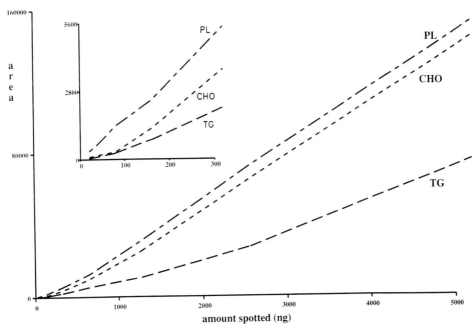

Figure 4.1. Calibration curves for cholesterol (Chol.), triglycerides (TG), and phospholipids (PL). Adapted from Sebedio and Juaneda (11).

mixture one should determine the FID response factor of each compound with a similar mixture and similar quantities spotted on the rods. Therefore, it is not of great importance that the FID response is not linear over the total sample range because it is possible to work in different linear sections of the curve.

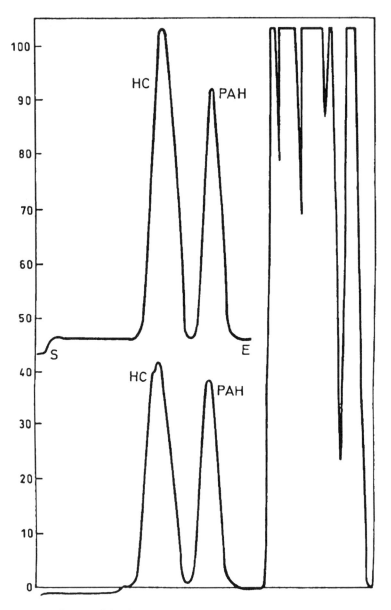

Figure 4.2. Full scan of the first chromarod of the set and (inset) a partial scan of the second chromarod. From Ackman et al. (3).

Major Applications

As already mentioned, many papers were published on the application of the TLC-FID technique to the lipid field. Two reviews were also recently published, one by Ackman et al. (3), followed by one by Shantha (4). The purpose of this section therefore will not be to review all the applications, but just to give a few examples showing the potential of this technique.

One of the major applications of the TLC-FID technique still remains the quantitative analysis of lipid classes of biological samples. Data using this technique were published as early as 1978 by Sipos and Ackman (6). For this type of analysis, great care must be taken in selecting the solvent system, as a small change in the solvent proportions may induce drastic changes in the elution order of some components, for example TG and FFA (15).

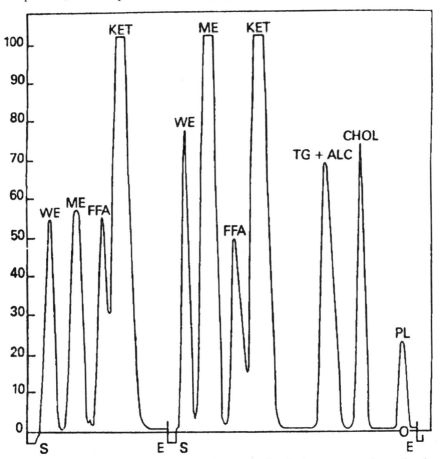

Figure 4.3. Right; full scan of the partially scanned rod of Figure 4.2, after redevelopment. Left; partial scan of a third rod. From Ackman et al. (3).

Due to the possibility of using multiple developments and partial scanning, this technique was applied successfully to the analysis of marine lipids, or the analysis of lipids recovered from seawater (9,13,16–19). Such an analysis with model compounds selected to represent lipid classes expected to occur in marine dissolved or particulate matter is reported in Figures 4.2, 4.3, and 4.4 where KET, represents a long-chain methyl ketone which is used as an internal standard.

The first development in a mixture of hexane-formic acid (99:1, v/v [Figure 4.2]) followed by a partial burn permitted the separation of the aliphatic and aromatic hydrocarbons from the other lipids which were left around the origin. In a second development effected in a slightly more polar solvent (hexane-diethyl ether-formic acid, 99:1:0.1, v/v), WE, ME, FFA, and KET were moved away from the other lipid classes (Figure 4.3). Again the rods were partially burned leaving the other lipid classes near the origin (TG, ALC, C, and PL). This procedure was

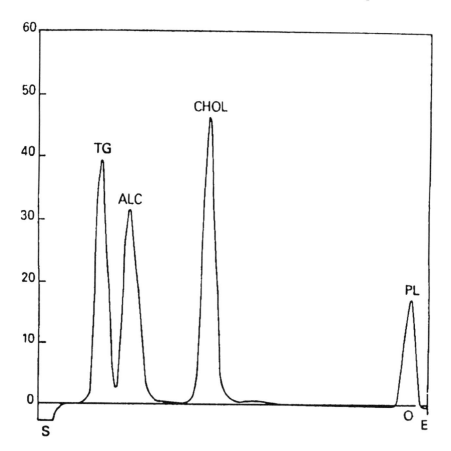

Figure 4.4. Full scan of the partially scanned (third) rod of Figure 4.3 after redevelopment. From Ackman et al. (3).

repeated (Figure 4.4) until all the neutral lipids were resolved and quantified, leaving the phospholipids at the origin. The phospholipids could also be fractionated, but this is a very tedious process, which has been the subject of many papers (8,20–23).

A complex procedure (Figure 4.5) has recently been proposed by De Scrijver and Vermeulen (23) to fractionate most of the phospholipid classes. This procedure involved oxalic-acid-impregnated rods and multiple development. The two major drawbacks of this technique are that it is time consuming and that the rods need to be impregnated each time before use, and oxalic acid is expected to form a complex with the inorganic phospholipid moiety.

However, comparable methods using high-performance liquid chromatography and a light-scattering detector have been shown to give excellent phospholipid separations (24,25). The major advantage of HPLC over the TLC-FID method is that phospholipid fractions can be collected for further analysis, although the TLC-FID technique would be useful for routine analysis considering that 10 different samples can be scanned at once.

Another interesting application is the analysis of the oxysterol oxides which could be present in foodstuffs (26–28). This is a real problem considering that the analysis of such compounds is very tedious and that oxysterols have been shown to present some biological properties (29). A good separation of the major oxysterols formed from cholesterol (Table 4.7) has been described by Bascoul et al. (Figure 4.6) (30). Unfortunately, nothing was published for the other oxysterols which could be formed from the major phytosterols.

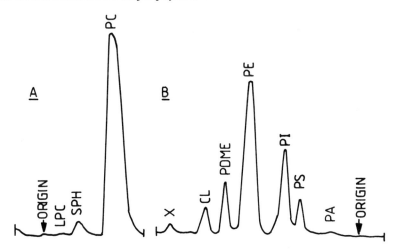

Figure 4.5. TLC/FID separation of chicken liver phospholipids on SII Chromarods impregnated with 0.01 M oxalic acid in acetone. (a) First partial phospholipid scan after double development in $CHCl_3/MeOH/CH_3COOH/HCOOH/H_2O$ (80:35:2:1:3 v/v). (b) Second partial phospholipid scan after double development in $CHCl_3/MeOH/30\% NH_4OH$ (60:35:0.9, v/v). In (a) and (b), scanning direction was from left to right. From De Schrijver and Vermeulen (23).

TABLE 4.7
Predominant Oxysterols Formed During Autoxidation of Cholesterol

Oxysterol	Number in Figure 4.6
3β-5-6β-trihydroxy-5α-cholestane	1
3β-7α-dihydroxy-cholesta-5-ene	2
3β-7β-dihydroxy-cholesta-5-ene	3
3β-hydroxy-7-oxo-cholesta-5-ene	4
5-6α-epoxy-5α-cholesta-3β-ol	5a
5-6β-epoxy-5β-cholesta-3β-ol	5b
3β-20-dihydroxy-(20S)-cholesta-5-ene	6
3β-25-dihydroxy-cholesta-5-ene	7
Cholesterol	8
7-oxo-cholesta-3-5-diene	9

As shown in Figure 4.7, quantification and separation can be improved by carrying a total hydrogenation before the TLC-FID analysis (31). This is especially important for animal tissues or for marine lipids where a great number of fatty acids varying in chain length and degree of unsaturation have to be quantified. Hydrogenation also produces a stable material where oxidation problems can be avoided.

Rods can be used as manufactured but can be modified by impregnation with different substances, such as boric acid (32), oxalic acid (23,33), silver nitrate (34–36), or copper sulfate (37–38). Oxalic acid impregnation of rods has been used

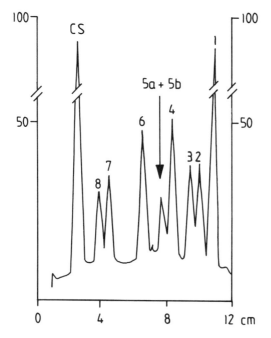

Figure 4.6. TLC-FID separation of a mixture of cholesterol oxides. See Table 4.7 for peak identification. From Bascoul et al. (30).

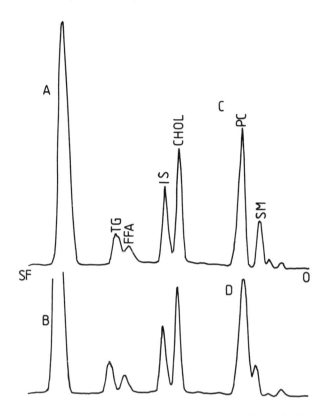

Figure 4.7. TLC-FID showing the effect of hydrogenation on the peak shape and separation of human plasma lipids on SIII Chromarods in the solvent system hexane-diethylether-formic acid (90:10:1, v/v), partial scan followed by complete redevelopment for analysis of polar lipids in the solvent system chloroform-methanol-water (70:30:3.5, v/v). (a) Neutral lipids, unhydrogenated. (b) Neutral lipids, hydrogenated. (c) Polar lipids, unhydrogenated. (d) Polar lipids, hydrogenated. Abbreviations: SE, steryl esters; FFA, free fatty acids; IS, internal standard, fatty alcohol; SM, sphingomyelin; PC, phosphatidylcholine; TG, triacylglycerol; O, origin and SF, solvent front. From Shantha (4).

to obtain an improved separation of phospholipids, especially between phosphatidylinositol and phosphatidylserine. Boric acid has been used for a complete separation of the triglyceride hydrolysis mixture, while copper sulfate impregnation was utilized to obtain a better response of the samples, a better reproducibility, and a lower rod-to-rod variability. Impregnation with silver nitrate has been used to separate the geometrical mono- and diethylenic fatty acid isomers as shown in Figure 4.8. Determination of *trans* monoethylenic isomers in margarine was also achieved.

The TLC-FID technique is an attractive method for qualitative and quantitative analysis of lipids of different sources. However, great care must be taken for quan-

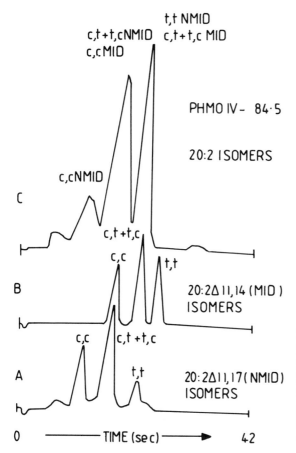

Figure 4.8. TLC-FID separation on AgNO$_3$ rods of a mixture of *cis, trans* isomers of 20:2 (a and b) and of the (c) 20:2 isomers isolated from a partially hydrogenated menhaden oil. From Sebedio and Ackman (35).

titative analysis (amount of sample spotted on the rod, the method of treating the rod, and the utilization of response factors) as the quality and reproducibility of the analyses depend primarily on the accurate setting up and calibration of the instrument.

References

1. Ackman, R.G. (1981) In *Methods in Enzymology.* Lowenstein, J.M., Academic Press, Vol. 72, pp. 205–52.
2. Ranny, M. (1987) *Thin Layer Chromatography with Flame Ionization Detection.* D. Reidel Publishing Company, Dordrecht, Holland.
3. Ackman, R.G., Meleod, C.A., and Banerjee, A. (1990) *J. Planar Chromatogr. 3:*450–490.
4. Shantha, N.C. (1992) *J. Chromatogr. 624:*21–35.
5. Crane, R.T., Goheen, S.C., Larkin, E.C., and Rao, G.A. (1983) *Lipids 18:*74–80.
6. Sipos, J.C., and Ackman, R.G. (1978) *J. Chromatogr. Sci.* 16:443–447.
7. Kramer, J.K.G.; Thompson, B.K. and Farnworth, E.R. (1986) *J. Chromatogr. 355:*221–228.

8. Murray, D.K. (1985) *J. Chromatogr. 331:*303–312.
9. Delmas, R.P., Parrish, C.C., and Ackman, R.G. (1984) *Anal. Chem. 55:*1272–1277.
10. Petersson, B. (1982) *J. Chromatogr. 242:*313–322.
11. Sebedio, J.L., and Juaneda, P. (1991) *J. Planar Chromatogr. 4:*35–41.
12. Fraser, A.J., Tocher, D.R., and Sargent, J.R. (1985) *J. Exp. Mar. Biol. Eco. 88:*91–99.
13. Parrish, C.C., and Ackman, R.G. (1985) *Lipids 20:*521–530.
14. Sebedio, J.L., Astorg, P.O., Septier, C., and Grandgirard, A. (1987) *J. Chromatogr. 405:*371–378.
15. Kramer, J.K.G., Fouchard, R.C., and Farnworth, E.R. (1980) *J. Chromatogr. 198:*279–285.
16. Volkman, J.K., Everitt, D.A., and Allen, D.I. (1986) *J. Chromatogr. 356:*147–162.
17. Parrish, C.C. (1986) *Can. J. Fish. Aquat. Sci. 44:*722–731.
18. Ohshima, T., Ratnayake, W.M.N., and Ackman, R.G. (1987) *J. Am. Oil Chem. Soc. 64:*219–223.
19. Ackman, R.G. (1981) *Chem. Industry,* 716–722.
20. Kaitaranta, J.K., and Nicolaides, N. (1981) *J. Chromatogr. 205:*339–347.
21. Foot, M., and Clandinin, M.T. (1982) *J. Chromatogr. 241:*428–431.
22. Kramer, J.K.G., Farnworth, E.R., and Thompson, B.K. (1985) *Lipids 20:*536–541.
23. De Schrijver, R., and Vermeulen, D. (1991) *Lipids 26:*74–76.
24. Christie, W.W. (1987) *High-Performance Liquid Chromatography and Lipids,* Pergamon Press, Oxford, pp. 106–132.
25. Juaneda, P., Rocquelin, G., and Astorg, P. (1990) *Lipids 25:*756–759.
26. Finocchiaro, E.T., and Richardson, T. (1983) *J. Food Protection 46:*917–925.
27. Addis, P.B. (1986) *Fd Chem. Toxic. 24:*1021–1030.
28. Maerker, G. (1987) *J. Am. Oil Chem. Soc. 64:*388–391.
29. Hwang, P.L. (1991) *BioEssays 13:*583–589.
30. Bascoul, J., Domergue, N., Olle, M., and Crastes de Paulet, A. (1986) *Lipids 21:*383–386.
31. Ackman, R.G., and Ratnayake, W.M.N. (1989) *J. Planar Chromatogr. 2:*219–223.
32. Tatara, T., Fujii, T., Kawase, T., and Minagawa, M. (1983) *Lipids 18:*732–736.
33. Banerjee, A.K., Ratnayake, W.M.N., and Ackman, R.G. (1985) *Lipids 20:*121–125.
34. Tanaka, M., Itoh, T., and Kaneto, H. (1979) *Yukagaku 28:*22–25.
35. Sebedio, J.L., and Ackman, R.G. (1981) *J. Chromatogr. Sci. 19:*552–557.
36. Sebedio, J.L., Farquharson, T.E., and Ackman, R.G. (1985) *Lipids 20:*555–560.
37. Kaimal, T.N.B., and Shantha, N.C. (1984) *J. Chromatogr. 288:*177–186.
38. Kramer, J.K.G., Fouchard, R.C., and Farnworth, E.R. (1986) *J. Chromatogr. 351:*571–573.

Chapter 5

High-Performance Liquid Chromatography: Normal-Phase, Reverse-Phase Detection Methodology

Vijai K.S. Shukla

International Food Science Centre A/S, P.O. Box 44, Sønderskovvej 7, DK-8520 Lystrup, Denmark.

Introduction

Development of high-performance liquid chromatography (HPLC) (or high-pressure, high-price, high-speed, and modern liquid chromatography) began in the late 1960s. The major advantages of HPLC include high resolution, speed, versatility, sensitivity, and automatic operation, which led to its acceptance both as a research tool and for routine analysis. The superiority of HPLC over other existing analytical techniques for analyzing lipids is attributed to the following:

1. High-performance liquid chromatography is the method of choice for separating the nonvolatile, thermally labile high molecular weight compounds at ambient temperatures.
2. High-performance liquid chromatography provides a means for the determination of multiple components in a single analysis.
3. Both aqueous and nonaqueous samples can be analyzed without pretreatment.
4. The availability of various solvents and column-packing materials provides a high degree of selectivity for specific analyses.
5. Separation times are short. Normally analyses are completed in a few minutes (high speed) or seconds (super speed) with excellent precision and accuracy.
6. The most important feature is that the separated components can be collected and recovered easily from the mobile phase for further analysis or characterization by complementary techniques, such as mass spectrometry, nuclear magnetic resonance, or infrared spectroscopy.

High-performance liquid chromatography has grown enormously during the past decade and is now accepted as a major tool for the analysis of lipids (1). It is still one of the fastest developing areas of analytical chemistry. The driving force leading to this growth has come from drastic improvements in HPLC-column technology and instrumentation.

High-performance liquid chromatography can be divided into various categories. The most popular of these for lipids are adsorption chromatography (2,3) on silica, and partition chromatography, mostly reversed phase partition. Today a large proportion of all work in HPLC is done using chemically bonded stationary phases.

Reversed phase chromatography on octadecylsilane-bonded phases continues to dominate HPLC modes, accounting for more than 65% of the present-day applications. Since reversed phase columns are much easier to handle and are considered better for the separation of a homologous series of compounds, these have been extensively employed for the separation of lipids in the author's laboratory.

The most common method of preparing reversed phase packing employs a surface reaction between a silica support and an appropriate organochlorosilane (or organoalkoxy) modifier. The extent to which the bonding reaction approaches completeness (maximum coverage) is an important determinant of column quality. The steric hindrance ensures that the bonding reaction is always incomplete. Some manufacturers apply a second silanization using trimethylchlorosilane as a silylation agent to remove residual silanols. The presence of unreacted, accessible silanols on the silica surface affects column stability, retention, reproducibility, and peak asymmetry. Because the reaction of water or other reagents in the mobile phase leads to the dissolution of the underlying silica, leading to reduced column lifetime (chemical instability), the accessible silanols give rise to mixed retention mechanisms. Separations on a column filled with such packing will occur through an undesirable mixture of normal phase and reversed phase processes. This phenomenon creates a major problem when comparing separations from one laboratory to another, leading to serious misinterpretations. Although considerable advances have been made for the reproducibility of the column chemistry and the radial compression technology, perfection has yet to be achieved.

Although columns with particles as small as 5 µm have been in routine use for several years, research continues towards the development of smaller particles. Theoretically a particle diameter, d_p, of about 2 µm will be optimal for many separations. Recent commercial availability of 3 µm particles led to the development of high-speed liquid chromatography (HSLC), which demands improved instrumentation to reduce extracolumn band-broadening effects. Employing this high-speed system, performance levels up to 450 plates/sec are attainable, thus allowing many isocratic separations to be performed in 1–2 min with over 10,000 theoretical plates. Besides offering high-speed/high-resolution analysis, the HSLC system also yields lower solvent consumption and reduced gradient cycle times. Several reports from the author's laboratory demonstrating the uses of 3 µm-silica and octadecylsilane (ODS) particles for lipid analyses have appeared.

During the course of the separation of triglycerides using 3 µm ODS-2 particles, the need for high efficiency of the column was realized, because many of the triglyceride components are coeluted related and need high plate numbers for efficient separations. The approach of generating very high plate numbers by serially coupling two standard columns was used. Thus, it was possible to achieve baseline separation of all the triglycerides of cocoa-butter by extending the resolving power of the coupled system.

Usually HPLC columns packed with 3 µm particles can achieve better separations than those obtained for large particle columns in less than one-half the total time, requiring very low dead volume liquid chromatographic systems. It is also worth noting that substantial time savings can be obtained during method development due to quick changeovers between mobile phases, in addition to the shorter analysis time and better efficiency. This opens up new capabilities for the routine HPLC laboratory. The lifetime of 3 µm columns is apparently the same as 5 µm, provided special care is taken in carefully clarifying the mobile phase and the samples.

Throughout the history of chromatography, chromatographers have striven to increase HPLC performance either by increasing the efficiency of the separation, reducing the time required for the separation, or by a combination of both simultaneously. Recently, very high speed liquid chromatographic principles have been applied to separate triglycerides in less than 10 min. This system is similar to a conventional LC system in concept, except that the band-broadening or extracolumn effects of the system have been reduced drastically.

Selecting an HPLC pump is not much different from selecting any other piece of analytical instrumentation. The usual parameters of capability are needed; versatility, accuracy, precision, ruggedness, size, and cost are all parameters to be considered seriously. Recommended pump parameters for various applications are listed in Table 5.1.

Significant advances have been made in HPLC instrumentation based on microprocessor technology. The evolution of integrated HPLC systems which control factors affecting performance saves considerable time and minimizes error in routine analyses. Sample preparation is the rate-limiting step in the automation of HPLC systems. Recent advances in sample-handling robotics have addressed this limitation. Robotic systems are available to reliably and precisely perform extractions, filtration, and other cleanup procedures prior to injection. Occasionally, it is necessary to modify the available HPLC-integrated systems to optimize separations

TABLE 5.1
Recommended Pump Parameters

Parameter	Mode of Operation		
	Research	Quality Control	Preparative
Pressure, (psi)	5000	3000	1500
Flow			
Range (mL/min)	10	10	20+
Accuracy	±5%	±5%	±10%
Reproducibility	±1%	±1%	±5%
Solvent storage	Unlimited	200–500 mL	Unlimited
Gradient elution capability	Necessary	Not critical	Not critical
Pulse-free delivery	Necessary	Necessary	Not critical

in lipid analyses. Often it is desirable during lipid analyses to operate at below ambient temperatures. There are few if any HPLC systems available that take into account operating below ambient temperatures while minimizing the extracolumn band-broadening effect from the injector and detector.

Detection Systems

The progress of HPLC for lipid analysis has been restricted due to the lack of a sensitive universal HPLC detector. Although there has been considerable growth in the development of various detection systems, much remains to be done with respect to improving detection technology for lipid analysis.

The general characteristics for proper selection of the detector are sensitivity, specificity, detectability, linearity, repeatability, and dependability. The detector-cell dead volume should be sufficiently small to prevent excessive peak broadening. It should not be affected by its laboratory location (drafts and room-temperature fluctuations). It should be insensitive to mobile-phase temperature and flow variations, nondestructive where fraction collection may be considered, and rugged and easy to use and maintain. Typical specification of the most commonly used detectors are listed in Table 5.2.

The ultraviolet (UV) detector was one of the earliest and most popular HPLC detectors used for analyzing lipids. Although most lipids generally lack chromophores facilitating UV detection and absorb in the 190–210 nm range, shorter wavelength UV detection is more sensitive and permits the use of gradients. However, it precludes the use of certain common lipid solvents, such as chloroform or acetone, which are opaque in the UV region of interest. Shukla et al. (4) have demonstrated the use of short wavelength UV detection at 220 nm for the quantitative analysis of triglycerides. At this wavelength, UV absorption of triglycerides is basically due to the ester C=O function of these molecules.

Refractive index (RI) detectors (differential refractometers) are probably the next most common HPLC detectors and are far less sensitive than UV. This detector is not suitable for gradient elution and is very sensitive to temperature changes and pressure fluctuations. Thus, it is impossible to achieve optimal separations.

Transport flame ionization or the moving wire detector devised by Maggs and Young (5) proved effective for the detection of lipids where the use of UV and RI was not feasible. Therefore, it is often termed a "universal detector." The eluents are deposited on a moving wire or belt and the solutes are removed from the volatile solvents in an intermediate furnace region and the nonvolatile lipid is carried through a flame ionization detector (FID) where it is combusted and detected. The main advantages of this type of detector are: it could be used with any volatile solvent using gradient-elution techniques, its universal application for lipids, and the detector response is rectilinearly related to the amount of the eluting lipid.

Phillips et al. (6,7) have successfully demonstrated the use of their moving belt-FID for quantitative analyses of lipid classes and triglyceride species of vegetable

TABLE 5.2
Typical Specifications for Commonly-Used Liquid Chromatograph Detectors

Parameter (units)	UV (absorbance)	RI (RI units)	Radioactivity	Electrochemical (μ amp)	Infrared (absorbance)	Fluorometer	Conductivity (μMho)
Type	Selective	General	Selective	Selective	Selective	Selective	Selective
Useful with gradients	Yes	No	Yes	No	Yes	Yes	No
Upper limit of linear dynamic range	2–3	10^{-3}	N.A.[a]	2×10^{-5}	1	N.A.	1000
Linear range (max)	10^5	10^4	Large	10^6	10^4	10^3	2×10^4
Sensitivity at ±1% noise, full-scale	0.002	2×10^{-6}	N.A.	2×10^{-9}	0.01	0.005	0.05
Sensitivity to favorable sample	2×10^{-10} g/mL	1×10^{-7} g/mL	50 cpm ^{14}C/mL	10^{-12} g/mL	10^{-6} g/mL	10^{-11} g/mL	10^{-8} g/mL
Inherent flow sensitivity[b]	No	No	No	Yes	No	No	Yes
Temperature sensitivity	Low	10^{-4} °C	Negligible	1.5%/°C	Low	Low	2%/°C

[a] N.A., not available.
[b] Because of sensitivity to temperature changes, some detectors appear to be flow sensitive.

oils. The author studied this detector in Privett's laboratory and found it very efficient with gradient elution, permitting the analysis of lipids in the low nanogram range by HPLC. However, there are two drawbacks: samples are destroyed and cannot be used for further analyses by complementary techniques and it is difficult to operate.

Tracor Instruments (Austin, Texas) launched the Tracor model 945 detector, incorporating a rotating disc with a fibrous quartz belt and FID detection system. Although this system was robust and easy to operate, it lacked sensitivity and flexibility in solvent selection due to the restriction of using only volatile solvents during triglyceride analysis.

The Mars detector, based on the principle of light scattering, may revolutionize lipid analysis (1). Gradient elution can be successfully applied providing the flexibility previously restricted to other detection methodologies. This provides minimal baseline drift when nonaqueous solvents are used. A range of volatile solvents are compatible with this detector provided that the sample to be analyzed is less volatile than the solvent. Although the response from the mass detector is nonlinear, it is highly reproducible and related to sample mass, permitting quantitation of eluting species. This detector is highly sensitive for all eluting triglycerides.

Mowery (8) applied a dielectric constant (DC) detector to evaluate corn and cottonseed oils. This detector is generally more sensitive than an RI detector and is ideally suited for use with nonaqueous mobile phases. This is a purely research approach, and the detector is not available commercially.

Parris was the first to report the utility of the infrared (IR) detector with gradient elution for the separation of triglycerides. The sensitivity of the IR detector is similar to that of the RI monitor. There are serious problems in finding the solvent window to monitor and excessive baseline drift with changes in solvent composition. Perhaps Fourier Transform Infrared (FTIR) will broaden the use of this detector. Compared with classical IR spectroscopy, FTIR spectroscopy has two main advantages: the signal-to-noise ratio is about 150 times greater, and the energy per unit time is 100–200 times higher.

Fluorescence labelling has been used in lipids, resulting in higher sensitivity of fluorescence detection during rapid scanning of HPLC effluents. Liquid chromatography-videofluorometry will be a powerful tool for lipid analysis in the future.

Over the past decade, considerable efforts have been made to produce HPLC-mass spectrometer (LC-MS) systems. Although LC-MS has enormous potential for lipid analysis, its implementation has been restricted by practical problems. The complexity and the high cost precludes its use in routine HPLC.

The best results are obtained in HPLC methodology if the following rules are followed strictly:

1. Keep air out while degassing the mobile phase and prevent air leaks.

2. Filter the mobile phase as well as the samples.

3. Flush the system regularly.

4. It is the volume, and not the time of flushing, that is important:

Vm = 0.1 L
Vm = volume of mobile phase in the column (mL).

Conclusions

HPLC has developed as a most powerful tool in lipid analysis. Further development of a universal HPLC detector will open new avenues of lipid research.

References

1. Shukla, V.K.S., (1988) *Prog. Lipid Res. 27:*5–38.
2. Hamilton, J.G., and Comai K. (1988) *Lipids 23(12):*1150–53.
3. Rhodes, S.H., and Netting, A.G. (1988) *Journ. of Chromatography 448:*135–43.
4. Shukla, V.K.S., Schiøtz, N.W., and Batsberg, W. (1983) *Fette Seifen Anstrichm. 85:*274–78.
5. Young, T.E., and Maggs, R.J. (1967) *J. Anal. Chim. Acta 38:*105–12.
6. Phillips, F.C., Erdahl, W.L., and Privett, O.S. (1982) *Lipids 17:*992–97.
7. Phillips, F.C., Erdahl, W.L., Nedenicek, J.D., Nutter, L.J., Schmit, J.A., and Privett, O.S. (1984) *Lipids 19:*142–50.
8. Mowery, R.A., Jr. (1982) *J. Chromatogr. Sci. 20:*551.

Chapter 6

Lipid Class Separations Using High-Performance Liquid Chromatography

William W. Christie

The Scottish Crop Research Institute, Invergowrie, Dundee, Scotland DD2 5DA, UK.

Introduction

In recent years, high-performance liquid chromatography (HPLC) has become the most widely used method for the separation of lipid classes, for phospholipids especially. Adsorption chromatography (normal or straight-phase) on silica gel has been used in most instances. The silica gel used for this purpose is a porous solid with a surface area inversely related to the size of the pores, 6 nm being standard. The adsorptive properties are due to silanol (hydroxyl) groups, which are attached to the surface and can be free or hydrogen bonded. In addition, there is water of hydration, which exists first in a strongly bound layer and then in one or more loosely bound layers on the surface. The latter can have a marked effect on reproducibility of separations, especially of nonpolar lipids, as it is readily removed, inadvertently or otherwise, by elution with dry solvents. To ensure reproducibility in retention times and resolution, it is better to arrange that only the strongly bound water layer remains. While this can be achieved by devising solvent-elution schemes to remove as much as is required to produce the desired adsorptivity in the column in gradient runs, it is harder with isocratic elution. Here the answer may lie in using hexane partly saturated with water as part of the mobile phase. Silica gel then presents a rather heterogeneous surface to an analyte, and control of its properties requires some skill and experience.

A number of stationary phases have therefore been manufactured with organic moieties bonded chemically to the silica gel with the intention of providing a more uniform surface to act as an adsorbent. Those with diol, nitrile, amino, and sulfonic acid residues appear to be of special value, and some interesting applications to lipid-class separations have been described.

The other important component of a chromatographic system is the mobile phase, but the composition may be restricted by the type of detector available to the analyst. Indeed, this has generally been the dominant factor governing the approach to the problem, since lipids in general lack chromophores that might facilitate spectrophotometric detection. For example, refractive index, ultraviolet (200–210 nm), and to a lesser extent refractive detectors are suited to small-scale isolation of particular lipid classes for analysis by other procedures, since they are nondestructive (although the refractive index detector can only be used with isocratic elution). They tend to be less useful for quantitative analysis of lipid classes, although some workers have employed them in this way after careful calibration. Some of the more successful quantitative analyses of simple lipids by means of HPLC have made use of detectors operating on the evaporative light-scattering principle, and these and other "universal" or "mass" detectors are now being used in large numbers of laboratories. The literature on this subject prior to 1987 has been reviewed elsewhere by the author (1), while the properties of detectors, especially those operating on the evaporative lightscattering principle, have been reviewed more recently (2). In this review, no attempt is made to cover the topic exhaustively. Instead, papers have been cited which appear to best illustrate the principles governing lipid-class separations and the various types of detection.

Separation of Simple Lipid Classes

In most of the published separations of simple lipid classes, adsorption chromatography with columns of silica gel has been used with a variety of different elution/detection systems. Applications of the Pye Unicam (Cambridge, U.K.) LCM2 transport detector, which is no longer manufactured, were among the first reports of separations of this kind. A silica-gel column with a gradient of ethanol into hexane-chloroform (9:1 v/v) were employed to effect good separations of triacylglycerols, diacylglycerols, sterols, free fatty acids, and monoacylglycerols in about 15 min, although a lengthy reequilibration period was necessary before the next sample could be analyzed (3). The method was applied to the simple lipids of soybean preparations, resulting in rectilinear detector responses to the standards, permitting excellent reproducibility in quantitative analyses. This work is now nearly 20 years old and has seen little improvement or utilization in that time, mainly because of detector problems. The same detection system was utilized in the analysis of the waxes of human sebum and similar lipids (4). Later, the method was applied to the analysis of the partial hydrolysis products of triacylglycerols (5), and a similar separation, making use of an evaporative light-scattering detector, has been described briefly (6).

A refractive index detector was utilized with a silica-gel column and isocratic elution with isooctane-tetrahydrofuran-formic acid (90:10:0.5 v/v) to separate most of the common simple lipid classes encountered in animal-tissue extracts, such as those from the liver (5). It was noteworthy that cholesterol esters, triacylglycerols, and cholesterol all gave symmetrical peaks. The system could be of value for the isolation of specific lipid classes on a small scale (1–2 mg) for further analysis. Although an attempt was made to use the technique quantitatively, the results were not reproducible because negative solvent peaks tended to interfere, the linearity of the response for each lipid class was not determined rigorously, and some variation in response with fatty acid composition was observed. However, others showed that this elution system could give acceptable accuracy with relatively simple mixtures, such as those obtained by commercial glycerolysis of seed oils, if an internal standard (ricinoleic acid) was used and a careful calibration was performed (8). The detector response was found to be rectilinear for up to 1 mg of glyceride.

Relatively few applications of UV spectrophotometry at 200–210 nm in the separation of simple lipids have been described, possibly because this form of detection is of limited value for quantification purposes. However, detection at 206 nm and a column of Porasil™ silica gel were used in the separation of simple lipids from serum and liver extracts (9). Isocratic elution with hexane-propan-2-ol-acetic acid (100:0.5:0.01 v/v) or hexane-butyl chloride-acetonitrile-acetic acid (90:10:1.5:0.01 v/v) gave good separations of cholesterol esters, triacylglycerols, free fatty acids, cholesterol, and some unidentified compounds. Some partial fractionation according to the nature of the fatty acid constituents of each lipid class was observed with the second solvent system, possibly as a consequence of partition effects with a layer of solvent molecules adhering to the silica-gel adsorbent, and an even more marked effect of this kind was observed by others (10). Partial separations of molecular species have no real analytical value as they are not easy to reproduce; it is better to strive for a single sharp peak for each lipid class. Recovery of lipid classes from the column eluent was essentially complete, so fractions could be collected for quantification or for the analysis of the fatty acid constituents. Ultraviolet detection has also been used in the fractionation of leaf waxes (11).

Isocratic elution of a column of cyanopropyl-bonded phase with 0.05% propan-2-ol in hexane gave good resolution of simple lipids from sheep livers (12). As spectrophotometric detection at 210 nm was used, unsaturated terpenoid "lipids," such as retinol, vitamin E, dolichol, ubiquinone, and their esterified forms, showed up especially prominently. Much better reproducibility of retention times was observed in this work, than with the more usual silica-gel adsorbents, and this has also been the author's experience. Unfortunately, the free fatty acid fraction and other acidic lipids appear to be adsorbed rather strongly. Excellent results were obtained in the author's laboratory with this type of column (Spherisorb™ S3CN) and gradient elution with light-scattering detection, removing the adsorption effects by adding a little acetic acid to the injection solvent, as illustrated in Figure 6.1 (13).

A great deal of work has been done to establish conditions for particular simple lipid separations using HPLC, but inevitably many compromises have been made because of detector limitations. It is to be hoped that more widespread availability of mass-selective detectors will lead to further developments.

Analysis of Phospholipids

To overcome detection difficulties, two basic solvent systems transparent at UV wavelengths in the range of 200–210 nm were developed for phospholipid separations, hexane-propan-2-ol-water and acetonitrile-water (sometimes with added methanol) mixtures, and these still find almost universal application today. The use of the latter was first described in 1976, acetonitrile-methanol-water [61:21:4 v/v (14)]. In this system, phosphatidylethanolamine elutes before phosphatidylcholine which is before sphingomyelin. Indeed all the choline-containing phospholipids are well resolved. A further benefit is that acidic lipids, such as phosphatidylinositol

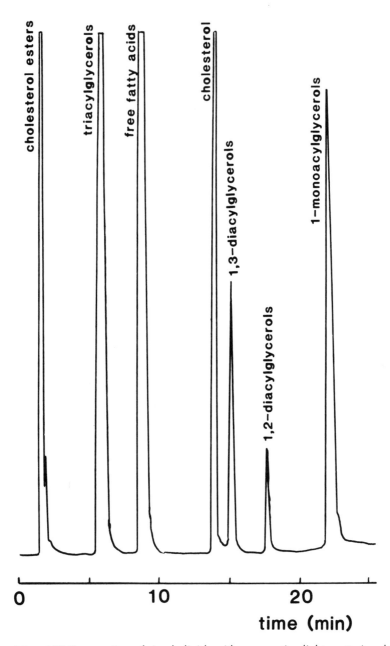

Figure 6.1. HPLC separation of simple lipids with evaporative light-scattering detection. The mobile phase was a gradient of hexane and methyl-*tert*-butyl ether with an HPLC column (100 × 4.6 mm) containing Spherisorb™ S3CN (13). Reproduced by permission of the *Journal of High Resolution Chromatography*.

and phosphatidylserine, are eluted with relative ease before phosphatidylethanolamine. For example, by isocratic elution with acetonitrile-methanol-sulfuric acid (100:2.1:0.05, v/v), phosphatidic acid, cardiolipin, phosphatidylinositol, and phosphatidylserine were clearly separated from each other before phosphatidylethanolamine emerged (15). Even better results should be attainable with gradient elution, although sulfuric acid is not really recommended as a constituent of a mobile phase.

Mobile phases based on hexane-propan-2-ol-water have also been used in many laboratories since their introduction in 1977 (16,17). Again, phosphatidylethanolamine elutes before phosphatidylcholine, but the other choline-containing lipids, such as sphingomyelin and phosphatidylcholine, tend to be less well resolved. The acidic lipids, such as phosphatidylinositol, phosphatidylserine, and phosphatidic acid, are separated from each other, but in this instance they emerge between phosphatidylethanolamine and phosphatidylcholine. By adding additional solvents to the basic mixture or by using gradients, it has been possible to improve the resolution of the choline-containing lipids. This system has proved easier to adapt to simultaneous separation of simple lipids and glycolipids than the system based on acetonitrile.

These results are summarized in Table 6.1. It is evident that the selectivities of the solvents used in the mobile phase can exert a marked effect on the separation of individual phospholipids, and in particular it can change the order of elution of specific components.

In addition, it is essential to note that phospholipids are ionic molecules and require a counter-ion in solution. If they are obtained from tissues by a conventional chloroform-methanol extraction, including a wash with sodium chloride solution, sodium will be the predominant counter-ion. However, special precautions may be necessary to ensure that small amounts of other ions do not remain, since different ionic forms of the acidic phospholipids may have different mobilities on chromatography.

TABLE 6.1
Elution Order of Phospholipids in Mobile Phases Based on Acetonitrile and Propan-2-ol[a]

Acetonitrile-Based	Propan-2-ol-Based
Phosphatidic acid	Cardiolipin
Cardiolipin	Phosphatidylethanolamine
Phosphatidylinositol	Phosphatidylinositol
Phosphatidylserine	Phosphatidylserine
Phosphatidylethanolamine	Phosphatidic acid
Phosphatidylcholine	Phosphatidylcholine
Sphingomyelin	Sphingomyelin
Lysophosphatidylcholine	Lysophosphatidylcholine

[a]Note: There may be some modification to the order given (especially of cardiolipin), depending on the nature of other solvents and any ionic species in the mobile phase.

One practical method to avoid difficulties during HPLC is to add counter-ions or acids to the mobile phase. Sulfuric and phosphoric acids often have been used, but apart from damaging HPLC equipment they will bring about complete destruction of any plasmalogens present. The author has used a serine buffer (0.5 mM, pH 7.5), which is relatively innocuous, and appeared to give well-shaped peaks with difficult analytes, such as phosphatidylserine and phosphatidylinositol (18), while others have used acetic acid (19). In spite of the presence of ionic species in the mobile phase, it was still possible to use evaporative light-scattering detection. This was also possible with mobile phases containing ammonia (20), when such addition affected the selectivity of the separation. On the other hand, column life was greatly reduced because silica dissolves at a high pH. There must be opportunities to test a range of organic and inorganic ionic species, at controlled pH values for their suitability as counter-ions. There may also be the potential to change the selectivity of the mobile phase to effect specific separations, especially of the acidic phospholipids.

Of the large number of procedures of this kind described, that of Patton and coworkers (21) appears particularly effective, and has been adopted by many others (Figure 6.2). It has the merit of employing isocratic elution, thus reducing the quantity of costly equipment. Hexane-propan-2-ol-25 mM phosphate buffer-ethanol-acetic acid was the mobile phase with silica gel as stationary phase, and detection was at 205 nm. With a rat liver extract, phosphatidylethanolamine eluted just after the neutral lipids, and was followed by each of the acidic lipids (i.e., phosphatidic acid, phosphatidylinositol, and phosphatidylserine), then by diphosphatidylglycerol, and the individual choline-containing phospholipids. Only phosphatidylcholine and sphingomyelin overlapped slightly. There was a fairly long elution time, but most of the important phospholipid classes did emerge eventually.

As each component was eluted, it was collected, washed to remove the buffer, and determined by phosphorus assay. In addition, the fatty acid composition of each lipid class was obtained with relative ease, by gas-liquid chromatographic (GLC) analysis after transmethylation. Direct quantification from the UV trace has been attempted by others (22–24), but is dependent on the degree of unsaturation of each lipid, and very careful calibration using standards identical to the lipids from the samples is essential.

In contrast, with eluents based on acetonitrile-methanol, the acidic lipids elute ahead of phosphatidylethanolamine and phosphatidylcholine and each of the choline-containing phospholipids is especially well resolved. A practical isocratic elution system of acetonitrile-methanol-water (50:45:6.5 v/v) was used to obtain an excellent resolution of phosphatidylethanolamine, lysophosphatidylethanolamine, phosphatidylcholine, sphingomyelin, and lysophosphatidylcholine, see Figure 6.3 (25). Others have preferred gradient elution, since this tends to give sharper peaks in relatively shorter times (26,27).

Both of these basic types of solvent systems were first described over 15 years ago. A few alternatives have been described since then, but the advent of light-scattering detectors has opened up many new opportunities, since the primary limita-

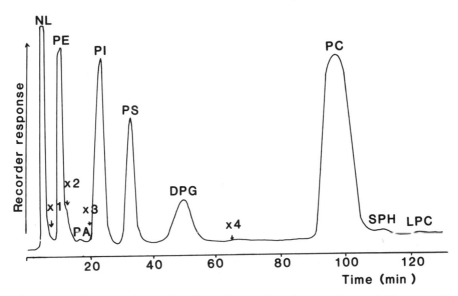

Figure 6.2. Isocratic elution of rat liver phospholipids from a column of silica gel with hexane-isopropanol-25 mM phosphate buffer-ethanol-acetic acid (367:490:62:100:0.6 v/v) as the mobile phase at a flow rate of 0.5 mL/min for the first 60 min then 1 mL/min, and with spectrophotometric detection at 205 nm (21). Reproduced by permission of Patton et al. and the *Journal of Lipid Research*. Redrawn from the original publication. Abbreviations: NL, neutral lipids; PE, phosphatidylethanolamine; PA, phosphatidic acid; PI, phosphatidylinositol; PS, phosphatidylserine; DPG, diphosphatidylglycerol; PC, phosphatidylcholine; SPH, sphingomyelin; LPC, lysophosphatidylcholine; X1, X2, X3, and X4, unidentified lipids.

tion now is the volatility of the solvent. There is a host of solvents—ethers, alcohols, ketones, aromatic, and halogen or other hetero-atom-containing compounds that should be tried. Potential remains to change the selectivity of separations further and to enhance the capability to isolate specific phospholipid components.

It is also possible to change the selectivity by utilizing a stationary phase other than silica gel. For example, one of several stationary phases with organic moieties bonded chemically to silica gel had benzene sulfonate residues as the functional group and has been used in phospholipid separations with isocratic elution (28). A column of Partisil™-SCX and eluted with acetonitrile-methanol-water at a flow rate of 2.5 mL/min was used to effect separation of the main ethanolamine- and choline-containing phospholipids of animal tissues. Phosphatidylinositol eluted at the solvent front. There was a particularly good separation of phosphatidylcholine and sphingomyelin, and no ions appear to be required in the mobile phase. While spectrophotometry at 203 nm was used to detect the components, a phosphorus assay was preferred for quantification purposes. Yet others used a bonded phase in

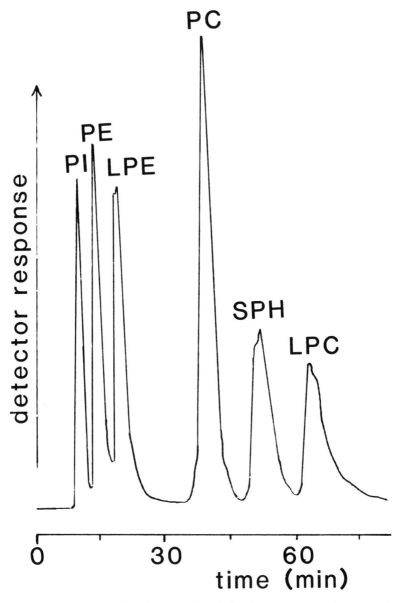

Figure 6.3. Separation of a reference phospholipid mixture on a silica-gel column, and by isocratic elution with acetonitrile-methanol-water (50:45:6.5 v/v) at a flow rate of 0.4 mL/min. Detection was by phosphate analysis (25). Reproduced by permission of Kaitaranta and Bessman and *Analytical Chemistry*. Redrawn from the original publication. Abbreviations: LPE, lysophosphatidylethanolamine; and see Figure 6.2.

which an aminopropyl group was the feature responsible for the separation (29–32). Thus, with acetonitrile-methanol-acetic acid as the mobile phase (with UV detection at 205 nm), choline-containing lipids eluted before those containing ethanolamine, while acidic phospholipids were retained very strongly (32). Others have used cyanopropyl-bonded stationary (33) or diol (19) phases to good effect. Again, there must be further opportunities to improve the separation of phospholipids by using such stationary phases with a wider range of mobile phases.

It should be noted that analysts do not always require a perfect analytical system that resolves every single lipid class. For example, if the interest is in the properties of a specific phospholipid class, it may simply be necessary to optimize the elution scheme so that the compound of interest is isolated in a relatively pure state; resolution of other components can be ignored. Then knowledge of factors controlling the separations is invaluable. However, an ideal analytical system should give sharp well-resolved peaks for all the main phospholipids in tissue extracts, especially the acidic and choline-containing components, and it should be stable and reproducible for months of continuous use. Ideally, it should be adaptable to the simultaneous analysis of simple lipids and of glycolipids.

Separation of Simple and Complex Lipids in a Single Chromatographic Run

It is possible to separate individual simple lipids followed by each of the main phospholipid classes in a single run, as was first demonstrated in the laboratory of the late Orville Privett [reviewed by Christie (1)]. In a sense, this work was ahead of its time, since a homemade detector of the transport-flame ionization type was employed with columns of silica gel inferior to those presently available.

The author obtained a more practical separation of this type by using a complex ternary gradient system and evaporative light-scattering detection (34). With this detector, there was the capacity to use complicated gradients, greatly increasing the available options. For example, it was possible to take the lipids of a sample, such as rat kidney, and separate components ranging in polarity from cholesterol esters to lysophosphatidylcholine in a single run on a short column of 3 µm silica gel (Figure 6.4). All the main components were resolved (on the 0.2–0.4 mg scale) in only 20 min. The baseline was absolutely steady, although there were abrupt changes of solvent at some points. In essence, isooctane (or hexane) was used to elute the cholesterol esters and triacylglycerols separately, then propan-2-ol was introduced to remove cholesterol and other simple lipids before water was introduced to elute each of the phospholipids in turn. The addition of chloroform improved the resolution of the choline-containing phospholipids. Finally, the polarity of the gradient was reversed to remove all of the water and to restore the column to its original activity over an additional 10 min. We were able to run 15 samples per 8-hour day, and to inject more than 1000 samples onto a column with no loss of resolution.

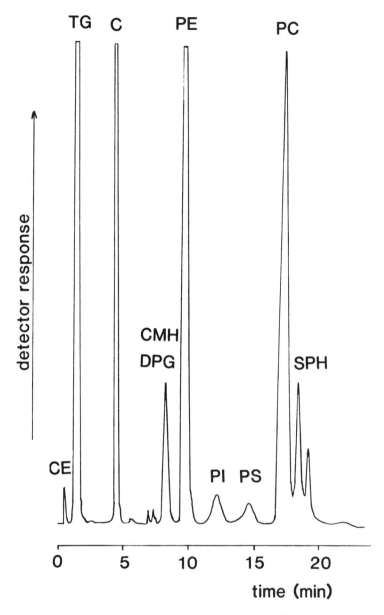

Figure 6.4. Separation of rat kidney lipids (0.35 mg) by HPLC on a column (5 × 100 mm) of Spherisorb™ silica gel (3 μm particles) with evaporative light-scattering detection, with a gradient elution scheme based on hexane, propan-2-ol and water (18). Reproduced by permission of *Journal of Chromatography*. Abbreviations: CE, cholesterol esters; TG, triacylglycerols; C, cholesterol; CMH, ceramidemonohexoside; and see Figure 6.2.

Quantification can be a problem, since the sensitivity of the evaporative light-scattering detector tends to drop rapidly for small components. When the lipid droplets in the airstream are too small, they no longer reflect or refract the light. However, the author's experience is that careful calibration of the mass detector can give results as good as most alternative methods. The quantification problem can be partially resolved by using phosphatidyldimethylethanolamine as an internal standard, as described in an analysis of the phospholipids of milk (35).

It was also determined that longer column life and sharper peaks for the acidic components were obtained if ionic species, for example, serine, are buffered with ethylamine to the correct pH (7.50) (18). A concentration of 0.5 mM in water had virtually no effect on the baseline during light-scattering detection.

This basic separation procedure has been adapted for fluorescence detection (36). Others have modified the gradient-elution procedure to improve the resolution of phospholipid and cerebroside components (37). In this work, the Varex ELSD II light scattering detector, a much more sensitive detector than the ACS model, was employed and produced calibration curves that were similar in shape to those from the ACS model, although they did differ appreciably otherwise. In this instance, N-oleoylethanolamine was utilized as an internal standard. The Varex detector and the basic solvent system gave excellent results in the analysis of lipids from human biopsy specimens (38). It was not found necessary to add ionic species to the mobile phase, but silica-saturation columns were incorporated into each of the solvent lines and this must have helped. A further modification of the basic procedure enabled automatic injection and computer handling of the data to be employed in routine analyses of lipid extracts (39). The original procedure has also been adapted for the analysis of phospholipids alone (40), with cholesterol being employed as an internal standard. The response of the detector for each phospholipid relative to the cholesterol standard was linear over a wide range.

Plant lipids present added difficulties to analysts because of their high content of glycolipids of various kinds in addition to phospholipids. Total neutral lipids and individual phospholipid classes from plant lecithins have been well separated on a diol phase, Lichrospher™ 100 diol (propanediol chemically bonded to silica), by gradient elution with hexane-propan-2-ol-water-acetic acid mixtures of increasing polarity (19). This elution system has recently been combined with plasma spray tandem mass spectrometry for analysis of lipids (41). However, progress has been made in developing methods for resolving individual simple lipids, glycolipids, and phospholipids from plant sources.

For example, Christie and Morrison developed an altered elution scheme to separate the glycolipids in cereal lipids on a column of silica gel (42). By eluting with hexane-butanone-acetic acid before starting the phospholipid gradient, both groups of compounds were successfully resolved. Others adapted the ternary elution system, developed in the author's laboratory, for comprehensive analysis of cereal lipids, including the galactosyldiacylglycerols and *N*-acylphosphatidylethanolamine (43). In essence, the modification involved using acetic acid in the mobile phase and 10 μm silica in the column, permitting a more rapid and complex gradient.

Moreau and colleagues developed a different modification of the basic procedure with a similar goal, involving a longer HPLC column, extended gradients, and reduced flow rates. Initially they used a transport flame-ionization detector (44), but subsequently it was shown that much better results were possible with evaporative light-scattering detection [Figure 6.5 (45)]. In this instance, not only were the main

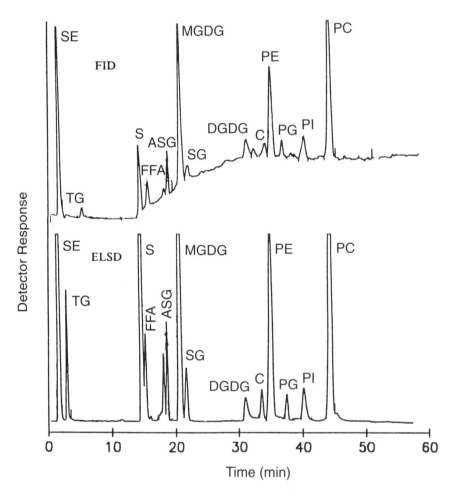

Figure 6.5. Comparison of the analysis of lipid classes from corn coleoptiles by HPLC with a Tracor transport-flame ionization detection (FID) and Varex evaporative light-scattering detection (ELSD) (B), each with a total of 125 µg of lipid (45). Reproduced by permission of Moreau and Portland Press. Abbreviations: SE, sterol esters; S, sterols; FFA, free fatty acids; ASG, acylated sterol glycosides; MGDG, monogalactosyldiacylglycerols; SG, sterol glycosides; DGDG, digalactosyldiacylglycerols; C, cardiolipin; PG, phosphatidylglycerol; and see Figures 6.2 and 6.4.

simple lipids, galactosyldiacylglycerols, and phospholipids separated, but also the sterolglycosides and cerebrosides. The value of this approach has now been confirmed in another laboratory with analyses of wheat flour lipids (46). Hammond (47) has published brief details of a very different ternary gradient-elution scheme for the analysis of wheat flour lipids, and further details would be valuable.

Conclusions

A great deal of work has now been published on the separation of lipid classes from plant and animal tissues by using HPLC. The wider availability of evaporative light-scattering detectors means that the opportunity to test novel solvent systems has been greatly increased. Similarly, there are many chemically bonded stationary phases that should be investigated.

The major problem remaining appears to be in reproducing the analytical results of acidic phospholipids, especially phosphatidylserine, phosphatidylinositol, and phosphatidic acid. The answer will probably be to find a suitable noncorrosive ionic species to add to the mobile phase. Presently, solvents containing ammonia appear to give the best results but result in reduced column life. An alternative to this is required urgently.

Acknowledgment

This paper is published as part of a program funded by the Scottish Office Agriculture and Fisheries Department.

References

1. Christie, W.W. (1987) *High-Performance Liquid Chromatography and Lipids*, Pergamon Press, Oxford.
2. Christie, W.W. (1992) In *Advances in Lipid Methodology—One*. Edited by W.W. Christie, Oily Press, Ayr. pp. 239–71.
3. Kiuchi, K., Ohta, T., and Ebine, H. (1975) *J. Chromatogr. Sci. 13:*461–66.
4. Aitzetmuller, K., and Koch, J. (1978) *J. Chromatogr. 145:*195–202.
5. Aitzetmuller, K. (1977) *J. Chromatogr. 139:*61–8.
6. Perrin, J-L., Prevot, A., Stolyhwo, A., and Guiochon, G. (1984) *Rev. Franc. Corps Gras 31:*495–501.
7. Greenspan, M.D., and Schroeder, E.A. (1982) *Anal. Biochem. 127:*441–48.
8. Ritchie, A.S., and Jee, M.H. (1985) *J. Chromatogr. 329:*273–80.
9. Hamilton, J.G., and Comai, K. (1984) *J. Lipid Res. 25:*1142–48.
10. Schlager, S.I., and Jordi, H. (1981) *Biochim. Biophys. Acta 665:*355–58.
11. Zabkiewicz, J.A., and Steele, K.D. (1982) *Chromatographia 16:*92–97.
12. Palmer, D.N., Anderson, M.A., and Jolly, R.D. (1984) *Anal. Biochem. 140:*315–19.
13. El-Hamdy, A.H., and Christie, W.W. (1993) *J. High Resolut. Chromatogr. 16:*55–57.
14. Jungalwala, F.B., Evans, J.E., and McCluer, R.H. (1976) *Biochem. J. 155:*55–60.
15. Islam, A., Smogorzewski, M., Pitts, T.O., and Massry, S.G. (1989) *Minor Electrolyte Metab. 15:*209–13.

16. Hax, W.M.A., and Geurts van Kessel, W.S.M. (1977) *J. Chromatogr. 142:*735–41.
17. Geurts van Kessel, W.S.M.; Hax, W.M.A.; Demel, R.A. and de Gier, J. 1977. *Biochim. Biophys. Acta 486:*524–30.
18. Christie, W.W. (1986) *J. Chromatogr. 361:*396–99.
19. Herslof, B., Olsson, U., and Tingvall, P. (1990) In *Phospholipids,* edited by I. Hanin, and G. Pepeu, Plenum Press, New York, pp. 295–98.
20. Abidi, S.L. (1991) *J. Chromatogr. 587:*193–203.
21. Patton, G.M., Fasulo, J.M., and Robins, S.J. (1982) *J. Lipid Res. 23:*190–96.
22. Briand, R.L., Harold, S., and Blass, K.G. (1981) *J. Chromatogr. 223:*277–84.
23. Heinze, T., Kynast, G., Dudenhausen, J.W., and Saling, E. (1988) *J. Perinat Med. 16:*53–60.
24. Heinze, T., Kynast, G., Dudenhausen, J.W., Schmitz, C., and Saling, E. (1988) *Chromatographia 25:*497–503.
25. Kaitaranta, J.K., and Bessman, S.P. (1981) *Anal. Chem. 53:*1232–35.
26. Nissen, H.P., and Kreysel, H.W. (1983) *J. Chromatogr. 276:*29–35.
27. Nissen, H.P., Topfer-Petersen, E., Schill, W.B, and Kreysel, H.W. (1983) *Fette Seifen Anstrichm. 85:*590–95.
28. Gross, R.W., and Sobel, B.E. (1980) *J. Chromatogr. 197:*79–85.
29. Kiuchi, K., Ohta, T., and Ebine, H. (1977) *J. Chromatogr. 133:*226–30.
30. Hanson, V.L., Park, J.Y., Osborn, T.W., and Kiral, R.M. (1981) *J. Chromatogr. 205:*393–400.
31. Chen, S.S-H., and Kou, A.Y. (1984) *J. Chromatogr. 307:*261–69.
32. Caruncio, V., Nicoletti, I., Frezza, L., and Sinibaldi, M. (1984) *Annal. Chim. 74:*331–39.
33. Samet, J.M., Friedman, M., and Henke, D.C. (1989) *Anal. Biochem. 182:*32–36.
34. Christie, W.W. (1985) *J. Lipid Res. 26:*507–12.
35. Christie, W.W., Noble, R.C., and Davies, G. (1987) *J. Soc. Dairy Technol. 40:*10–12.
36. Homan, R., and Pownall, H.J. (1989) *Anal. Biochem. 178:*166–71.
37. Lutzke, B.S,. and Braughler, J.M. (1990) *J. Lipid Res. 31:*2127–30.
38. Markello, T.C., Guo, J., and Gahl, W.A. (1991) *Anal. Biochem. 198:*368–74.
39. Redden, P.R., and Huang, Y.-S. (1991) *J. Chromatogr. 567:*21–27.
40. Juaneda, P., Rocquelin, G., and Astorg, P.O. (1990) *Lipids 25:*756–59.
41. Valeur, A., Michelsen, P., and Odham, G. (1993) *Lipids 28:*225–29.
42. Christie, W.W., and Morrison, W.R. (1988) *J. Chromatogr. 436:*510–13.
43. Carr, N.O., Daniels, N.W.R., and Frazier, P.J. (1989) In *Wheat End-Use Properties,* Proceedings of ICC Meeting, Helsinki. pp. 151–72.
44. Moreau, R.A., Asmann, P.T., and Norman, H.A. (1990) *Phytochemistry 29:*2461–66.
45. Moreau, R.A. (1990) In *Plant Lipid Biochemistry, Structure, and Utilization,* edited by P.J. Quinn, and J.L. Harwood, Portland Press, London. pp. 20–22.
46. Conforti, F.D., Harris, C.H., and Rinehart, J.H. (1993) *J. Chromatogr. 645:*83–88.
47. Hammond, E.W. (1989) *Trends Anal. Chem. 8:*308–13.

Chapter 7
Silver-Ion High-Performance Liquid Chromatography

William W. Christie

The Scottish Crop Research Institute, Invergowrie, Dundee, Scotland DD2 5DA, UK.

Introduction

The principle of silver-ion (or argentation) chromatography is very simple and is dependent on the fact that the π electrons of double bonds in lipid fatty acyl residues react reversibly with silver ions to form polar charge-transfer complexes; the greater the number of double bonds the stronger the complexation effect. The technique has been adapted to liquid-liquid partition, gas-liquid, column, and thin-layer chromatography (TLC), but most lipid researchers associate it with TLC. On the other hand, high-performance liquid chromatography (HPLC) recently has been shown to have much to offer in terms of the quality of resolution, convenience, and cleanliness, both of the sample and the laboratory environment. The mechanism, methodology, and applications to lipids have been reviewed by Nikolova-Damyanova (1).

In TLC, silver nitrate is simply incorporated into the aqueous slurry used to suspend the silica gel; the plates are spread and activated in the usual way, though some care is necessary to minimize exposure to light. With a suitable mobile phase, fully saturated lipids do not form complexes and will migrate to the top of the plate, those containing one monoenoic fatty acyl residue come next, and components of increasing unsaturation then follow. Many different lipid classes have been analyzed in this way, including methyl esters of fatty acids, sterol esters, and phospholipids and derivatives, but most industrial analysts will associate the technique with the fractionation of molecular species of triglycerides. With confectionery fats, it is possible to distinguish between monoenoic species in which the unsaturated fatty acid is in either the central or the outer positions of the glycerol moiety, and with care this can be extended to other molecular species (1).

There are many drawbacks to silver-ion TLC procedures, however, not the least being that silver ions are eluted from TLC plates and contaminate fractions in preparative applications. Silver nitrate leaves indelible stains on benches, equipment, and fingers. On the other hand, the equipment required is simple, inexpensive, and available in many laboratories, so undoubtedly the technique will continue to be used widely.

Because of the disadvantages of silver-ion TLC, in recent years analysts have looked to HPLC for the separation of molecular species of lipids, especially triglycerides. The technique has been employed in the reversed-phase mode primarily with bonded octadecylsilyl groups (ODS) as the stationary phase with mobile phases containing acetonitrile as the major component [reviewed elsewhere (2)]. Separations are then based on both the combined chainlengths and the number of double bonds in the fatty acid residues, one double bond reducing the effective chainlength of an acyl moiety by the equivalent of about two methylene groups. By using long columns and small particle sizes (3 µm), some remarkable separations of molecular fractions have been achieved from relatively simple oils and fats of commercial interest. Isocratic elution is often possible, so that a variety of detection and quantification procedures can be used.

Nonetheless, there are still some problems, and the most important is the identification of components. Because of the dual nature of the factors involved in the separation, intuitive identification of peaks on the recorder trace is generally not possible. This is not an insuperable problem with the relatively simple seed oils, but can lead to great difficulties with more complex traces, such as those from fish oils, milk fat, or any sample with a fatty acid composition out of the ordinary.

HPLC with Silver Nitrate/Silica Gel Adsorbent

In contrast to reversed-phase chromatography, silver-ion chromatography utilizes a single property, the degree of unsaturation, as the basis of separation. There have been many attempts to adapt silver-ion chromatography to HPLC and at first these met with only limited success. The approach of Hammond and co-workers was to impregnate an HPLC-grade silica gel with silver nitrate and only then pack it into columns [reviewed elsewhere (3)]. Both the grade of silica gel and the method of impregnation were found to be critical. However, isocratic and gradient elution procedures were developed with transport-flame ionization detection to obtain excellent resolution and direct quantification of the sample, for example confectionery fats. The chromatograms were relatively easy to interpret, in contrast to those from reversed-phase HPLC. Unfortunately, applications to highly unsaturated fats have not been published. Again, silver ions eluting continuously in the mobile phase were a major problem, although the life of the column could be prolonged by introducing a silver-saturation column into the solvent line. Nonetheless, this method does permit separation of some regiospecific isomers in addition to degree of unsaturation (4).

Silver-Resin Chromatography

As an alternative, macroreticular ion-exchange resins have been converted to the silver-ion form for use in column chromatography [reviewed in greater detail elsewhere (1)]. The silver ions are held in position by ionic bonds and are not easily removed from the support by simple elution with organic solvents. The first separations were relatively crude (5), but by grinding the ion-exchange particles (Amberlyst XN1010™) and selecting a cut in the region of 270–350 mesh, it proved possible to obtain good separations of methyl ester derivatives of fatty acids by degree of unsaturation (6,7). A mobile phase of methanol was preferred with an increasing amount of acetonitrile, since the latter has a high affinity for silver ions and can readily displace unsaturated solutes.

The virtue of this approach is that the columns can be used almost indefinitely without loss of resolution or sample capacity. They have a relatively high capacity (4–14 g) so are well suited to preparative-scale applications. On the other hand, the resin can only be used with a limited range of solvents, since it tends to expand and contract with changes in the mobile phase.

HPLC with Ion Exchangers in the Silver-Ion Mode

The approach to silver-ion HPLC adopted by the author was to load a silica-based ion-exchange medium (chemically bonded sulfonic acid groups) with silver ions (8). Preparation of the column involved a standard prepacked column with the appropriate stationary phase (Nucleosil™ 5SA) and introducing the silver ions via a Rheodyne injector while pumping water through the column. Finally, the aqueous phase was replaced with organic solvents. Only 50–80 mg of silver ions are bound to the stationary phase, but this is sufficient for many useful separations.

For much of the work with this column, an evaporative light-scattering detector (ELSD) was employed, since this permitted the use of complex gradients and mobile phases containing solvents such as dichloromethane and acetone. Applications of this type of detector in lipid analysis have been reviewed (9). It is also possible to use ultraviolet detection with appropriate fatty acid derivatives.

Some of the early experiments with related phases of the ion-exchange type linked to silver ions were disappointing, largely because methanol was incorporated into the mobile phase and a small proportion of residual free sulfonic acid moieties catalyzed *trans*esterification when lipids were on the column (10). However, if aprotic solvents are used, there are no problems of this kind. In the analysis of molecular species of triacylglycerols, the simplest elution scheme was a gradient of acetone into dichloroethane-dichloromethane, but this is suitable only for fats with a relatively small proportion of linoleic acid, such as sheep adipose tissue or milk fat (11). The trisaturated (SSS) species are eluted first and are followed by disaturated-monoenoic (SSM), saturated-dimonoenoic (SMM), and so forth. Indeed, it was possible to separate not only the usual fractions with saturated and *cis*-monoenoic residues but also those with *trans* double bonds. Therefore, the procedure may have some potential for the analysis of partially hydrogenated fats, such as those in margarines. A similar type of separation has recently been recorded for human-milk triacylglycerols, in which there are more molecular species (12).

Most samples of potential interest contain a higher proportion of linoleic acid, and this can be accommodated with the ternary gradient system simply by introducing acetonitrile into acetone after the first fractions are eluted, as was demonstrated with maize (corn), safflower seed oil, and rat adipose tissue; the last is illustrated in Figure 7.1. The retention time of one dienoic (D) acyl residue appears to be equivalent to about 2.5 monoenes. One triene [18:3(n-3)] is exactly equal to two

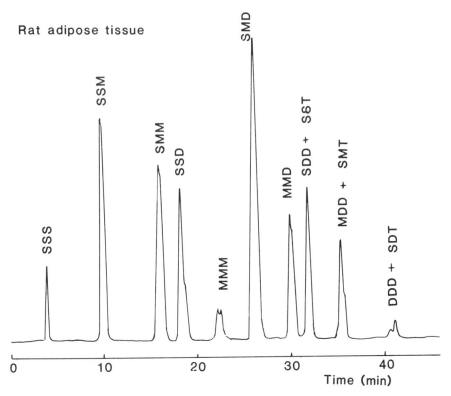

Figure 7.1. Separation of triacylglycerols from rat adipose tissue by HPLC on a Nucleosil™ 5SA column in the silver-ion form (11). Abbreviations: S, saturated; M, monoenoic; D, dienoic; T, trienoic fatty acyl residues. Reproduced by permission of the *Journal of Chromatography*.

dienes [18:2(n-6)], so there is some overlap of dienoic and trienoic fractions when α-linolenic acid is present in a sample. On the other hand, molecular species of triglycerides containing γ-linolenic acid, as in evening primrose oil, are retained less strongly, and the separation is quite distinctive (13). Thus, the order of elution of the triglyceride species with these systems is easy to understand, unlike HPLC in the reversed-phase mode.

A similar elution scheme was employed for palm oil and cocoa butter, and this type of analysis is well suited to confectionery fats where a separation of the important SSM species is obtained rapidly. Indeed, the resolution is good enough that there is ample opportunity to speed up the separation by using a shorter column, faster flow rate, or steeper gradient. Unfortunately, species of the type SMS are not separated from SSM according to the position of the acyl residue on the triglyceride molecule, although this can be accomplished on silica gel impregnated with silver nitrate as described previously.

In this work, quantification was accomplished by collecting fractions via a stream splitter, adding an internal standard, and *trans*esterifying for analysis of the methyl ester derivatives by gas chromatography. This permitted simultaneous identification and quantification of the fractions. Such procedures would be tedious in routine use, and after careful calibration with appropriate standards, there is no reason why the response of the mass detector should not be used directly. Of course, detectors of the transport-flame ionization type could also be utilized.

A more exacting test of the column was the seed oil of the meadowfoam plant (*Limnanthes alba*), which is unusual in that the triacylglycerols contain mainly C_{20} and C_{22} fatty acids with double bonds in positions 5 and 13, that is 5*c*-20:1, 5*c*-22:1, 13*c*-22:1, and 5*c*,13*c*-22:2, in addition to small amounts of the more common fatty acids. Triacylglycerols of this oil were resolved by HPLC in the silver-ion and reversed-phase modes, and then by the two techniques used in a complementary fashion (14). Silver-ion chromatography gave a distinctive resolution in which fractions differing solely in the position and chain length of a single monoenoic fatty acyl group were resolved (Figure 7.2). Reversed-phase chromatography also gave fractions containing single positional isomers, but the pattern was more difficult to discern since fractions containing 22:2 tended to overlap with those containing 20:1. Used in concert, more fractions were seen than when either technique was used on its own. This was also true for a recent analysis of human-milk triacylglycerols (12).

The methodology has been applied with considerable success to samples as highly unsaturated as linseed oil and fish oils. With the former [Figure 7.3 (11)], the most abundant single fraction is trilinolenin, with nine double bonds, and a simple progression of fractions having increasing numbers of double bonds are eluted until this species is reached. The ultimate challenge for a silver-ion column is a fish oil. Triacylglycerols from Atlantic herring (*Clupea harengus*) and other species have been fractionated using the author's silver-ion HPLC technique by extending the polarity range of the gradient (15). Resolution was excellent at first, when the least unsaturated molecules eluted; it was initially surprising to find appreciable amounts of SSS and SMM fractions, for example. Baseline resolution was achieved up to the MMD species, but it could no longer be sustained when molecules containing trienoic and more highly unsaturated fatty acids began to elute, because the wide range of positional isomers caused components to overlap. Nonetheless, the elution pattern was reproducible. Some valuable separations of species containing two saturated and/or monoenoic fatty acids and one polyenoic fatty acid were achieved, and fractions with up to 14 double bonds were resolved. Double-bond indices (average number of double bonds in each triacylglycerol molecule) were calculated to estimate the separations possible. Comparable results have now been published by others (16).

Many more fractions from the Atlantic herring oil were obtained when each of the fractions from the silver-ion column were subjected to HPLC in the reversed-phase mode, again illustrating again the resolving power of the techniques when used in concert (17). On the other hand, each fraction from the reversed-phase column still contained a large number of distinct molecular species. High-performance

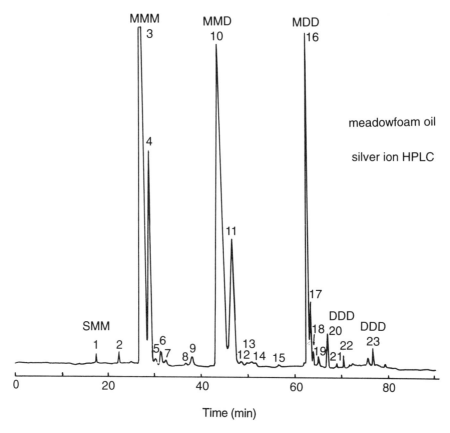

Figure 7.2. Separation of triacylglycerols from meadowfoam oil (*Limnanthes alba*) by HPLC on a Nucleosil™ 5SA column in the silver-ion form (11). Peak 3 is tri-5-20:1, for example; in peaks 4 to 9, one of the 5-20:1 residues is replaced by a monoenoic fatty acid with the double bond in a different position. Abbreviations: see Figure 7.1. Reproduced by permission of the *Journal of the American Oil Chemists' Society*.

liquid chromatography in combination with mass spectrometry would undoubtedly be extremely valuable in identifying individual components. The technique has yet to be applied to phospholipids or their diacylglycerol moieties, but it has given excellent results with cholesterol esters (18).

This methodology is also very useful for the resolution of simple fatty acid esters into discrete fractions. For example, simple gradient-elution schemes were devised to separate methyl ester derivatives of fatty acids into fractions with 0–6 double bonds (8,19). This particular separation has proved to be useful for simplifying complex mixtures of fatty acids for structural analysis by gas chromatography-mass spectrometry (GC–MS). For example, fatty acids from marine invertebrates (19–21), algae (22), animal tissues (23), cheese [saturated and unsaturated oxo fatty

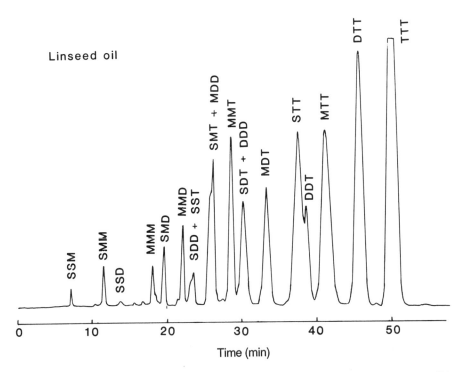

Figure 7.3. Separation of triacylglycerols from linseed oil by HPLC on a Nucleosil™ 5SA column in the silver-ion form (11). Abbreviations: see Figure 7.1. Reproduced by permission of the *Journal of Chromatography*.

acids (24,25)], seed oils (14,26–28), and heated frying oils [cyclic fatty acids (29)]. Perhaps the most powerful demonstration of the value of this combination of techniques was an application to the fatty acids of *Dysidea fragilis*, a sponge from the Black Sea, in which over a hundred distinct fatty acids were fully characterized (30). The silver-ion chromatogram is illustrated in Figure 7.4. The fatty acids found included mono- and multi-methyl-branched isomers (some with double bonds), cyclic, mono-, di-, and polyunsaturated (methylene-interrupted) fatty acids of many different biosynthetic families, and di- and trienoic acids with several methylene groups between the double bonds; they ranged in chain length from C_{13} to C_{27}.

Much better resolution of positional and geometrical isomers of fatty acid derivatives was achieved by isocratic elution with dichloromethane-dichloroethane [1:1 v/v (31)]. Phenacyl as opposed to methyl esters appeared to give better resolution and could be detected by UV absorption. For example, phenacyl esters of oleic and vaccenic acids and their *trans* analogues (Figure 7.5), three of the four possible geometrical isomers of linoleic acid, and six of the eight possible linoleate isomers were separable by this means. When the method was applied to the monoenoic esters from partially hydrogenated soybean oil, a substantial, if incomplete, resolu-

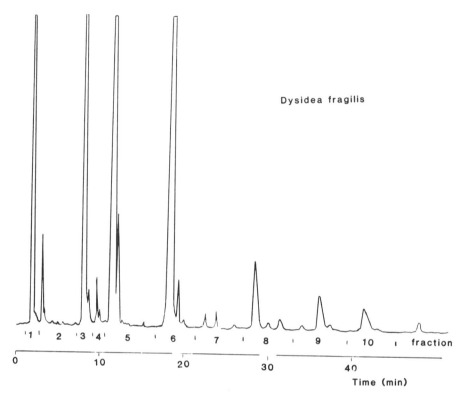

Figure 7.4. Separation of methyl esters of fatty acids from the sponge *Dysidea fragilis* by HPLC on a Nucleosil™ 5SA column in the silver-ion form (30). Fraction 1, saturated components; Fraction 2, *trans*-monoenes; Fraction 3, *cis*-monoenes; Fraction 5, dienes; Fraction 6, nonmethylene-interrupted trienes; Fraction 7, methylene-interrupted trienes; Fractions 8, 9, and 10 = tetra-, penta-, and hexaenoic components, respectively. Reproduced by permission of *Lipids*.

tion of the complex mixture of *cis* and *trans* positional isomers was possible. The potential for quantitative analysis of *trans* isomers was explored, and while reasonable results were obtained, the methodology appears more suited to separation of specific isomers. On the other hand, it was possible to use a short HPLC column (5 cm × 4.6 mm i.d.) of Nucleosil™ 5SA in the silver-ion form for rapid (5–10 min) separation of *cis* and *trans* isomers of methyl esters as a relatively simple and accurate method of determining the *trans*-monoenoic acid content of fats and oils (32). The preferred technique was to analyze the samples by GC first to determine the relative proportions of saturated and monoenoic components, and then to isolate the saturated esters with the *trans*-monoenes as a fraction for reanalysis by GC. The precision was better than by Fourier transform infrared spectroscopy. At present, there is no definitive method for determining the *trans* content of fats and oils, but

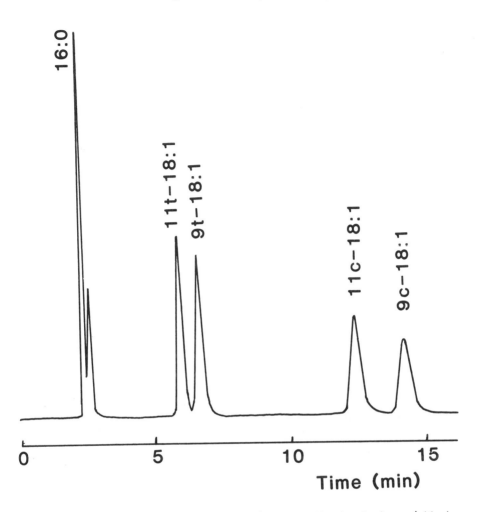

Figure 7.5. Separation of phenacyl esters of 9-*trans*-, 11-*trans*, 9-*cis*-, and 11-*cis*-octadecenoic acids by HPLC on a Nucleosil™ 5SA column in the silver-ion form (31). Reproduced by permission of the *Journal of Chromatography*.

most analysts utilize some form of silver-ion chromatography when great accuracy is required (33). Silver-ion HPLC in combination with GC has the potential to be a suitable reference method.

The chromatographic behavior of simple derivatives of series of isomeric fatty acids has also been investigated systematically (34). Baseline separation of the three common naturally occurring octadecenoic acids, the 6-, 9-, and 11-isomers (as phenacyl esters), was readily obtained with a mobile phase of 1,2-dichloroethane-dichloromethane (1:1 v/v) with acetonitrile (0.025%) as a polar modifier. Fixed

TABLE 7.1
Relative Capacity Factor Ratios, k''^a of Methyl and Phenacyl Esters of Isomeric cis-Monoenoic Fatty Acids on a Silver-Ion Column

Fatty Acid	k'' Methyl Ester	Phenacyl Ester
3-18:1	0.7	1.0
4-18:1	1.3	2.7
5-18:1	3.2	3.7
6-18:1	2.7	4.9
7-18:1	1.8	5.1
8-18:1	2.9	4.6
9-18:1	2.9	3.8
10-18:1	2.9	3.7
11-18:1	2.6	3.2
12-18:1	2.6	3.1
13-18:1	2.5	3.0
14-18:1	2.4	3.2
15-18:1	2.5	3.3
16-18:1	1.9	2.6
17-18:1	2.0	3.0

[a] $k'' \dfrac{k'(\text{analyte})}{k'(\text{octadecene})}$.

Source: Nikolova-Damyanova et al. (34).

temperatures were essential for reproducible retention times, but contrary to earlier views, lowering the temperature did not have a direct effect on improving the resolution; it did so indirectly simply by increasing the time of elution. With isomeric cis-octadecenoates, it was confirmed that better separations were possible with methyl as opposed to phenacyl esters.

In methyl esters when the double bond was close to the carboxyl group, the retention time was lower, reaching a maximum for 5- and 6-18:1, falling for 7-18:1, and almost plateauing for the 8–14-isomers; a further small maximum was evident for the 15-isomer. In contrast, there tended to be much greater differences between the chromatographic behavior of isomers when they were in phenacyl ester form especially when the double bonds were relatively close to the carboxyl group (Table 7.1). There was a maximum in this instance at the 7-18:1 derivative, and in general there was a better resolution of isomers in which the double bond was in positions 3–11, although not beyond that point. Both the proximal and terminal part of a monoenoic molecule have some potential to influence the interaction between silver ions and the double bond, but it was confirmed that the distance between the carboxyl group and the double bond was the more important.

Supercritical Fluid Chromatography in the Silver-Ion Mode

Supercritical fluid chromatography (SFC) is only beginning to make an impact in the analysis of lipids [reviewed elsewhere (35)], but it has been successfully used for silver-ion chromatography by adapting the HPLC methodology described previously. Thus, a fused silica column packed with Nucleosil™ 5SA was converted to the silver form in situ, and this was used for separation of molecular species of triacylglycerols (36–38) and of methyl ester derivatives (39). The mobile phase was carbon dioxide with acetonitrile as a polar modifier (with a little 2-propanol to ensure homogeneity) and with UV detection. In essence, the chromatograms resembled those published for HPLC, although the nature of the detector response was very different from that of ELSD. A miniaturized version of the latter has now been developed, however, and this affords better baseline stability and improved chromatograms (40). Separation of lipids by degree of unsaturation recently has been accomplished by means of SFC with permanganate-impregnated anion-exchange columns, suggesting that other ligands which form charge-transfer complexes might be worthy of investigation (41,42).

Solid-phase Extraction in the Silver-Ion Mode

Although it can by no means be termed high-performance liquid chromatography, some interesting separations of lipids have been achieved by solid-phase extraction (SPE) using methodology analogous to the HPLC procedures described previously [reviewed elsewhere (43)]. Christie (44) showed that Bond Elut™ columns packed with a silica-based benzenesulfonic acid medium could be converted to the silver-ion form and used to achieve excellent separations of methyl ester derivatives of fatty acids. The column was loaded with silver ions in situ by eluting with a silver nitrate solution, before it was flushed with organic solvents. Dichloromethane, acetone, and acetonitrile in various proportions were employed in the optimal elution scheme to obtain satisfactory resolution of components with 0–6 double bonds. Substantial changes in solvent composition were utilized at each step, and there was very little cross-contamination especially with the early fractions. The procedure has been applied to the analysis of fatty acids in soil lipids (45). A similar method was described independently by Ulberth and Achs (46).

The procedure was adapted for the analysis of molecular species of triacylglycerols from relatively saturated fats (47) and for cholesterol esters (18). With the former, dichloromethane-methyl acetate mixtures gave good resolution of species (47), while cholesterol esters were eluted under the same conditions as methyl esters, since the double bond in the sterol moiety did not appear to influence the separation (18).

Christie (unpublished) also tried metal salts other than silver in such SPE columns with methyl esters of fatty acids. No detectable resolution was obtained with rhodium. In contrast, platinum and palladium gave clean fractions containing saturated esters, but no conditions could be devised to recover any of the unsaturated components.

Silver-Ion Chromatography with Reversed-Phase HPLC Columns

While is possible to obtain some distinctive separations of lipids by employing HPLC with a reversed-phase column and silver ions in the mobile phase (48–51), the disadvantages of using a corrosive solvent mixture of this kind are such that the method can only be recommended when no alternative is available.

The Mechanism of Silver-Ion Chromatography

The mechanism of the interaction between silver ions and double bonds involves formation of a σ bond by overlap of the filled π orbital of the olefin with the free *s* orbital of silver(I), and a π bond formed by overlap of the vacant antibonding π orbitals of the olefin with filled *d* orbitals of silver(II) (52). It is evident that the strength of the complex is determined by accessibility of the electrons in the filled orbitals and by steric inhibition of the orbitals. However, the mechanism of the interaction between unsaturated centers and silver ions in practical chromatographic systems is poorly understood.

In most of the published work with TLC and silver nitrate incorporated into a silica-gel *G* layer, R_f values of different components are highly variable. They are dependent on factors such as the proportion of silver nitrate in the layer and its degree of hydration, atmospheric humidity and temperature, as well as the nature of the mobile phase, and such factors are not easily controlled. In contrast, the stable silver-ion column for HPLC described previously, with silver ions linked via ionic bonds to phenylsulfonic acid moieties which in turn are bound to a silica matrix, has relatively well-defined properties and chemistry. It is possible to control many of the chromatographic parameters, especially the composition and flow rate of the mobile phase, as well as the column temperature, with a high degree of accuracy. Therefore, this system was employed to obtain reproducible numerical values to describe retention characteristics in order to obtain a better understanding of the quantitative nature of the interactions between the double bonds in lipids and silver ions (34).

For example, the effect of double-bond position in isomeric monoenes on silver-ion retention was investigated quantitatively as described previously. Analogous results were obtained with a partial series of methylene-interrupted *cis*, *cis* dienes. Relative capacity factors values (k") were obtained for some nonmethylene-interrupted octadecenoic acids and they are listed in Table 7.2. The allenic compound, 9,10-18:2 and a conjugated diene (9-*cis*,11-*trans*-18:2) were retained to a lesser extent than oleate, presumably because of a very weak interaction between silver ions and the delocalized π electrons of the *bis*-double-bond systems. In contrast, isomers with more than one methylene group between the double bonds were very strongly retained, those with two methylene groups between the double bonds being held most strongly of all, followed by those with three and then those with

TABLE 7.2
Capacity Factor Ratios, $k''{}^a$ for Methyl and Phenacyl Esters of Isomeric Octadecadienoates.

Fatty Acid	k''	
	Methyl Ester	Phenacyl Ester
9,10-18:2	—	0.7
9-cis,11-trans-18:2	—	2.2
9-cis,12-cis-18:2	10.3	10.9
6-cis,10-cis-18:2	45.0	—
8-cis,12-cis-18:2	42.5	—
6-cis,11-cis-18:2	29.0	—
7-cis,12-cis-18:2	26.6	—
5-cis,12-cis-18:2	17.7	—

[a] $k'' = \dfrac{k'(\text{analyte})}{k'(\text{octadecene})}$.

Source: Nikolova-Damyanova et al. (34).

five. It was not possible to find isocratic elution conditions to permit the separation of a wide variety of polyunsaturated fatty acid derivatives in a reasonable time. To obtain quantitative information on the degree of retention of such compounds, therefore, capacity factors for a range of fatty acids with 3–6 double bonds were determined relative to that of the corresponding linoleate derivative, which was in turn related to that of the oleate derivatives (taken as 1.0). The results are listed in Table 7.3. It appeared that the capacity factors of polyunsaturated esters on silver-ion HPLC rose in an almost exponential manner with the number of double bonds, so that docosahexaenoate was retained up to 30 times as strongly as oleate. Again, phenacyl esters were held more strongly than methyl esters.

The results were explained in terms of a dual interaction with a silver-ion and both the π electrons of two double bonds, or those of one double bonds and the free electrons of the carbonyl moiety of methyl esters, the benzene ring, or oxygenated moieties of phenacyl esters. Certainly electron-rich esters, such as the phenacyl

TABLE 7.3
Ratio of the Capacity Factors k', of Methyl and Phenacyl Esters of Some Polyunsaturated Fatty Acids to That of Oleate

Fatty Acid	$k'_{\text{PUFA}}/k'_{18:1}$	
	Methyl Ester	Phenacyl Ester
9-18:1	1.0	1.0
9,12-18:2	2.9	3.5
6,9,12-18:3	7.1	9.4
9,12,15-18:3	7.9	9.9
5,8,11,14-20:4	10.3	14.2
5,8,11,14,17-20:5	16.7	19.7
4,7,10,13,16,19-22:6	21.5	30.1

Source: Nikolova-Damyanova et al. (34).

derivatives, were held much more strongly than were methyl esters when the double bond was within about eight carbons of the carboxyl group, and the elution patterns of series of isomers were very different. From 9–18:1 onwards, when the possibility of such a simultaneous interaction would seem to be less likely, there was no significant difference between methyl and phenacyl esters. The finding that esters with electron-withdrawing substituents, such as dinitrophenyl groups, were poorly resolved supported this hypothesis, as did the observation that monoenoic alcohols were also very strongly retained by the silver-ion HPLC column, and positional isomers were completely resolved.

X-ray crystallographic studies have shown that one silver-ion can interact with two unsaturated molecules simultaneously (53,54), or with two double bonds in a single molecule (55). An interaction between one silver-ion and two double bonds at the same time may explain the chromatographic behavior of dienoic derivatives with silver-ion HPLC. When the distance between the double bonds was optimal, as in a 1,5-*cis,cis*-diene system, fatty acids were very strongly retained, and the effect diminished as the number of methylene groups between the double bonds was varied. If the double bonds interacted singly with silver ions, it might have been anticipated that the kinetics of the system would have retentions comparable in magnitude to the sum of the individual parts. This theory of complexation predicts that a triene would be held twice as strongly as a diene, a tetraene three times as strongly, and so forth. Such simple relationships were not found, possibly because interactions with the ester moiety have to be taken into consideration and because the conformations of polyenes may permit some interactions between silver ions and double bonds that are remote from each other, via the formation of pseudocyclic structures, such as that recently proposed for arachidonic acid (56).

In contrast, in an experimental reversed-phase HPLC system with a mobile phase containing a silver salt, it was possible to compare the retention characteristics of a solute in the presence and absence of silver ions in the mobile phase, and simple ratios for the interactions of double bonds with silver ions were obtained (57). It was possible to calculate equivalent chain-length and fractional chain-length values, which appeared to indicate that double bonds and silver ions formed simple 1:1 complexes in the mobile phase. While this system had little practical value, as discussed previously, it did permit determination of the equilibrium constant for the formation of phenacyl oleate (or argentation constant), which was calculated to be in the range 0.059–0.067, depending on the experimental conditions.

Conclusions

Silver-ion chromatography is a mature technology that has gained fresh impetus from the development of stable columns for HPLC based on ion-exchange media. A large number of practical applications have now been described, and others will certainly follow. In addition, this methodology allows new insights into the mechanism of silver-ion chromatography.

Acknowledgment

This paper is published as part of a program funded by the Scottish Office Agriculture and Fisheries Department.

References

1. Nikolova-Damyanova, B. (1992). In *Advances in Lipid Methodology—One*. Edited by W.W. Christie, Oily Press, Ayr. pp. 181–237.
2. Christie, W.W. (1987). *High-Performance Liquid Chromatography and Lipids*, Pergamon Press, Oxford.
3. Hammond, E.W., and Irwin, J.W. (1988). In *HPLC in Food Analysis*. Edited by R. Macrae, Academic Press, London. pp. 95–132.
4. Jeffrey, B.S.J. (1991). *J. Am. Oil Chem. Soc. 68:*289–93.
5. Emken, E.A., Scholfield, C.R., and Dutton, H.J. (1964). *J. Am. Oil Chem. Soc. 41:*388–90.
6. Adlof, R.O. (1988). *J. Am. Oil Chem. Soc. 65:*1541–42.
7. Adlof, R.O. (1991). *J. Chromatogr. 538:*468–573.
8. Christie, W.W. (1987). *J. High Res. Chromatogr. Chromatogr. Commun. 10:*148–50.
9. Christie, W.W. (1992). In *Advances in Lipid Methodology—One*. Edited by W.W. Christie, Oily Press, Ayr. pp. 239–71.
10. Adlof, R.O., and Emken, E.A. (1980). *J. Am. Oil Chem. Soc. 57:*276–78.
11. Christie, W.W. (1988). *J. Chromatogr. 454:*273–84.
12. Winter, C.H., Hoving, E.B., and Muskiet, F.A.J. (1993). *J. Chromatogr. 616:*9–24.
13. Christie, W.W. (1991). *Fat Sci. Technol. 93:*65–6.
14. Nikolova-Damyanova, B., Christie, W.W., and Herslof, B. (1990). *J. Am. Oil Chem. Soc. 67:*503–07.
15. Laakso, P., Christie, W.W., and Pettersen, J.W. (1990). *Lipids 25:*284–91.
16. McGill, A.S., and Moffat, C.F. (1992). *Lipids 27:*360–70.
17. Laakso, P., and Christie, W.W. (1991). *J. Am. Oil Chem. Soc. 68:*213–23.
18. Hoving, E.B., Muskiet, F.A.J., and Christie, W.W. (1991). *J. Chromatogr. 565:*103–10.
19. Christie, W.W., Brechany, E.Y., and Stefanov, K. (1989). *Chem. Phys. Lipids 24:*116–20.
20. Stefanov, K.L., Christie, W.W., Brechany, E.Y., Popov, S.S., and Andreev, S.N. (1992). *Comp. Biochem. Physiol. 103B:*687–90.
21. Stefanov, K., Seizova, K., Brechany, E.Y., and Christie, W.W. (1992). *J. Nat. Products—Lloydia 55:*979–81.
22. Stefanov, K., Konaklieva, M., Brechany, E.Y., and Christie, W.W. (1988). *Phytochemistry 27:*3495–497.
23. Adkisson, H.D., Risener, F.S., Zarrinkar, P.P., Walla, M.D., Christie, W.W., and Wuthier, R.E. (1991). *Fed. Am. Soc. Exp. Biol. J. 5:*344–53.
24. Brechany, E.Y., and Christie, W.W. (1992). *J. Dairy Res. 59:*57–64.
25. Brechany, E.Y., and Christie, W.W. *J. Dairy Res. 61:*111–115.
26. Christie, W.W., Brechany, E.Y., and Shukla, V.K.S. (1989). *Lipids 24:*116–20.
27. Wretsenjo, I., Svensson, L., and Christie, W.W. (1990). *J. Chromatogr. 521:*89–98.
28. Griffiths, G., Brechany, E.Y., Christie, W.W., Stymne, S., and Stobart, K. (1989). In *Biological Role of Plant Lipids*. Edited by P.A. Biacs, K. Gruiz, and T. Kremmer, Plenum Press, New York. pp. 151–53.

29. Christie, W.W., Brechany, E.Y., Sebedio, J.L., and Le Quéré, J.L. *Chem. Phys. Lipids,* in press.
30. Christie, W.W., Brechany, E.Y., Stefanov, K., and Popov, S. (1992). *Lipids 27:*640–44.
31. Christie, W.W., and Breckenridge, G.H.M. (1989). *J. Chromatogr. 469:*261–69.
32. Toschi, T.G., Capella, P., Holt, C., and Christie, W.W. 1993. *J. Sci. Food Agric. 61:*261–66.
33. Firestone, D., and Sheppard, A. (1992). In *Advances in Lipid Methodology—One.* Edited by W.W. Christie, Oily Press, Ayr. pp. 273–322.
34. Nikolova-Damyanova, B., Herslof, B.G., and Christie, W.W. (1992). *J. Chromatogr. 609:*133–40.
35. Laakso, P. (1992). In *Advances in Lipid Methodology—One.* Edited by W.W. Christie, Oily Press, Ayr. pp. 81–119.
36. Demirbuker, M., and Blomberg, L.G. (1990). *J. Chromatogr. Sci. 28:*67–72.
37. Demirbuker, M., and Blomberg, L.G. (1991). *J. Chromatogr. 550:*765–69.
38. Demirbuker, M., Blomberg, L.G., Olsson, N.U., Bergqvist, M., Herslof, B.G., and Jacobs, F.A. (1992). *Lipids 27:*436–41.
39. Demirbuker, M., Hagglund, I., and Blomberg, L.G. (1992). *J. Chromatogr. 605:*263–68.
40. Demirbuker, M., Anderson, P.E., and Blomberg, L.G. (1993). *J. Microcol. Sep. 5:*141–47.
41. Demirbuker, M., and Blomberg, L.G. (1992). *J. Chromatogr. 600:*358–63.
42. Demirbuker, M., Hagglund, I. and Blomberg, L.G. (1992). In *Contemporary Lipid Analysis, 2nd Symposium Proceedings,* edited by N.U. Olsson, and B.G. Herslof, LipidTeknik, Stockholm. pp. 30–47.
43. Christie, W.W. (1992). In *Advances in Lipid Methodology—One.* Edited by W.W. Christie, Oily Press, Ayr. pp. 1–17.
44. Christie, W.W. (1989). *J. Lipid Res. 30:*1471–73.
45. Zelles, L., and Bai, Q.Y. (1993). *Soil Biol. Biochem. 25:*495–507.
46. Ulberth, F., and Achs, E. (1990). *J. Chromatogr. 504:*202–206.
47. Christie, W.W. (1990). *J. Sci. Food Agric. 52:*573–77.
48. Schomburg, G., and Zegarski, K. (1975). *J. Chromatogr. 114:*174–78.
49. Chan, H.W.S., and Levett, G. (1978). *Chem. Ind. (London):*578–79.
50. Takano, S., and Kondoh, Y. (1987). *J. Am. Oil Chem. Soc. 64:*380–83.
51. Baillet, A., Corbeau, L., Rafidison, P., and Ferrier, D. (1993). *J. Chromatogr. 634:*251–56.
52. Dewar, M.S.J. (1951). *Bull. Soc. Chim. France 18:*C71–79.
53. Kasai, P.H., McLeod, D., and Watanabe, T. (1980). *J. Am. Chem. Soc. 102:*179–90.
54. Ganis, P., and Dunitz, J.D. (1967). *Helv. Chim. Acta 50:*2379–86.
55. *Gmelin's Handbuch der Anorganische Chemie.* (1975). Vol. 61. Tl B5 Springer, Berlin. p. 26.
56. Rich, M.R. (1993). *Biochim. Biophys. Acta 1178:*87–96.
57. Nikolova-Damyanova, B., Christie, W.W., and Herslof, B.G. *J. Chromatogr. 653:*15–23.

Chapter 8

Utilization of Silver-Ion High-Performance Liquid Chromatography for Separation of the Geometrical Isomers of α-Linolenic Acid

Pierre Juanéda

Institut National de la Recherche Agronomique, Station de Recherches sur la Qualité des Aliments de l'Homme - Unité de Nutrition Lipidique, 17 rue Sully, 21034 Dijon, France.

Introduction

Geometrical isomers of α-linolenic acid have been detected in oils after deodorization (1) or frying operations (2). Upon heat treatment, α-linolenic acid (18:3 Δ9c,12c,15c) can be isomerized into seven other isomers containing one, two, or three *trans* ethylenic bonds. These isomers are 18:3 Δ9c,12c,15t; 18:3 Δ9c,12t,15c; 18:3 Δ9t,12c,15c; 18:3 Δ9c,12t,15t; 18:3 Δ9t,12t,15c; 18:3 Δ9t,12c,15t; and 18:3 Δ9t,12t,15t. Unfortunately even using two columns of different polarity (2,3), gas chromatography (GC) does not allow the total separation of the eight isomers as either methyl ester derivatives, or isopropyl esters on a CPSil 88 column as described by Wolff (4 and Chapter 13). On the other hand, thin-layer chromatography (TLC) on silver nitrate coated plates is a common technique which allows the separation of fatty acids differing in the degree of unsaturation (5). However, the separation of geometrical isomers containing more than two double bonds is not easy. The development of HPLC in the silver-ion mode (Chapter 7) has enabled progress in the separation of the *trans* geometrical isomers of unsaturated fatty acids, in particular for mono- and diunsaturated fatty acids (5). Separations of geometrical isomers of linoleic, linolenic, and arachidonic fatty acid methyl esters on a commercial column containing silver ions were also described by Adlof (6). Unfortunately, the different isomers were not fully identified. The aim of the present work was to completely resolve the eight geometrical isomers of α-linolenic acid, which could be formed during processing vegetable oils. Moreover, it was necessary to collect these isomers in as pure a form as possible for further structural studies.

Materials

All the liquid chromatographic separations were effected using a ternary solvent HPLC instrument. Fatty acid methyl esters were detected with a light-scattering detector, whereas an ultraviolet detector (238 nm) was used for the phenacyl derivatives of fatty acids. The column (25 cm × 4.8 mm i.d.) was packed with Nucleosil® SSSA. It was loaded with silver nitrate as described by Christie (7). Briefly, the column was washed first with a 1% ammonium nitrate solution for 1 h at a flow rate of 1 mL/min, followed by distilled water at the same flow rate for 1 h. Silver nitrate (200 mg) in distilled water (1 mL) was injected onto the column via a Valco valve with a 100 µL loop at 1-min intervals. The column was rinsed with distilled water for 20 min, then with methanol for 1 h at a flow rate of 1 mL/min, and finally with the eluting solvent.

Fatty acid methyl esters were analyzed by GC on a chromatograph fitted with a split/splitless injector at 250°C and a flame-ionization detector (FID) at 250°C. Three fused silica columns of different polarities were used. Helium was used as the carrier gas.

Separation

The geometrical isomers of 18:3 fatty acid were derived from methyl esters according to Morrison and Smith (8). A light-scattering detector was used at 25°C, with filtered air as nebulized gas. The solvents used for the eluting gradient were 1,2-dichloroethane (DCE), dichloromethane (DCM), and acetonitrile (ACN), as described by Christie and Breckenridge (9). The mobile phase changed over 30 min from a mixture of DCE/DCM (50:50 v/v) to a mixture of DCE/DCM/ACN (50:50:3, v/v). This composition was maintained until the elution of the last compound and then reverted to the starting composition. The flow rate was 0.7 mL/min, and the column was maintained at 38°C. As showed in Figure 8.1b, the eight isomers were eluted in three peaks. Changing the composition of the solvents and/or the temperature did not improved the separation.

Consequently, phenacyl derivatives of the 18:3 geometrical isomers were preferred, since these derivatives were reported to be particularly effective for silver-ion chromatography (10). The fatty acids were converted into phenacyl derivatives as described by Wood and Lee (11). The same chromatographic conditions were tested with the phenacyl derivatives (Figure 8.1a). The 18:3 Δ9t,12t,15t and Δ9c,12c,15c isomers were well separated, but the resolution of the mono- and ditrans isomers was unsatisfactory. Unlike the methyl esters, the phenacyl derivatives of the eight isomers were separated according the number of *trans* double bounds (mono-, di-, and tritrans).

Changing the temperature from 38°C to 10°C increased the retention time and improved the resolution of the di- and monotrans (12). The eight isomers were partially separated, but complete separation of the monotrans isomers was not obtained. Changing the DCM/DCE ratio, while keeping the ACN level constant, did not change the order of elution nor the resolution of the di- and monotrans isomers. However, it was noticed that DCE was not necessary when the analyses were performed at 10°C. Modification of the DCM/ACN ratio did not improve the separation. When ACN was replaced by methanol (MeOH) in the eluting solvent the separation was improved and the eight isomers were well separated as shown in Figure 8.2. The mobile phase changed over 30 min from 95% DCM: 5% MeOH to

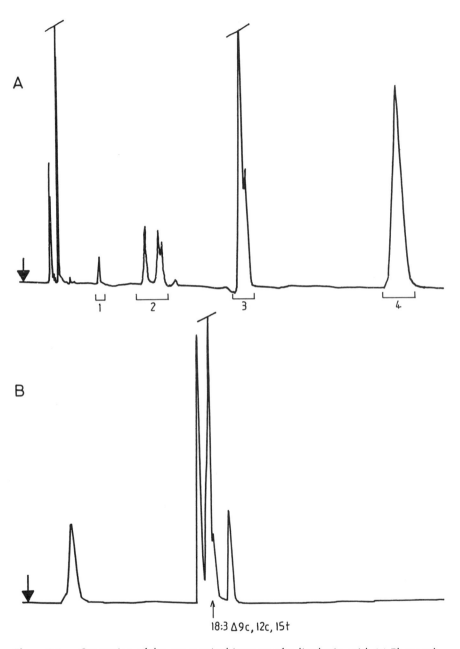

Figure 8.1. Separation of the geometrical isomers of α-linolenic acid. (a) Phenacyl derivatives and UV detection; (1) tri*trans*, (2) di*trans*, (3) mono*trans*, (4) tri*cis*. (b) Methyl ester derivatives and light scattering detection.

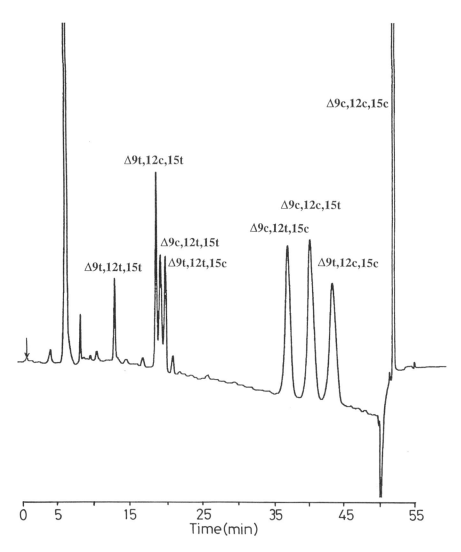

Figure 8.2. Separation of the eight geometrical isomers of α-linolenic acid using silver-ion HPLC.

75% DCM: 25% MeOH. After a total of 45 min, the mobile phase was changed to DCM/ACN (100:5 v/v) until 18:3 Δ9c,12c,15c eluted, before reverting to the original mobile phase composition. The flow rate was maintained at 0.7 mL/min. The solvents and the column were maintained at 10°C, using a refrigerated circulating bath.

The utilization of DCM/ACN (100:5 v/v) in the gradient allowed a rapid elution of the all *cis* 18:3 (in about 55 min). Maintaining the mixture of DCM and MeOH increased the elution time of this isomer to about 2 h.

Identification

All the peaks were collected in order to identify all the isomers by GC. The phenacyl derivatives were hydrolyzed with 0.1 M potassium hydroxide in 98% ethanol at 50°C for 3 h. After acidification and extraction, the free fatty acids were converted into methyl esters as described by Morrison and Smith (8). The identification was effected by comparison of the elution order and the equivalent chain length (ECL) values of the 18:3 isomers from published experimental data (2,4,13).

Grandgirard et al. (3) reported that the utilization of columns of two different polarities (Carbowax and Silar 10C) allowed the identification of most of the isomers. For the present work, three columns were used a DBwax column (30 m × 0.25 mm i.d.), a CPSil 84 (50 m × 0.25 mm i.d.) and a BPX70 (30 m × 0.25 mm i.d.). The order of elution of the eight isomers is shown in Figure 8.3. Each HPLC fraction was injected on the 3 columns. The different isomers were 99–100% pure for tri- and each mono*trans* and 95–99% for each di*trans*. The GC chromatograms showed only one major peak and permitted determination of the order of elution of the isomers.

To complete this identification, the ECL values were determined. The ECL values of the 18:3 isomers were calculated according to the method described by

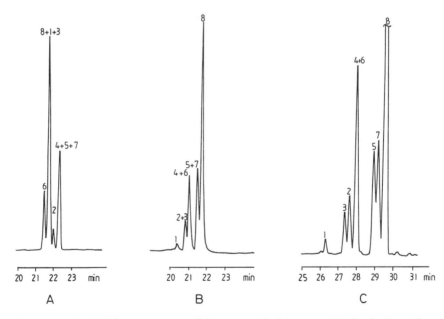

Figure 8.3. Partial chromatograms of the geometrical isomers of α-linolenic acid methyl esters on DBWax (a), CPSil 84 (b), and BPX70 (c). 1. Δ9*trans*,12*trans*,15*trans*; 2. Δ9*trans*,12*cis*,15*trans*; 3. Δ9*cis*,12*trans*,15*trans*; 4. Δ9*trans*,12*trans*,15*cis*; 5. Δ9*cis*,12*trans*,15*cis*; 6. Δ9*cis*,12*cis*,15*trans*; 7. Δ9*trans*,12*cis*,15*cis*; 8. Δ9*cis*,12*cis*,15*cis*.

TABLE 8.1
Gas Chromatography Identification of the 18:3 Isomers by Comparison of the ECL Values Between the Standard Mixture and the Collected HPLC Fractions

Collected Peaks (Figure 8.2)	18:3 Isomers[a]	DBWax (60°C to 190°C at 20°C/min)	CPSil 84 (60°C to 190°C at 20°C/min)	BPX70 (60°C to 170°C at 20°C/min)
Peak 1	Δ9t,12t,15t	19.28	19.88	19.07
Peak 2	Δ9t,12c,15t	19.34	20.12	19.32
Peak 3	Δ9c,12t,15t	19.30	20.10	19.28
Peak 4	Δ9t,12t,15c	19.42	20.21	19.39
Peak 5	Δ9c,12t,15c	19.42	20.41	19.56
Peak 6	Δ9c,12c,15t	19.21	20.19	19.39
Peak 7	Δ9t,12c,15c	19.42	20.43	19.61
Peak 8	Δ9c,12c,15c	19.30	20.54	19.69

[a] c = cis; t = trans

Ackman (14) using $C_{18:0}$ and $C_{21:0}$ fatty acid methyl esters as an internal standard. The ECL values are presented in Table 8.1. A difference of 0.03–0.04 carbon units should be sufficient to show a resolution between two isomers. On the BPX70, the coefficient of resolution between the 18:3 Δ9c,12t,15t (ECL = 19.28) and 18:3 Δ9t,12c,15t (ECL = 19.32) is 1.56. In order to confirm the identification 18:3 Δ9c,12c,15t and 18:3 Δ9t,12c,15c synthetic molecules supplied by J.M. Vatele (ESCIL, Villeurbanne, France) were used. In conclusion, the development of the silver-ion liquid chromatography allows the separation of complex fatty acid mixtures, particularly the geometrical isomers of polyunsaturated fatty acids.

References

1. Ackman, R.G., Hooper, S.N., and Hooper, D.L. (1974) *J. Am. Oil Chem. Soc. 51:*42–9.
2. Grandgirard, A., Sebedio, J.L., and Fleury, J. (1984) *J. Am. Oil Chem. Soc. 61:*1563–1568.
3. Grandgirard, A., Julliard, F., Prevost, J., and Sebedio, J.L. (1987) *J. Am. Oil Chem. Soc. 64:*1434–1440.
4. Wolff, R.L. (1992) *J. Chromatogr. Sci. 30:*17–22.
5. Nikolova-Damyanova, B. (1992) In *Advances in Lipid Methodology.* Vol. 1. Edited by W.W. Christie, The Oily Press, Ayr. pp. 181–237.
6. Adlof, R.O. (1994) *J. Chromatogr. A 659:*95–99.
7. Christie, W.W. (1987) *J. High Resol. Chromatogr. Chromatogr. Commun. 10:*148–150.
8. Morrison, W.R., and Smith, L.M. (1964) *J. Lipid Res. 5:*600–608.
9. Christie, W.W., and Breckenridge, G.H. (1989) *J. Chromatogr. 469:*261–269.
10. Nikolova-Damyanova, B., Herslof, B., and Christie, W.W. (1992) *J. Chromatogr. 609:*133–140.
11. Wood, R., and Lee, T. (1987) *J. Chromatogr. 254:*237–246.
12. Juanéda, P., Sebedio, J.L., and Christie, W.W. (1994) *J. High Res. Chromatogr. 17:*321–324.
13. Rakoff, H., and Emken, E.A. (1982) *Chem. Phys. Lipids 31:*215–225.
14. Ackman, R.G. (1972) *Prog. Chem. Fats Other Lipids 12:*167–284

Chapter 9

High-Performance Size-Exclusion Chromatography Applied to the Analysis of Edible Fats

M. Carmen Dobarganes and Gloria Márquez-Ruiz

Instituto de la Grasa (C.S.I.C.), Avda. Padre García Tejero, 4, 41012 Sevilla, Spain.

Introduction

Among chromatographic techniques, exclusion chromatography is unique in that separation is based on the relative molecular size of the compounds. Typically, the stationary phase consists on macromolecules that have been cross-linked to form a three-dimensional network characterized by a specific pore size. Larger solute molecules will be excluded and emerge first, while the smaller ones will diffuse into the pores of the gel partially or completely.

The first applications of this technique date from the late 1950s, although it has been during the last decade when high-performance stationary phases have been extensively used. Efforts focusing on optimizing the mechanical stability of stationary phases, packing methodology, and particle-size distribution have permitted increased flow rates and improved resolution.

Although certain initial applications took advantage of complementary effects, such as partition or adsorption, improvements of the technique were directed to minimize or avoid such parallel effects so that the migration of molecules between the stationary phase and the mobile phase was driven only by diffusion. Assuming that only diffusion occurs, the order of elution of solute molecules will depend exclusively on the molecular size, which is closely related to the molecular weight (MW) provided that molecules have the same shape. It is possible to estimate the MW of an unknown molecule by plotting retention volume versus the logarithm of MW for a series of known MW standards. These plots can provide accurate determinations of MW for molecules which adopt a similar conformation in solution to that of the standard (1).

High-performance size-exclusion chromatography (HPSEC) has found its main application in the range of high-MW molecules, for the analysis of polymers and biomolecules. However, development of lipophilic gels extended its application to the separation of organic compounds of low MW. Particularly for fats and oils, normally comprised of over 95% triglycerides (TG) which are satisfactorily analyzed by other chromatographic techniques, HPSEC has found only limited application since this technique is not appropriate to separate groups of lipid compounds differing in MW only slightly as a result of variable fatty acid (FA) composition or unsaturation degree.

Nevertheless, recent advances of HPSEC offer lipid researchers a powerful chromatographic technique for certain applications, with distinct advantages over other forms of chromatography, that is, excellent recovery of the sample, simplicity of the mobile phase, predictable elution order, and high reproducibility. Application of low- and high-performance size-exclusion chromatography in lipid analysis has been recently reviewed, both in organic and aqueous media (2). The present paper will focus on specific applications in the field of edible fat analysis, in clear connection with the use of high-performance supports able to separate molecules in the low-MW range.

General Characteristics of HPSEC in the Analysis of Edible Fats

Considering the principle of size-exclusion chromatography, HPSEC analyses are characterized by their simplicity, regardless of the application. Once decided that the technique is useful for a given purpose, selection of conditions involves a limited number of possibilities, mainly those related to the stationary phase, solvent used as the mobile phase, and detection system.

Stationary phase

For HPSEC in organic solvents, silica-based phases can be used, although copolymers of styrene divinyl benzene are practically the only stationary phase used for fat and oil analyses. The most important parameters influencing resolution are the pore volume, pore-size distribution, and particle size. Copolymers of styrene divinyl benzene are available over a wide range of pore sizes, but 50, 100, and 500 Å are essential porosities for low-MW separations (100–20,000 MW). Size-exclusion columns are generally larger than those used in the other chromatographic modes, so that the amount of stationary phase and thus the effective pore volume available is increased. Packed columns are normally around 30 cm × 0.8 cm i.d. and they are used in series often. Thus, effective selection within a broad range is accomplished by the first column and fractionation within a more defined range is achieved on the second or third column (3).

Given that resolution in size-exclusion chromatography is dependent on diffusion, optimization of stationary phase particles has received special attention in order to enhance mass transfer. The development of spherically shaped monosized particles has contributed considerably to improving particle-size and pore-size distribution and hence efficiency and separation capacity (4). Particle sizes of 5 and 10 µm are the most commonly used.

Mobile Phase

In HPSEC, a single solvent is used, and its selection is limited only by the solubility of the sample. Columns are normally filled with the same solvent used to dissolve the sample.

Supports of copolymer styrene divinyl benzene have been designed to operate across a wide spectrum of solvents. Columns can even be transferred easily and rapidly between solvents of differing polarity without damage to the packed bed. Tetrahydrofuran (THF) is the most commonly used solvent by far, although toluene and dichloromethane have been selected for certain applications.

Flow rates between 0.5 and 1.5 mL/min are usually selected, which permits analyses to be accomplished in less than 30 min.

Detection Systems

Selection of a detection system is dependent on the analytical objective for each application. Thus, infrared detectors have been used to register absorbance of specific functional groups, such as the ester carbonyl or free alcohol groups (5). Ultraviolet (UV) detection has been used to obtain high sensitivity for products with conjugated dienes in their structure (6). In both cases, only qualitative results can be obtained.

For quantitative purposes in HPSEC applications, the universal detectors most commonly used are the refractive index detector (RID) and the evaporative light-scattering (mass) detector (ELSD). Refractive index detection is the simplest since a single solvent is used and linear responses are normally obtained in the ranges of interest. Also, ELSD has been recently described for the analysis of lipid classes, including polymers (7,8). In a recently published paper, Hopia and Ollilainen compared the properties of RID and ELSD for HPSEC applications (9). In the analysis of glyceridic compounds, not only were similar response factors found for RID, but the response was linear in the ranges studied. For ELSD, the sensitivity was slightly higher, but response factors depended on the amounts injected. Table 9.1 summarizes the conditions normally used for HPSEC in edible fat analyses, where those most commonly applied are specified.

Frying Fats

Frying fats, used to prepare a large variety of foods. are normally exposed to high temperature in the presence of air and moisture. Under these conditions, they may

TABLE 9.1
Conditions Normally Used for HPSEC in the Analysis of Edible Fats

Stationary phase: Copolymers of styrene-divinyl benzene[a]
                    Particle size: 3, 5, 10 μm
                      Pore size: 50, 100[a], 500[a], 1000 Å

Mobile Phase: Tetrahydrofuran[a], toluene, dichloromethane

Detection systems: Refractive index detector[a]
                     Evaporative light-scattering detector

[a]Most commonly used.

undergo important changes due to hydrolytic, oxidative, and thermal reactions. Among the main products found are polymers, dimers, oxidized TG, as well as diglycerides (DG) and FA. It is important to note that while hydrolysis involves breaking the ester bond and releasing FA, monoglycerides (MG), and DG—the normal compounds originating in the stage previous to fat absorption—oxidative and thermal degradation takes place in the unsaturated acyl groups of the TG and modifies the nutritional properties of the fat through formation of oxidized TG, dimers, and polymers. This means that the compounds present in frying fats differ in nutritional significance, therefore it is of great interest to know the contribution of the main groups of compounds to the total alteration.

The most accepted technique for the analysis of frying fats is polar compound determination based on adsorption chromatography, which has advantages due to its simplicity, reproducibility, accuracy of the gravimetric determination, and the possibility of checking the efficiency of separation (10). However, by determining the level of polar compounds exclusively, it is not possible to distinguish the main groups of compounds originated. In this sense, HPSEC offers an optimal analytical tool, considering the differences in MW existing between the compounds involved. Thus, hydrolytic compounds have significantly lower MW than the TG, while MW of dimers and polymers are significantly higher. Satisfactory results were obtained for the analysis of polymeric compounds prior to the development of high-performance chromatographic supports (11–13). Perrin et al. (14) were the first to propose high-performance columns of polystyrene divinyl benzene for the analysis of polymers. Three columns of different pore size, 50, 100, and $50Å–10^6Å$, were used separately and in series. Tetrahydrofuran was selected as the solvent and a refractometer as the detector. The potential of the technique was illustrated by Perrin et al. and later supported by Kupranycz et al. (15), for the analysis of oils with increasing alteration levels. At the same time, White and Wang (6) evaluated different heated soybean oils using a variable wavelength UV detector. Two columns of 500 and 1000 Å were employed, and methylene chloride served as the mobile phase. In spite of the lack of exact quantitative data, the authors concluded that some estimate of the relative quantities of high-MW compounds formed could be given and suggested this method as a rapid usage test in the oil industry to compare stabilities of various frying oils.

Sample concentrations of 30–50 mg/mL and loops of 10–20 µL have generally resulted in good resolution in polymer analysis in all the laboratories. The method stands out for its simplicity, as it is only necessary to dilute the fat in the appropriate solvent before the chromatographic determination, which is performed with a single solvent in 10–30 min and depends on the flow rate and number of columns used. Because of the complexity of the polymeric fractions, quantitative results are usually obtained from the calculation Peak Area % divided by Total Area.

The IUPAC Commission on Oils, Fats, and Derivatives has recently published the results of two interlaboratory tests on analyses of polymerized TG, concluding that repeatability was generally high but reproducibility was rather poor for samples with low polymer content. Hence the Commission finally decided to adopt a general methodology limiting application to samples containing 3% or more of polymerized TG (16). The method proposed a single column of 30 cm × 0.77 cm i.d. packed with a high-performance spherical gel made of copolystyrene divinyl benzene of 5 μm, THF as the mobile phase, and a refractive index detector with a sensitivity at full scale at least 1×10^{-4} of refractive index. The sample concentration suggested was 50 mg/mL for an injection valve with a 10 μL loop. The analysis time was about 10 min with a flow rate of 1 mL/min.

A different analytical approach has been suggested to increase sensitivity and widen the application of polymer determination to any kind of fat or oil, heated or not (17), by conveniently combining the two methods proposed by IUPAC, namely the determination of polar compounds by adsorption chromatography (10) and the determination of polymerized triglycerides by HPSEC (16) which is applied to the polar fraction obtained by the former method. Advantages result from this scheme in comparison with the analysis of the total fat sample. This is illustrated in Figure 9.1, which shows high-performance size-exclusion chromatograms of nonheated cottonseed oil (a), frying cottonseed oil (b), and the polar compounds obtained from the latter (c). Application of HPSEC to the polar compounds separated by adsorption chromatography permits, first, a substantial increase in quantitation possibilities of all the groups of alteration compounds because of the concentration effects; second, an independent determination of oxidized TG monomers as a measurement of oxidative alteration (since nonaltered TG would elute in the first separation stage); and third, simultaneous evaluation of DG as an indication of hydrolytic alteration, avoiding any overlap with TG in the whole sample.

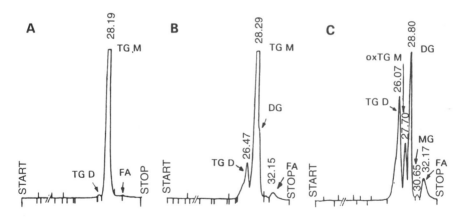

Figure 9.1. High-performance size-exclusion chromatograms of nonheated cottonseed oil (a), frying cottonseed oil (b), and the polar compounds obtained from frying cottonseed oil (c). Abbreviations: TGD, triglyceride dimers; TGM, triglyceride monomers (nonaltered and oxidized); oxTGM, oxidized triglyceride monomers; DG, diglycerides; MG, monoglycerides; and FA, fatty acids.

This methodology has been revealed as a promising alternative for frying oil evaluation (18,19), although it may be time-consuming for routine quality-control laboratories as the determination of polar compound takes approximately 2 h. Thus, it is suggested as a complementary analysis for laboratories where polar compound determination is already undertaken following the present regulation, and not only for analyses of polymer content but with the broader aim of obtaining fairly complete information on the type of degradation. Silica-column chromatography used for the initial separation by polarity may be substituted for methods which require lower amounts of solvent and less time, such as solid phase extraction using NH_2 columns (8) or silica-gel cartridges (20).

Previous methodologies for polymer analysis in frying fats based on urea non-adduct formation, solvent partition, or distillation have been substituted for HPSEC, where hydrolytic and oxidized compounds can be simultaneously quantitated after elimination of the nonpolar triglycerides, thus significantly improving the possibilities for frying fat evaluation.

Also, of particular mention is the HPSEC use to evaluate polymerized fatty acids or fatty acid methyl esters of proven interest in early studies in the area of oleochemicals. Christopoulou and Perkins (21) presented an interesting systematic study of this application and emphasized its possibilities for the analysis of thermally oxidized oils. The system used was composed of two styrene/divinylbenzene copolymer columns with toluene as the mobile phase and refractometry as the mode of detection. Quantitation of monomer, dimer, and trimer content in various samples correlated well with results obtained by gas-liquid chromatography (GLC) analysis and gravimetrically by size-exclusion chromatography (SEC). Additionally, comparisons of SEC with other chromatographic techniques available for analysis of monomer, dimer, and trimer FA, such as GLC and thin-layer chromatography with flame ionization detection (TLC-FID), have been also carried out (22) and no significant differences were found between the quantitative results given by any of the methods tested.

Similar to the previous comments on frying oil evaluation, analysis of FA monomers, dimers, and polymers exclusively by HPSEC, and therefore based on molecular size only, may be rather poor when the major fraction of the sample is composed of nonpolar FA. Again, it is possible to enhance the determination possibilities by separating two fractions by polarity prior to the HPSEC analysis (23). The method proposed allows quantitation of unaltered FA in addition to four groups of degradation compounds, nonpolar FA dimers, oxidized FA monomers, polar FA dimers, and FA polymers. This approach is of great value to examine the nutritional effects of frying oil consumption as it is focused on the specific evaluation of the fatty acyl groups of triglycerides, which are the products of fat digestion that are ultimately absorbed (24). Additionally, the analysis of frying fats after transesterification when combined with direct evaluation of the fat may be useful for acquiring a better understanding of fat alteration.

Oxidized Fats

Lipid oxidation significantly affects the useful storage life of foods, but the efficacy of antioxidants and the oxidative stability of edible oils are difficult to evaluate in view of the questionable conditions and methodologies currently used. At present, recommended testing protocols for oxidative stability state the necessity of determining the degree of oxidation at suitable time intervals and by more than one method, thus measuring different types of lipid oxidation (25).

Along with the well-established methods which provide information on primary or secondary oxidation products, HPSEC offers a complementary technique for the evaluation of oxidized compounds originating at low temperatures. High-performance size-exclusion chromatography has been applied satisfactorily to the area of fish oils. Polyunsaturated fatty acids such as eicosapentaenoic and docosahexaenoic acids are particularly abundant in fish oils. On one hand, they may account for a lower incidence of coronary heart disease but, on the other, they are readily oxidized, and their ingestion may entail a risk of exposure to potentially harmful products. Therefore, an increasing number of general studies on autoxidation of fish oils at low temperatures have been carried out, with special emphasis on determining possibly deleterious compounds. As for HPSEC applications in this area, it is worth noting those evaluating oxidation materials in encapsulated fish oils, a popular health supplement to reduce the risk of cardiovascular disease. A series of connected polystyrene-divinyl benzene columns (500, 100, and 100 Å) and THF as the mobile phase were used. The results supported the need for a close control of commercial fish-oil capsules and the utility of HPSEC in this particular regard (26,27).

In the same direction, Burkow and Henderson (28) proposed HPSEC for the routine assessment of fish-oil quality. They used an Ultrastyragel 500 Å column with dichloromethane (0.8 mL/min) as the mobile phase. In experiments carried out at 35°C in artificial daylight, the oils oxidized rapidly, with polymer content values reaching as high as 35–45% after 4 d. Later, these authors improved the separation by coupling three columns in series with effective fractionation ranges from 100–30,000 daltons. The new combination was applied to both TG and hydrolyzed samples. The detection system chosen was an evaporative mass detector, and glycerol was used as internal standard. The difficulties found in quantitation led to the use of curvilinear calibration graphs to give the response-weight relationship between the polymers and the internal standard (7).

Oxidized TG have also been proposed as a measurement of the level of oxidation in fats and fatty foods other than fish oils (29,30). Following examination of HPSEC profiles and data of the polar compounds originating after storing of various fatty foods at different temperatures, it was evident that oxidized TG was the group of compounds which experienced the largest increase. The results obtained in oils extracted from roasted sunflower seeds and peanuts after storage for 30, 60, and 90 d are presented in Table 9.2. A significant increase of oxidized TG (polymers plus monomers) could be observed, most notably in sunflower seeds, in contrast, DG and FA coming from hydrolysis remained at the initial levels. Therefore,

TABLE 9.2
Total Level and Distribution of Polar Compounds (mg/g oil) in Oils from Roasted Sunflower Seeds and Peanuts after Storage at Room Temperature for 30, 60, and 90 Days

	Time (d)	Total PC	PC Distribution			
			TGP	oxTGM	DG	FA
Sunflower seeds	0	53	0.8	33.9	9.6	8.7
	30	86	9.2	60.8	8.9	7.1
	60	139	25.2	93.5	9.9	10.4
	90	184	63.0	101.7	9.8	9.5
Peanuts	0	30	0.2	18.0	8.0	3.8
	30	31	0.4	19.2	8.2	3.2
	60	35	0.8	21.1	9.3	3.8
	90	48	0.8	33.7	9.8	3.7

Abbreviations: PC, polar compounds; TGP, triglyceride polymers; oxTGM, oxidized triglyceride monomers; DG, diglycerides; and FA, fatty acids.

the analytical method was suggested as an objective measurement of the total oxidation level in fats, as quantitation included both the primary and secondary compounds formed. Similar results were found by Hopia et al. (31,32) in studies in which TG mixtures and various edible oils were autoxidized and polar compounds analyzed through HPSEC with a light-scattering detector, following preliminary elimination of nonpolar TG by solid phase extraction.

Quality and Characterization of Fats

An exceptional set of HPSEC applications of promising utility takes advantage of specific aspects which can help characterize or evaluate the quality of certain fats and oils. That is the case of mixtures of oils differing substantially in triglyceride MW. Thus, Husain et al. (33) reported recently the use of HPSEC for accurate determination of MW averages in a number of oils, fats, and their binary mixtures. The authors discussed the conditions under which the method might be suitable for the determination of adulteration in certain oils and fats.

High-performance size-exclusion chromatography has also been proposed for characterization of virgin and refined oils. Good results were obtained even through direct analysis of the total oil (34–36). Thus, Gomes et al. found dimer contents higher than 0.5% in refined olive oils, while no dimers were detected in virgin oil samples. However, improved evaluation of minor compounds can be achieved by previous elimination of nonpolar triglycerides, especially for application to oils containing very low levels of polar glyceridic compounds.

By using this methodology, the authors found that the absence of dimers is the most useful parameter for the characterization of virgin oils, while the occurrence of dimers plus a high ratio of DG/FA are those for refined oils (37,38). The procedure was later followed to quantitate oxidized triglycerides and diglycerides in different types of olive oils of certified origin (39). Furthermore, the authors evaluat-

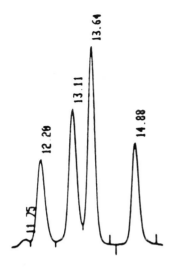

Figure 9.2. Significant part of a chromatogram of polar compounds from an olive oil after refining. Retention times (min): 12.2, triglyceride dimers; 13.1, oxidized triglyceride monomers; 13.6, diglycerides; and 14.9, fatty acids and polar unsaponifiable fraction.

ed the possibility of HPSEC for quality assessment of refined oils in a laboratory-scale refining process (40). Quantitation of polar compounds in crude oils and in samples taken at different stages of the refining process indicated that the quality of an initial crude oil could be deduced from the resulting refined oil by virtue of certain markers of oxidative and hydrolytic alterations. An illustrative example is presented in Figure 9.2, which shows the polar fraction from an olive oil after refining. On one hand, compounds of higher MW than that of oxidized TG monomers are formed, mainly as a consequence of the high temperature in the deodorization step. On the other hand, oxidized TG and DG remain after refining, so their level in the different samples is a measure of oxidation and hydrolysis, respectively, regardless of whether the oil is a crude or refined. Finally, the peak including FA and the polar unsaponifiable fraction decreases due to the neutralization step. Recently, this approach has also been used to monitor oils during industrial refining (41).

Low-Caloric Fats

High-performance size-exclusion chromatography evaluation has been extended to lipid samples other than classical fats and oils. A great deal of work is currently done on low-digestibility fats that can be heated at high temperature and hence used in baked and fried low caloric fatty foods. Among the fat substitutes proposed, sucrose polyesters (SPE), sucrose molecules esterified by 5–8 fatty acids, probably stands out and is pending approval for use in foods by the U.S. Food and Drug Administration. Birch and Crow (42) relied on HPSEC to determine SPE in faeces and diets and proposed the method for absorption measurement of such materials, thus improving previous methods based on radioactive markers and hence not suitable for human subjects. The authors established the conditions best suited to quantitate SPE as a single peak eluting before other components of the extracts, by using two µStyragel columns (500 and 1000 Å) in series, THF as eluant at a flow rate of 1.5 mL/min, and a refractive index detector.

More recently, studies on the polymers formed at high temperature from SPE were carried out with a single 500 Å PLGel column (60 cm), and indicated that dimers, trimers and mixed dimers were present when SPE was subjected to this thermal treatment in mixtures with TG (43). Susceptibility of SPE to thermal, oxidative, and hydrolytic modification is a current subject of study in the authors' laboratory. Starting from purified samples derived from olive oil and following an analytical procedure based on adsorption chromatography and HPSEC, it has been reported that oxidation of sucrose octaesters takes place more slowly than that of TG with a similar FA composition. The methodology was applied to the methyl ester derivatives in order to comparatively evaluate oxidized and polymeric fatty acyls in sucrose octaesters and TG, which differ markedly in molecular structure, with 8 and 3 acyl groups respectively. Basically, monomeric and, in lower amounts, dimeric compounds, were obtained at 100°C after 12 h (44).

Recently, valid analytical methods applicable to the evaluation of SPE/fat blends have been claimed (45). They are of tremendous importance for production and quality control, and in terms of the nutritional labeling which would be required for products containing SPE/lipid blends. For this purpose, reverse phase HPLC can be used (46) but HPSEC seems to be the simplest, most general, and most repro-

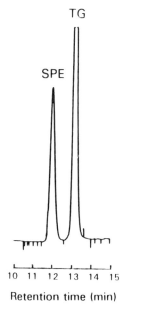

Figure 9.3. Efficacy of the separation obtained by HPSEC for a mixture olive oil SPE/sunflower oil (26.7% SPE) using the chromatographic conditions listed. Abbreviations: SPE, sucrose polyesters; and TG, triglycerides.

ducible analytical method. Ríos et al. recently suggested this approach to characterize SPE/TG mixtures (47) and later obtained satisfactory results for SPE/natural oil blends. Figure 9.3 illustrates the efficacy of the separation by HPSEC for an olive oil SPE/sunflower oil mixture. The validity of the method relies on the elution of SPE as a single peak and the excellent separation achieved between SPE and TG peaks (48). Similarly, the utility of HPSEC can be extended to the analysis of mixtures of natural oils with other fat substitutes provided that they differ significantly in MW, as occurs for sorbitol polyesters, trehalose polyesters, or raffinose polyesters.

In conclusion, the simplicity and high reproducibility of HPSEC support its validity for analysis of edible fats, not only to evaluate alteration due to frying or oxidative conditions but also to gain information on their quality and characterization. Furthermore, the potential of HPSEC seems to anticipate future applications in other areas of edible oils.

Acknowledgments

This work was supported in part by CICYT (Project ALI 91-0544).

References

1. Stellwagen, E. (1990) *Methods Enzymol. 182:*317–28.
2. Dobarganes, M.C., and Márquez-Ruiz, G. (1993) In *Advances in Lipid Methodology.* Vol. 2. Edited by W.W. Christie, The Oily Press, Dundee. pp. 113–137.
3. Fallon, A., Booth, R.F.G., and Bell, L.D. (1987) In *Laboratory Techniques in Biochemistry and Molecular Biology: Applications of HPLC in Biochemistry.* Vol. 17. Edited by R.H. Burdon and P.H. Knippenberg, Elsevier, Amsterdam. pp. 56–64.
4. Kulin, L.I., Flodin, P., Ellingsen, T., and Ugelstad, J. (1990) *J. Chromatogr. 514:*1–9.
5. Aitzetmüller, K. 1988. *Chem. Ind.* 452–464.
6. White, P.J., and Wang, Y. (1986) *J. Am. Oil Chem. Soc. 63:*914–20.
7. Burkow, I.C., and Henderson, R.J. (1991) *J. Chromatogr. 552:*501–506.
8. Hopia, A.I., Piironen, V.I., Koivistoinen, P.E., and Hyvonen, L.E.T. (1992) *J. Am. Oil Chem. Soc. 69:*772–776.
9. Hopia, A.I., and Ollilainen, V.-M. (1993) *J. Liquid Chromatogr. 16:*2469–2482.
10. Waltking, A.E., and Wessels (1981) *J. Assoc. Off. Anal. Chem. 64:*1329–1330.
11. Aitzetmüller, K. (1972) *J. Chromatogr. 71:*355–360.
12. Perkins, E.G., Taubold, R., and Hsieh, A. (1973) *J. Am. Oil Chem. Soc. 50:*223–225.
13. Unbehend, V.M., Scharmann, H., Strauss, H.J., and Billek, G. (1973) *Fette Seifen Anstrichmittel 75:*689–696.
14. Perrin, J.L., Redero, F., and Prevot, A. (1984) *Rev. Franc. Corps Gras 31:*131–133.
15. Kupranycz, D.B., Amer, M.A., and Baker, B.E. (1986) *J. Am. Oil Chem. Soc. 63:*332–337.
16. Wolff, J.P., Mordret, F.X., and Dieffenbacher, A. (1991) *Pure Appl. Chem. 63:*1163–1171.
17. Dobarganes, M.C., Pérez-Camino, M.C., and Márquez-Ruiz, G. (1988) *Fat Science Technol. 90:*308–311.
18. Arroyo, R., Cuesta, C., Garrido-Polonio, C., López-Varela, S., and Sánchez-Muñiz, F.J. (1992) *J. Am. Oil Chem. Soc. 69:*557–563.
19. Cuesta, C., Sánchez-Muñiz, F.J., Garrido-Polonio, C., López-Varela, S., and Arroyo, R. (1993) *J. Am. Oil Chem. Soc. 70:*1069–1073.

20. Sebedio, J.L., Septier, C., and Grandgirard, A. (1986) *J. Am. Oil Chem. Soc.* *63:*1541–1543.
21. Christopoulou, C.N., and Perkins, E.G. (1989) *J. Am. Oil Chem. Soc. 66:*1338–1343.
22. Chandrasekhara Rao T., Kale, V., Vijayalakshmi, P., Gangadhar, A.; Subbarao, R., and Lakshminarayama, G. (1989) *J. Chromatogr. 466:*403–406.
23. Márquez-Ruiz, G., Pérez-Camino, M.C., and Dobarganes, M.C. (1990) *J. Chromatogr. 514:*37–44.
24. Márquez-Ruiz, G., Pérez-Camino, M.C., and Dobarganes, M.C. (1992) *J. Am. Oil Chem. Soc. 69:*930–934.
25. Frankel, E.N. (1993) *Trends Food Sci. Techn. 4:*220–325.
26. Shukla, V.K.S., and Perkins, E.G. (1991) *Lipids 26:*23–26.
27. Sagredos, A.N. (1992) *Fat Sci. Technol. 94:*101–111.
28. Burkow, I.C., and Henderson, R.J. (1991) *Lipids 26:*227–231.
29. Pérez-Camino, M.C., Márquez-Ruiz, G., Ruiz-Mendez, M.V., and Dobarganes, M.C. (1990) *Grasas y Aceites 41:*366–370.
30. Pérez-Camino, M.C., Márquez-Ruiz, G., Ruiz-Mendez, M.V., and Dobarganes, M.C. (1991) In *Proceedings of Euro. Food Chem. VI.* Vol. 2. Edited by W. Baltes, T. Eklund, R. Fenwick, W. Pfannhauser, A. Ruiter, and H.P. Thier, Lebensmittelchemische Gesellschaft, Frankfurt. pp. 569–574.
31. Hopia, A.I., Lampi, A.-M., Piirönen, V.I., Hyvonen, L.E.T., and Koivistoinen, P.E., (1993) *J. Am. Oil Chem. Soc. 70:*779–784.
32. Hopia, A.I. (1993) *Food Sci. Technol. 26:*563–567.
33. Husain, S., Sastry, G.S.R., Raju, N.P., and Narasimha, R. (1988) *J. Chromatogr. 454:*317–326.
34. Gomes, T. (1988) *Riv. Ital. Sostanze Grasse 65:*433–438.
35. Gomes, T., and Catalano, M. (1988) *Riv. Ital. Sostanze Grasse 65:*125–127.
36. Gomes, T. (1989) In *Actes du Congrès International "Chevreul" pour l'étude des corps gras.* Vol. 3. ETIG, Paris. pp. 1169–1175.
37. Dobarganes, M.C., Pérez-Camino, M.C., and Márquez-Ruiz, G. (1989) In *Actes du Congrès International "Chevreul" pour l'etude des corps gras.* Vol. 2. ETIG, Paris. pp. 578–584.
38. Pérez-Camino, M.C., Ruiz-Mendez, M.V., Márquez-Ruiz, G., and Dobarganes, M.C. (1993) *Grasas y Aceites 44:*91–96.
39. Gomes, T. (1992) *J. Am. Oil Chem. Soc. 69:*1219–1223.
40. Dobarganes, M.C., Pérez-Camino, M.C., Márquez-Ruiz, G., and Ruiz-Mendez, M.V. (1990) In *Edible Fats and Oils Processing: Basic Principles and Modern Practices.* Edited by Erickson, The American Oil Chemists' Society, Champaign. pp. 427–429.
41. Hopia, A.I. (1993) *Food Sci. Technol.* in press.
42. Birch, C.G., and Sanders, R.A. (1990) *J. Am. Oil Chem. Soc. 53:*581–583.
43. Gardner, D.R., and Crowe, F.E. (1976) *J. Am. Oil Chem. Soc. 67:*788–795.
44. Ríos, J.J., Pérez-Camino, M.C., Márquez-Ruiz, G., and Dobarganes, M.C. (1992) *Food Chem. 44:*357–362.
45. Haumann, B.F. (1993) *INFORM 4:*1226–1235.
46. Tallmadge, D.H., and Lin, P.Y.T. (1993) *J. AOAC Internat. 76:*1396–1400.
47. Ríos, J.J., Pérez-Camino, M.C., Márquez-Ruiz, G., and Dobarganes, M.C. (1994) *J. Am. Oil Chem. Soc., 71:*385–390.
49. Márquez-Ruiz, G., Pérez-Camino, M.C., Ríos, G.G., and Dobarganes, M.C. (1994) *J. Am. Oil Chem. Soc. 71:*1017–1020.

Chapter 10
Stereospecific Analysis of Triacyl-sn-Glycerols

W.W. Christie

The Scottish Crop Research Institute, Invergowrie, Dundee, Scotland DD2 5DA, UK.

Introduction

Triacylglycerols consist of the trihydric alcohol, glycerol, each position of which is esterified to long-chain fatty acids. Glycerol itself has a plane of symmetry. On the other hand, when the two primary hydroxyl groups are esterified with different fatty acids, the resulting triacylglycerol is asymmetric and may display optical activity. The enantiomers can be designated without ambiguity by the conventional *D/L* or *R/S* systems, but problems arise in application to the complex mixtures of molecular species of triacylglycerols found in nature. A "stereospecific numbering" (*sn*) system was therefore recommended by an IUPAC-IUB commission on the nomenclature of glycerolipids (1). In a Fischer projection of a natural *L*-glycerol derivative, the secondary hydroxyl group is illustrated to the left of C-2; the carbon atom above this is then C-1, that below is C-3, and the prefix *sn* is placed before the stem name of the compound (Figure 10.1).

The main biosynthetic mechanism for triacylglycerol biosynthesis in plant and animal tissues is the *sn*-glycerol-3-phosphate pathway (2) in which *sn*-glycerol-3-phosphate, produced by the catabolism of glucose, is acylated in turn at positions *sn*-1 and *sn*-2 by specific transferases to form phosphatidic acid. The enzyme phosphatidate phosphatase removes the phosphate group, and the resulting 1,2-diacyl-*sn*-glycerol is acylated by a further acyltransferase to form a triacyl-*sn*-glycerol. Since the precursor has a defined stereochemistry, and each of the enzymes catalyzing the various steps in the process is distinct and can have preferences for specific fatty acids or combinations of these in the partially acylated intermediates, it is inevitable that natural triacylglycerols exist in enantiomeric forms. In effect, this implies that each position of the *sn*-glycerol moiety can have a characteristic fatty acid composition. Practical and theoretical aspects of the topic were reviewed in some detail (3).

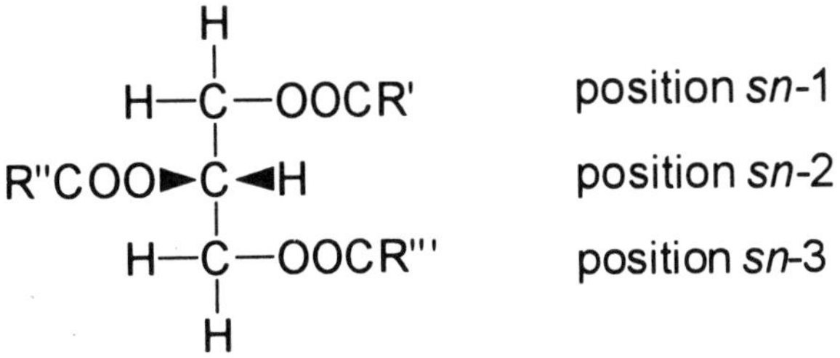

```
              H
              |
        H—C—OOCR'          position sn-1
              |
    R"COO►C◄H              position sn-2
              |
        H—C—OOCR'''        position sn-3
              |
              H
```

Figure 10.1. Fischer projection of a triacyl-*sn*-glycerol.

In the first experiments with synthetic triacylglycerols of high stereochemical purity, no optical activity could be detected, because it was too low to be measured by the equipment then available. Although modern instruments are much more sensitive, the only natural triacyl-*sn*-glycerols to show any observable optical activity as a result of glyceride chirality are the highly asymmetric ones, such as those in cow's milk (4). In addition, proton magnetic resonance spectroscopy with chiral shift reagents has been used to resolve the ester groups at the center of chirality of triacylglycerols, permitting a measure of enantiomeric purity and confirming the specific location of short-chain fatty acids in position *sn*-3 of certain natural fats (5). Neither of these techniques has any wider analytical value.

There was a sudden burst of interest in stereospecific analysis methodology in the mid-1960s, but this tended to fall off over a number of years. The opportunities opened up by developments in high-performance liquid chromatography (HPLC) have rejuvenated the topic.

Regiospecific Analysis with Lipases

Differences in the positional distributions of fatty acids on the glycerol moiety of natural triacylglycerols were first demonstrated systematically by means of hydrolysis with the enzyme pancreatic lipase, which permits the determination of the composition of position *sn*-2. This should be termed regiospecific (not stereospecific) analysis. The enzyme hydrolyzes the fatty acids from both primary positions of triacylglycerols, leaving 2-monoacyl-*sn*-glycerols as illustrated in Figure 10.2, and these can be isolated by chromatographic means for fatty acid analysis. Pig pancreatin is the most widely used source of the enzyme, and commercial preparations are inexpensive and are stable for long periods of time. Those triacylglycerols containing the normal range of saturated and mono-, di-, and trienoic fatty acids are apparently hydrolyzed at about the same rate. Glycerol esters of fatty acids such as 22:6(*n*-3) from fish oils, *trans*-3-hexadecanoic acid from some plant sources, phytanic acid and even γ-linolenic acid are hydrolyzed more slowly, because of steric hindrance between substituent groups and the ester bonds. In addition, pancreatic lipase hydrol-

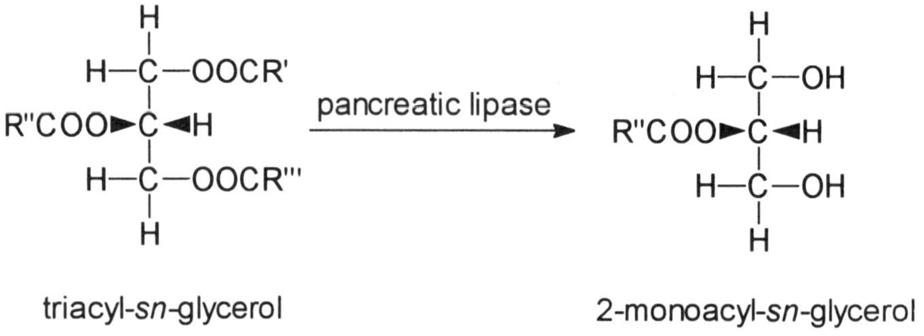

Figure 10.2. Pancreatic lipase hydrolysis of a triacyl-*sn*-glycerol.

yses triacylglycerol molecules that contain short-chain fatty acids more rapidly than those with only longer chain components; with a triacylglycerol such as 1-butyro-2,3-dipalmitin, the fatty acids on both primary positions are hydrolyzed at about the same rate, but faster than for tripalmitin. These problems do not arise with molecules containing a more normal range of fatty acids, where the 2-monoacylglycerols formed are representative of those in the native triacylglycerols.

In practice, calcium ions are essential for the reaction, bile salts are helpful, and the triacylglycerols must be well dispersed by vigorous shaking so that they are in a micellar form for hydrolysis to occur. Methyl oleate or cyclohexane is sometimes added to relatively saturated fats with high melting points to act as a carrier, but preincubation at 42°C for 5 min may be preferable (6). For optimal reaction, the concentrations of the various cations, bile salts, and the enzyme, the pH of the buffer and the temperature must be adjusted so that appreciable hydrolysis (50–60% is sufficient) occurs in a short time (1–2 min), thus minimizing undesirable side reactions, such as acyl migration. A semimicromethod developed by Luddy et al. (7) is probably the best practical procedure.

Because of the simplicity of the experimental technique, there has been a tendency to assume that the composition of fatty acids on the sole secondary hydroxyl group must have greater importance than those on the two primary positions. Certainly the composition of position sn-2 is of great importance when natural oils and fats are digested by animals, since 2-monoacyl-sn-glycerols are then formed which can be absorbed by the intestines and utilized as such. However, true stereospecific analyses have shown that the composition of all three positions in a fat can be distinctive and can cast light on important aspects of its biosynthesis.

Stereospecific Analysis Utilizing Lipases

As no lipase capable of distinguishing between positions 1 and 3 of a triacyl-sn-glycerol was known, it was necessary to develop stereospecific analytical procedures which made use of the stereospecificity of other enzymes. Brockerhoff (8) devised the first generally applicable procedure, and although this has been modified somewhat by subsequent research, it still provides the standard against which others must be judged. It is illustrated schematically in Figure 10.3. An equimolar mixture of the 1,2- and 2,3-sn-diacylglycerols was first prepared via partial hydrolysis by chemical or enzymatic means for synthetic conversion to phospholipid derivatives, which were in turn hydrolyzed by the stereospecific phospholipase A of snake venom, that is, an enzyme which reacts only with the "natural" 1,2-diacyl-sn-glycerophosphatide. The products were a lysophosphatide which contained the fatty acids originally present in position sn-1, unesterified fatty acids released from position sn-2, and unchanged "unnatural" 2,3-diacyl-sn-phosphatide. After isolation and *trans*-methylation, the fatty acid composition of each product was determined by gas chromatography (GC). The composition of position sn-2 was determined separately by means of pancreatic lipolysis as a check. Only the fatty acid

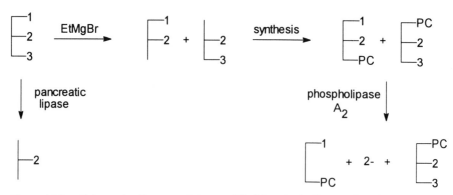

Figure 10.3. Schematic diagram of the modified Brockerhoff procedure for stereospecific analysis of triacyl-sn-glycerols (8). Abbreviations: PC, phosphorylcholine.

composition of position *sn*-3 was not determined directly by this method, but this could be calculated from the data for positions *sn*-1 and -2. The composition of position *sn*-3 could be determined directly by hydrolyzing the unchanged 2,3-diacyl-*sn*-phosphatide with the lipase from *Rhizopus arrhizus*, thus releasing the fatty acids from position *sn*-3 for analysis.

At first, hydrolysis with pancreatic lipase was used for the preparation of the intermediate diacylglycerols, but a Grignard reagent, ethyl magnesium bromide, was soon preferred because it had no fatty acid specificity and caused less acyl migration than other chemical methods. In the first Brockerhoff procedure (8), phosphatidylphenols were prepared from the diacylglycerols, but phosphatidylcholines, with better understood chromatographic properties, are easier to prepare and are now considered to be better for the purpose (9,10). For example, they can be hydrolyzed enzymatically by phospholipase C with some stereoselectivity, affording an additional analytical option (10).

Full experimental details for a recommended procedure incorporating features developed in several laboratories have been published by the author (3). Although this has many synthetic, enzymatic, and chromatographic steps, it gives good results in the hands of skilled workers.

An alternative Brockerhoff procedure (11), utilizing 1,3-diacylglycerols as the key intermediate, has proved to be of limited value. Similarly, a procedure developed by Lands et al. (12) has been little used, although re evaluation might be beneficial. Here, 1,2-/2,3-diacyl-*sn*-glycerols were prepared from the triacylglycerols, and they were phosphorylated by an enzyme, diacylglycerol kinase from *E. coli*, which is completely stereospecific for 1,2-diacyl-*sn*-glycerols. The fatty acid composition of the phosphatidic acid formed was determined, and that of position *sn*-2 was obtained separately by means of pancreatic lipolysis, so that the compositions of each of the primary positions could be calculated. One factor hindering development and use of the method has been the difficulty of preparing the enzyme, but this should no longer be a limitation as it is now available from commercial sources.

Chiral Chromatography via Diastereomeric Derivatives for Stereospecific Analysis

Most lipid analysts have an aversion to carrying out reactions for which aqueous media are required, such as enzymatic hydrolyses. In the last 2 years, alternative approaches to stereospecific analysis have been described that use simple chemical degradative and derivatization steps and the methodology of chiral chromatography [reviewed comprehensively elsewhere (13)]. One such method was developed in the author's laboratory and is shown schematically in Figure 10.4 (14–16). The principle is dependent on the fact that diastereomeric compounds can be resolved by adsorption chromatography on silica gel, because they have different physical and chemical properties. Thus a chiral derivatizing agent (R)-X will react with a racemic substance (R,S)-Y:

$$(R)\text{-}X + (R,S)\text{-}Y \longrightarrow (R)\text{-}X\text{-}(R)\text{-}Y + (R)\text{-}X\text{-}(S)\text{-}Y$$

The degree of separation of the two diastereomers in a chromatographic system will depend on the chiral structures of X and Y and the manner of their interactions with the mobile and stationary phases. In addition, it is noteworthy that the order of elution of the diastereomeric derivatives is reversed if the other enantiomer of the reagent, that is (S)-X, can be employed. With an HPLC column packed with silica gel, it appears that the presence of the hydrogen atom on the nitrogen atom between the chiral centers in the preferred derivatives is essential to the separation process and is presumed to be a primary site for hydrogen bonding to silanols on the adsorbent surface.

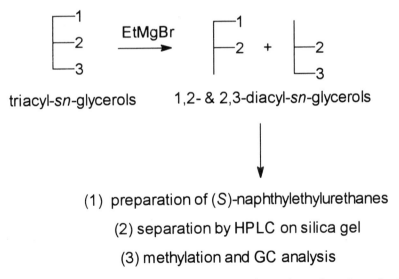

Figure 10.4. Schematic diagram for stereospecific analysis of triacyl-sn-glycerols via resolution of diastereomeric (S)-naphthylethyl urethane derivatives of diacyl-sn-glycerols by HPLC on silica gel (14,15).

The first step in the stereospecific analysis procedure is identical to that in most other methods, that is, partial hydrolysis of the triglycerides with ethyl magnesium bromide, giving among other products, a mixture of sn-1, 2-, 2,3-, and 1,3-diacylglycerols. While Grignard reagents do appear to cause less acyl migration during hydrolysis than occurs with other chemical reagents, this is still far from perfect and there is a need for a better method. Allyl magnesium bromide is reported to cause much less acyl migration and use of this hydrolytic reagent should be explored (17).

The second step involves reacting the products with a chiral derivatizing agent, (S)-(+)-1-(1-naphthyl)ethyl isocyanate, and isolation of the diacyl-sn-glycerol urethane derivatives (Figure 10.5) by chromatography on solid-phase extraction columns containing an octadecylsilyl phase [also reviewed elsewhere (18)]. In essence, the separation in this application is on the basis of molecular weight so that low molecular weight by-products and excess derivatizing reagents are eluted first and discarded, before the required diacylglycerol urethane derivatives are recovered.

The third and most important step involves resolution of the diacylglycerol urethanes by HPLC, simply on a column of silica gel. A simple isocratic mobile phase is employed, and since the derivatives absorb strongly in the ultraviolet, detection was straightforward. 1,3-Diacylglycerol urethanes elute early and this fraction is easily recovered; however, they are not used further as they are more susceptible to acyl migration than the other products of interest. As the derivatizing agent is chiral and a single enantiomer, the 1,2- and 2,3-diacyl-sn-glycerol urethanes formed from it are now diastereomers, so they are separable in a nonchiral environment. In practice, the 1,2-diacyl-sn-glycerol derivatives elute ahead of the 2,3-diastereomers, and the two distinct fractions can be collected.

Some separation of molecular species occurs within each diastereomeric fraction, and while this might be considered an advantage in some circumstances, for this purpose it is something of a nuisance because it restricts the range of fatty acid components in the triacylglycerols that can be investigated. However, most of the common fats with C_{16} and C_{18} fatty acids are in the practical range, and it may be possible with further development to extend this. For example, the HPLC resolution of the diastereomeric diacylglycerol urethane derivatives prepared from egg triacyl-sn-glycerols are illustrated in Figure 10.6. When the heights of the analogous peaks in the 1,2- and 2,3-diacyl-sn-glycerol groups are compared, it is evident that the tri-

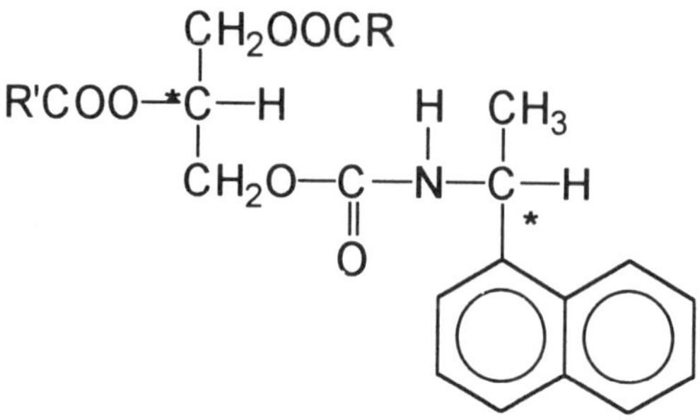

Figure 10.5. (S)-(+)-1-(1-Naphthyl)ethyl urethane (or carbamate) derivative of a 1,2-diacylglycerol.

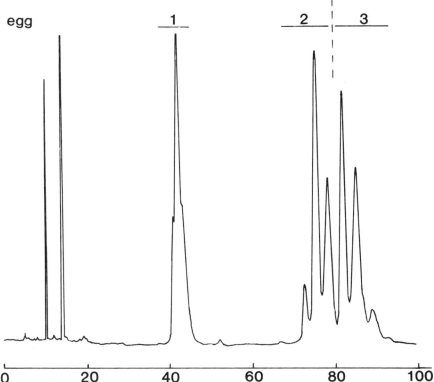

Figure 10.6. HPLC resolution of (S)-(+)-1-(1-naphthyl)ethyl urethane derivatives of diacyl-sn-glycerols prepared from egg triacyl-sn-glycerols. The mobile phase was isooctane with 0.33% 1-propanol (containing 2% water). The numbers 1, 2, and 3 above peaks refer to the 1,3-, 1,2-, and 2,3-sn-diacylglycerol derivatives respectively (15).

acyl-sn-glycerols are indeed highly asymmetric. The main component of the former is the palmitoyloleoyl species, while in the latter it is the dioleoyl species.

The final step involves methylation of each of the fractions for analysis by GC with the highest precision possible (19). Then, the results for the positional distributions are simply a matter of calculation. For example, as the fatty acid composition of the intact triacylglycerols is known, and that of the 1,2-diacyl-sn-glycerol derivatives has been determined, it is simple arithmetic to calculate the composition of position sn-3. Similarly, the composition of position sn-1 can be calculated by difference once that of the 2,3-diacylglycerols is known. That of position sn-2 can then be calculated by difference. Thus, the composition of all three positions is determined, without resorting to enzymes, by using standard chromatography columns and derivatizing agents that are readily available from commercial sources.

Racemic diacylglycerols in the form of R-(+)-1-phenylethyl urethane derivatives also have been resolved by HPLC on columns of silica gel in studies designed to determine the positional specificity of lipases in the hydrolysis of natural triacyl-sn-glycerols (20,21).

Stereospecific Analysis Utilizing HPLC with Chiral Phases

A different but comparable approach has been adopted by Professor Takagi and colleagues in Japan (22–28). They converted mono- and diacyl-sn-glycerols prepared from triacylglycerols to the 3,5-dinitrophenylurethane (DNPU) derivatives (Figure 10.7B) for resolution by HPLC on columns containing a stationary phase with chiral moieties bonded chemically to a base of silica gel, such as that illustrated (Sumipax™ OA-4100 in Figure 10.7A). The 3,5-dinitrophenyl moieties of the urethanes contributed to charge-transfer interactions with functional groups having π electrons on the stationary phase and thus aided the resolution.

In the preferred technique, 1 and 3-monoacyl-sn-glycerols were prepared from triacylglycerols by partial hydrolysis with ethyl magnesium bromide. Initially, these were isolated by TLC on silica gel impregnated with boric acid, separated into satu-

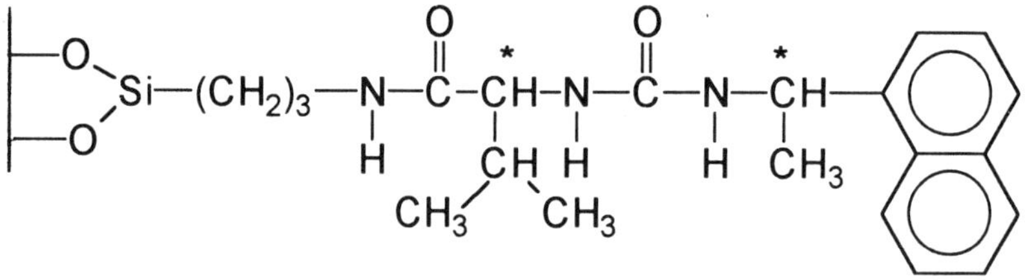

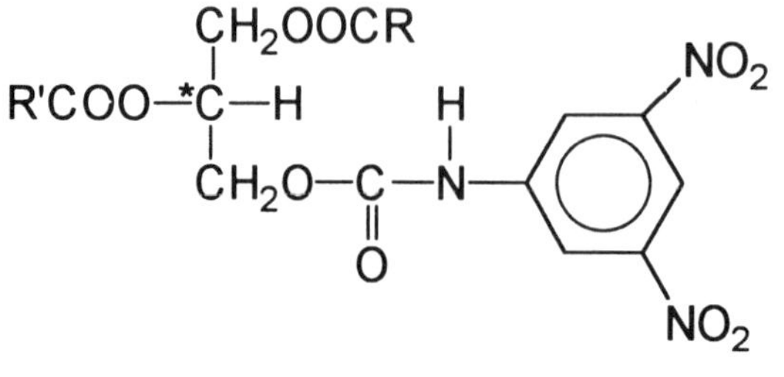

Figure 10.7. (a) Chiral moiety bonded chemically to a base of silica gel of Sumipax™ OA-4100. (b) DNPU derivative of a diacyl-sn-glycerol.

rated and unsaturated fractions by silver-ion chromatography, converted to the DNPU derivatives and then resolved on a chiral column. The distributions of fatty acids in the *sn*-1, -2, and -3 positions could be calculated from the data. Subsequently, an improved procedure was described in which a better chiral column was employed, so that there was no need for the step involving silver-ion chromatography. By lowering the column temperature and slowing down the flow rate, the method could even be applied to such complex triacyl-*sn*-glycerols as fish oils. It is illustrated in Figure 10.8 for 1 and 3-monoacyl-*sn*-glycerols, prepared from triacylglycerols of a fish oil by partial hydrolysis with ethyl magnesium bromide and separated as the DNPU derivatives by chiral-phase HPLC. The positional distributions in each of the *sn*-1, -2, and -3 positions could be calculated by analysis of the products.

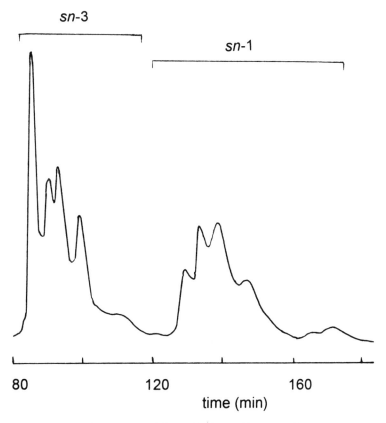

Figure 10.8. HPLC of 1-monoacylglycerols, formed by partial hydrolysis of herring oil triacyl-*sn*-glycerols with ethyl magnesium bromide, as their di-3,5-dinitrophenyl-urethanes derivatives on a chiral column OA-4000 (500 × 4 mm i.d.); mobile phase, hexane-1,2-dichloroethane-ethanol (40:12:3 v/v) at a flow rate of 0.5 mL/min at −7°C with detection at 254 nm (28). Reproduced by permission of the authors and the *Journal of the American Oil Chemists' Society*, redrawn from the original.

This procedure will certainly be widely used in the future, as the special chiral columns become less expensive and the required derivatizing agent becomes more readily available. Independent confirmation should be obtained that representative monoacylglycerols can indeed be produced without significant acyl migration by means of Grignard hydrolysis. As with the procedure developed in the author's laboratory, there is no need to use lipases.

Bezard and co-workers adapted related methodology to the analysis of molecular species of triglycerides in combination with stereospecific analysis (29–30). The procedure has recently been used by others for a seed oil (31).

Parallel resolutions of DNPU derivatives of natural chiral diacyl-*sn*-glycerols and structurally related compounds have also been accomplished by HPLC on chiral phases. For example, clean separations of single-acid 1,3-, 1,2-, and 2,3-diacyl-*sn*-glycerols (and analogous other lipids) was achieved and straight-line relationships between the logarithms of the retention volumes and carbon numbers were observed for homologous series of each form, especially with the Sumipax™ OA-4100 column (32–34). Molecular species differing by two carbon atoms were separated to the baseline and distinct peaks were obtained for homologues differing in total chain length by as many as six carbon atoms. As with the monoacylglycerol derivatives described previously, a long column (at a temperature as low as -30°C), slow flow rate, and a mobile phase of low polarity were utilized for optimum resolution.

A column in which a polymer of (*R*)-(+)-1-(1-naphthyl)ethylamine moieties was bonded chemically to silica gel (YMC-Pack A-K03™) was applied to the resolution of diacyl-*sn*-glycerol derivatives in a similar manner (35). Such compounds generated from natural triacyl-*sn*-glycerols by partial hydrolysis with a Grignard reagent were resolved, and the method was used for stereospecific analysis of the latter (36). In addition, it was utilized to isolate 1,2- and 2,3-*sn*-diacylglycerols so that the molecular species of each could be determined by capillary GC on a polar stationary phase (37). In this instance, the DNPU derivatives were resolved on the chiral column, before the parent diacylglycerol molecules were regenerated by reaction with trichlorosilane (without acyl migration). The methodology was used to study the specificity of enzymes involved in triacylglycerol biosynthesis (38,39).

Applications to Natural Fats and Oils

Much data was obtained by means of pancreatic lipase hydrolysis on the composition of position *sn*-2 of natural triacyl-*sn*-glycerols in the 1960s, and stereospecific analyses have provided much insight recently. As was well known from the early studies (40), position *sn*-2 of the triacylglycerols of seed oils tends to be greatly enriched in linoleic and linolenic acids, while saturated fatty acids are concentrated in the primary positions, and monoenoic acids are distributed among all three positions in similar proportions. There are exceptions to the rule, however, and cacao butter is an example where oleic acid is present largely in position *sn*-2. Although stereospecific analyses showed minor differences only in the distributions of saturated and monoenoic fatty acids between positions sn-1 and sn-3, it could be argued that too few samples have been analyzed for definitive comment. Longer chain fatty acids (C_{20}–C_{24}) are usually concentrated in the primary positions and may have a small preference for position sn-3. Coconut oil, which is relatively saturated, was reported to be more asymmetric than most other seed oils, with myristic and palmitic acids being concentrated in position sn-1, dodecanoic acid in position sn-2, and octanoic and decanoic acids in position sn-3 (41). In some seed oils containing unusual fatty acids, highly specific distributions have been found, for example an allenic estolide was found entirely in position sn-3 in triacylglycerols from *Sapium sebiferum* (42).

While there may appear to be little difference in the composition of positions sn-1 and sn-3 in seed oils, a close investigation has revealed greater underlying complexity in olive oil (16). The proportion of oleic acid in positions sn-1 and sn-3 was close to 70% in whole oil, tending to suggest that the triacylglycerols were not in fact asymmetric. When molecular species isolated from olive oil by silver-ion HPLC were subjected to the stereospecific analysis procedure, however, marked asymmetry was observed in some fractions. The trimonoenoic species (50% of the total) was symmetrical, but other species were not. In most fractions, oleic acid was predominantly in position sn-2, for example. In the disaturated-monoenoic fraction, there was a higher proportion of oleate in position sn-3 than in position sn-1, but the opposite was true of the saturated-dimonoenoic and saturated-monoene-diene fractions. Such results could not possibly have been predicted from the results of the stereospecific analysis of the whole triacyl-sn-glycerols. Therefore, it is evident that the structures of seed oils may be more complex than was originally believed, and this can only be revealed if a wide range of analytical methodology is applied to the problem.

In animal tissues, a high proportion of the triacylglycerols are located in adipose tissue. Substantial interspecies differences in the distributions of fatty acids among the three positions of the glycerol moiety have been found. For many species, saturated fatty acids are most abundant in position sn-1, although appreciable amounts of oleic acid may also be present. Position sn-2 tends to contain unsaturated fatty acids, especially linoleic acid, together with shorter chain fatty acids in some species; there is some preference for the longer chain fatty acids to be located in position sn-3. The main exception is the pig and related animals, where it is well known that palmitic comprise more than 70% of the fatty acids in position sn-2; it is less well known but just as interesting that most of the stearic acid is in position sn-1, while position sn-3 is occupied by more than 70% of oleic acid (43). Although the absolute fatty acid compositions of adipose tissue at different sites in an animal can vary somewhat, the relative distributions among the three positions tend to be similar.

The structures of milk lipids have been reviewed (44). In stereospecific analyses of bovine-milk triacylglycerols, the technical problems are substantial because of the presence of short-chain fatty acids, which give rise to difficulties in the isolation of the required diacylglycerol intermediates. It has proved possible to overcome these problems, and it is now well established that most of the butyric and hexanoic acids are in position *sn*-3 (45). Thus, the triacylglycerols of cows' milk are the most asymmetric to have been found in any animal. The fatty acid composition of the triacylglycerols of milk varies greatly for different species, depending both on the diet and the nature of the fatty acids synthesized in the mammary gland, but certain common structural features have been found with respect to the longer chain fatty acids. For nearly all species, much of the palmitic acid is located in position *sn*-2. Other than in the pig, distributions with a high palmitic acid content in position *sn*-2 are not common in animal triacylglycerols. Like palmitic acid, myristic and the medium-chain fatty acids are present in the greatest concentration in position *sn*-2 in cows' milk triacylglycerols, but stearic acid is concentrated in the primary positions, especially position *sn*-1, while the unsaturated fatty acids tend to be located in higher proportions in positions *sn*-1 and *sn*-3.

Structural analyses have been performed on triacylglycerols from a variety of animal tissues, in addition to those of commercial importance, as the information obtained is of great value to biochemists. To give one example, palmitic acid comprised over 70% of the fatty acids of position *sn*-1 of the plasma triacylglycerols in the chicken with relatively small amounts in positions *sn*-2 and *sn*-3, while oleic acid comprised 60% of the fatty acids in position *sn*-1 and more than 70% of that in position *sn*-3 (17). As virtually identical structures were found in the ovarian follicles (46) and in the egg (47), a common biosynthetic origin seemed to be indicated.

Conclusions

The complexity of the procedures for stereospecific analysis may have deterred many chemists and biochemists from exploring their potential. While the problems can be exaggerated, it is to be hoped that the improved methods involving chiral chromatography, will greatly simplify the task in future.

Acknowledgment

This paper is published as part of a program funded by the Scottish Office Agriculture and Fisheries Dept.

References

1. IUPAC-IUB Commission on Biochemical Nomenclature. (1967) *Biochem. J. 105*:897–902.
2. Weiss, S.B., and Kennedy, E.P. (1956) *J. Am. Oil Chem. Soc. 78*:3550–3551.
3. Christie, W.W. (1986) In *The Analysis of Oils and Fats*. Edited by R.J. Hamilton and J.B. Rossell, Elsevier Applied Science, London. pp. 313–339.
4. Anderson, B.A., Sutton, C.A., and Pallansch, M.J. (1970) *J. Am. Oil Chem. Soc. 47*:15–16.
5. Bus, J., Lok, C.M., and Groenewegen, A. (1976) *Chem. Phys. Lipids 16*:123–132.
6. Rossell, J.B., King, B., and Downes, M.J. (1983) *J. Am. Oil Chem. Soc. 60*:333–339.
7. Luddy, F.E., Barford, R.A., Herb, S.F., Magidman, P., and Riemenschneider, R.W. (1964) *J. Am. Oil Chem. Soc. 41*:693–696.

8. Brockerhoff, H. (1965) *J. Lipid Res. 6:*10–15.
9. Myher, J.J., and Kuksis, A. (1979) *Can. J. Biochem. 57:*117–124.
10. Christie, W.W., and Hunter, M.L. (1980) *Biochem. J. 191:*637–643.
11. Brockerhoff, H. (1967) *J. Lipid Res. 8:*167–169.
12. Lands, W.E.M., Pieringer, R.A., Slakey, P.M., and Zschocke, A. (1966) *Lipids 1:*444–448.
13. Christie, W.W. (1992) In *Advances in Lipid Methodology—One.* Edited by W.W. Christie, Oily Press, Ayr. pp. 121–148.
14. Laakso, P., and Christie, W.W. (1990) *Lipids 25:*349–353.
15. Christie, W.W., Nikolova-Damyanova, B., Laakso, P., and Herslof, B. (1991) *J. Am. Oil Chem. Soc. 68:*695–701.
16. Santinelli, F., Damiani, P., and Christie, W.W. (1992) *J. Am. Oil Chem. Soc. 69:*552–556.
17. Becker, C.C., Rosenquist, A., and Holmer, G. (1993) *Lipids 28:*147–149.
18. Christie, W.W. (1992) In *Advances in Lipid Methodology—One.* Edited by W.W. Christie, Oily Press, Ayr. pp. 1–17.
19. Christie, W.W. (1991) *Lipid Technology 3:*97–98.
20. Rogalska, E., Ransac, S., and Verger, R. (1990) *J. Biol. Chem. 265:*20271–276.
21. Rogalska, E., Cudrey, C., Ferrato, F., and Verger, R. (1993) *Chirality 5:*24–530.
22. Takagi, T., and Ando, Y. (1990) *Yukagaku 39:*622–628.
23. Takagi, T., and Ando, Y. (1991) *Lipids 26:*542–547.
24. Takagi, T., and Ando, Y. (1991) *Yukagaku 40:*288–292.
25. Suzuki, T., Ota, T., and Takagi, T. (1992) *J. Chromatogr. Sci. 30:*315–318.
26. Takagi, T., and Suzuki, T. (1992) *J. Chromatogr. 625:*163–168.
27. Takagi, T., and Suzuki, T. (1993) *Lipids 28:*251–253.
28. Ando, Y., Nishimura, K., Aoyanagi, N., and Takagi, T. (1992) *J. Am. Oil Chem. Soc. 69:*417–424.
29. Sempore, G., and Bezard, J. (1991) *J. Chromatogr. 557:*227–240.
30. Sempore, G., and Bezard, J. (1991) *J. Am. Oil Chem. Soc. 68:*702–709.
31. MacKenzie, S.L., Giblin, E.M., and Mazza, G. (1993) *J. Am. Oil Chem. Soc. 70:*629–631.
32. Takagi, T., and Itabashi, Y. (1987) *Lipids 22:*596–600.
33. Takagi, T., Okamoto, J., Ando, Y., and Itabashi, Y. (1990) *Lipids 25:*108–110.
34. Takagi, T., and Suzuki, T. (1990) *J. Chromatogr. 519:*237–243.
35. Itabashi, Y., Kuksis, A., Marai, L., and Takagi, T. (1990) *J. Lipid Res. 31:*1711–1717.
36. Yang, L.-Y., and Kuksis, A. (1991) *J. Lipid Res. 32:*1173–1186.
37. Itabashi, Y., Kuksis, A., and Myher, J.J. (1990) *J. Lipid Res. 31:*2119–2126.
38. Lehner, R., and Kuksis, A. (1993) *J. Biol. Chem. 268:*8781–6786.
39. Lehner, R., Kuksis, A., and Itabashi, Y. (1993) *Lipids 28:*29–34.
40. Brockerhoff, H., and Yurkowski, M. (1966) *J. Lipid Res. 7:*62–64.
41. Breckenridge, W.C. (1978) In *Handbook of Lipid Research.* Vol. 1. Edited by A. Kuksis, Plenum Press, New York. pp. 197–232.
42. Christie, W.W. (1969) *Biochim. Biophys. Acta 187:*1–5.
43. Christie, W.W., and Moore, J.H. (1970) *Biochim. Biophys. Acta 210:*46–56.
44. Christie, W.W. (1984) In *Developments in Dairy Chemistry: Lipids.* Edited by P.F. Fox, Applied Science Publishers, London. pp. 1–35.
45. Pitas, R.E., Sampugna, J., and Jensen, R.G. (1967) *J. Dairy Sci. 50:*1332–336.
46. Christie, W.W., and Moore, J.H. (1972) *Comp. Biochem. Physiol. 41B:*287–295.
47. Christie, W.W., and Moore, J.H. (1970) *Biochim. Biophys. Acta 218:*83–88.

Chapter 11
Structural Analysis of Peanut Oil Triacylglycerols

J.A. Bézard and B.G. Semporé

Université de Bourgogne, Unité de Nutrition Cellulaire et Métabolique, DRED EA 564, BP 138, 21004 F-Dijon, Cédex, France.

Introduction

Vegetable oils are extremely complex mixtures of molecular species of triacylglycerols. A triacylglycerol molecule consists of a trihydric alcohol, glycerol, each position of which is esterified to generally different long-chain fatty acids. If n represents the number of the different fatty acids detected in total triacylglycerols, the maximum of possible molecular species is n^3 (1). When only considering the major constituent fatty acids in the most simple seed oils, $n = 4$ and the number of molecular species amounts to 64. Obviously, the number of molecular species increases considerably when the minor fatty acids present in the oil triacylglycerols are taken into account. For example, in peanut oil the number of fatty acids that amount to more than 1 mol% of the total $n = 8$, and the maximum number ($N = n^3$) of possible molecular species of triacylglycerols amounts to 512 (2). Owing to the complexity of the natural oil triacylglycerol mixtures, analysis of the component triacylglycerol molecular species was carried out through several steps.

Step One

By taking into account only the nature of the three fatty acids esterified at the glycerol moiety without considering the position, the maximum number of possible triacylglycerols that can be formed from n different fatty acids can be calculated from the formula (1):

$$\frac{n^3 + 3n^2 + 2n}{6}$$

When only four major fatty acids are present, the maximum number of possible triacylglycerols is 20, instead of 64 molecular species, and when taking into account the eight peanut-oil fatty acids that amount to more than 1%, the number of possible molecular species drops from 512 to 120.

Analysis of the constituent triacylglycerols of seed oils without distinguishing positional isomers, is generally possible only after fractionation according to the two main characteristics of the constituent fatty acids unsaturation and carbon number. Fractionation according to unsaturation was first successfully performed by silver-ion thin-layer chromatography (TLC) (3). At present, silver-ion high-performance liquid chromatography (HPLC) (4) is preferred because it is less time consuming and because higher amounts of triacylglycerols can be fractionated for further analysis.

Peanut-oil triacylglycerols were first fractionated by silver-ion TLC in three steps to get better separations of the triacylglycerol classes (Figure 11.1 and Table 11.1) (2). Under these conditions 10 triacylglycerol classes were fractionated. Two fractions were poorly separated in conditions reported in B, and were rechro-

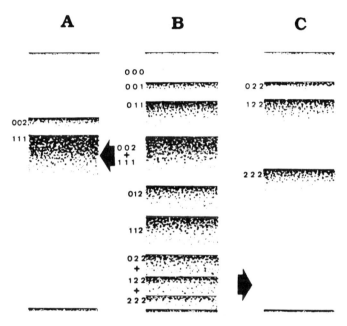

Figure 11.1. Separation of peanut oil triacylglycerols by TLC on silica impregnated with silver nitrate. Triacylglycerols of the band (002 + 111) and of the band (022 + 122 + 222) were rechromatographed in conditions A and C respectively, for better separation. Conditions for separation are listed in Table 11.1. *Source*: Semporé and Bézard (2).

TABLE 11.1
Conditions for Separation of Peanut-Oil Triglycerides

		A	B	C
Thickness (µm)		250	500	250
Silver nitrate%		20	10	20
Solvent	Hexane		158	155
(mL)	Ether		40	40
	Methanol	1	2	5
	Chloroform	199		

Source: Semporé and Bézard (2).

matographed under better conditions in A and C, respectively. Generally the triacylglycerol mixtures of each class are too complex for individual triacylglycerols to be identified and quantified and should be further fractionated according to carbon number by GLC (2) or by reverse-phase HPLC (5,6).

For peanut-oil triacylglycerols analysis (2), Figure 11.2 shows the chromatogram registered in the GLC analysis of the 011 triacylglycerols class. It shows six peaks corresponding to the six saturated fatty acids (0) present in the oil (16:0 to 26:0). The last peak was identified as containing 26:0. These six saturated fatty acids were associated with two positions containing molecules of monounsaturated fatty acids (1), and oleic acid represented more than 97% of the monounsaturated fatty acids of this fraction.

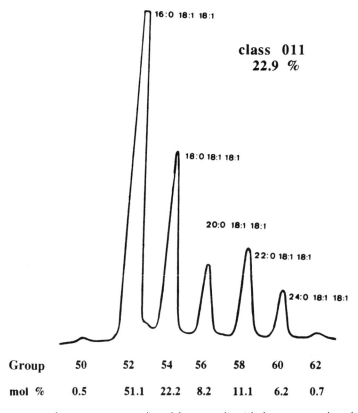

Figure 11.2. Chromatogram produced from gas-liquid chromatography of the class 011 triacylglycerols. Each peak corresponds to a group of triacylglycerols of same carbon number (total number of acyl carbon atoms). Group 62 corresponds to 26:0-18:1-18:1 later identified. *Source:* Semporé and Bézard (2).

TABLE 11.2
Major Triacylglycerols of Peanut Oil (> 1 mol%)

Triacylglycerols[a]		mol% Experimental	mol% 1,2,3-Random[b]
Triolein	18:1-18:1-18:1	24.6	21.9
Dioleolinolein	18:1-18:1-18:2	17.2	19.6
Palmitodiolein	16:0-18:1-18:1	11.7	11.4
Palmitooleolinolein	16:0-18:1-18:2	7.1	6.8
Oleodilinolein	18:1-18:2-18:2	5.5	5.9
Stearodiolein	18:0-18:1-18:1	4.8	3.8
Behenodiolein	22:0-18:1-18:1	2.6	3.3
Stearooleolinolein	18:0-18:1-18:2	2.5	2.3
Arachidodiolein	20:0-18:1-18:1	1.7	1.7
Behenooleolinolein	22:0-18:1-18:2	1.7	1.9
Palmitodilinolein	16:0-18:2-18:2	1.5	1.0
Gadoleodiolein	20:1-18:1-18:1	1.5	1.5
Lignocerodiolein	24:0-18:1-18:1	1.4	1.7
Dipalmitoolein	16:0-16:0-18:1	1.4	2.0
Arachidooleolinolein	20:0-18:1-18:2	1.2	1.0
Palmitostearoolein	16:0-18:0-18:1	1.2	1.3
Palmitobehenoolein	16:0-22:0-18:1	1.0	1.1
Lignocerooleolinolein	24:0-18:1-18:2	1.0	1.0
Oleogadoleolinolein	18:1-20:1-18:2	1.0	0.9

[a]Triacylglycerols with the same three component fatty acids but the position of which in the glycerol moiety is not known.
[b]Assuming that three different pools of fatty acids are separately distributed to the 1, 2, and 3 positions of all the glycerol molecules (1).
Source: Semporé and Bézard (2).

From data obtained by GLC fatty acid analysis of the different fractions isolated according to unsaturation and refractionated according to carbon number, it was possible to identify and quantify 62 triacylglycerols representing 99.9% of the total oil triacylglycerols (7,8). The 19 major triacylglycerols are listed in Table 11.2. Their percentages can be compared to those predicted by a random distribution (9,10). Data show that trioleoylglycerol (triolein) was the major triacylglycerol amounting to nearly one-fourth of the total triacylglycerols, an amount slightly higher than the random hypothesis predicted (nearly 22%).

Step Two

The next step in oil triacylglycerol analysis is the so-called regiospecific analysis, that is the determination of the fatty acids esterified at the internal (*sn*-2) position of the glycerol moiety. The number of triacylglycerols for which the fatty acid in the *sn*-2 position is known is given by the formula (1):

$$\frac{(n^3 + n^2)}{2}$$

for n different fatty acids.

The method is based on the regiospecific deacylation by mammalian pancreatic lipase (11–13), generally porcine lipase. The enzyme possesses near-absolute specificity for hydrolysis of primary *sn*-1 and *sn*-3 ester linkages in triacylglycerols, generating 2-monoacyl-*sn*-glycerols the fatty acids of which represent those esterifying the *sn*-2 position. To obtain more information on the fatty acid distribution at the *sn*-2 position of the triacylglycerols, partial deacylation by pancreatic lipase was carried out on triacylglycerols of the same degree of unsaturation that were fractionated by silver-ion chromatography.

This method was applied to the analysis of an African peanut oil (14). Nine classes of triacylglycerols were separated based on unsaturation by silver-ion TLC, and subjected to partial deacylation by rat pancreatic juice using the semimicromethod developed by Luddy et al. (15). Monoacylglycerols were analyzed by GLC for fatty acid composition. Diacylglycerols were isolated by TLC, acetylated and fractionated according to unsaturation into classes that were analyzed for fatty acid composition by GLC. The data obtained from both the monoacyl and diacylglycerols formed by hydrolysis of triacylglycerols of the same class, allowed quantification of the triacylglycerol isomers present. Those are defined as triacylglycerols comprising the same three fatty acids but differing by the fatty acid esterified at the *sn*-2 position.

A complete description of the calculation method was given elsewhere for the most complex class 012 (16). In this class, 18 isomers were identified and quantified. This method was applied to the 9 classes which allowed the proportion of 84 triacylglycerol isomers to be determined (14). The 17 isomers amounting to more than 1% of the total are listed in Table 11.3. The percentages thus calculated can be compared to those predicted by a 1,3-random-2-random distribution (10). In these 17 isomers the *sn*-2 position was occupied by an unsaturated fatty acid; in 11 isomers oleic acid was the major fatty acid of the oil, at 60% of the total. The more unsaturated linoleic acid was present in the remaining six isomers, even though it only composed 18% of the oil. The first isomer with a saturated fatty acid in the internal position was 2-palmitoyl-1(3)-dioleoyl-*sn*-glycerol (18:1-16:0-18:1) It only represented 0.53%, despite the relatively high proportion of 16:0 (10.5%) in the peanut oil analyzed. The preferential incorporation of unsaturated fatty acids, especially linoleic acid, at the *sn*-2 position of vegetable oils has been reported frequently (10).

Step Three

The last step in the analysis of the fatty acid distribution in triacylglycerol molecules is to determine the fatty acids present at the stereospecifically distinct, external *sn*-1 and *sn*-3 positions. The first methods used in such analyses were based on the stereospecificity of enzymes, especially phospholipases. In the last 3 years alternative methods based on chiral chromatography have been proposed.

Use of Lipases

Except for gastric lipase, which hydrolyzes ester linkages at the *sn*-3 position preferentially over the *sn*-1 position (17), no lipase is known that can distinguish

TABLE 11.3
Component Isomers (mol%) of Triacylglycerols of Peanut Oil

Isomers[a]	mol%	
	Experimental	Random[b]
Percentage > 1		
18:1-18:1-18:1	24.7	21.7
16:0-18:1-18:1	11.2	11.3
18:1-18:2-18:1	9.5	10.5
18:1-18:1-18:2	7.6	8.4
16:0-18:2-18:1	5.2	5.5
18:1-18:2-18:2	4.8	4.1
18:0-18:1-18:1	4.7	4.0
22:0-18:1-18:1	2.5	3.4
18:0-18:2-18:1	1.8	1.9
20:0-18:1-18:1	1.7	1.8
16:0-18:1-18:2	1.7	2.2
20:1-18:1-18:1	1.5	1.5
16:0-18:2-18:2	1.4	1.1
24:0-18:1-18:1	1.4	1.8
16:0-18:1-16:0	1.2	1.5
16:0-18:1-18:0	1.1	1.0
22:0-18:2-18:1	1.0	1.6

[a]The three component fatty acids and the fatty acid in the sn-2 position are known.
[b]Assuming that two different pools of fatty acids are separately and randomly distributed to the combined sn-1,3 positions and the sn-2 position of all glycerol molecules.
Source: Semporé and Bézard (14).

between the two external positions of a triacyl-sn-glycerol. However, phospholipases can differentiate the two positions. Brockerhoff (18) was the first to develop a stereospecific analytical procedure by using the stereospecificity of snake venom phospholipase A for phospholipid-like molecules from a mixture of 1,2- and 2,3-diacyl-sn-glycerols. The diacylglycerols were produced by a partial chemical or enzymatic deacylation. The enzyme only reacts with the 1,2-diacyl-sn-glycerophosphatides, generating lysophosphatides with the fatty acids from the sn-1 position of the deacylated triacylglycerols only. The composition of the sn-2 position was determined by means of pancreatic lipase, and the composition of the sn-3 position was deduced from data for the sn-1 and sn-2 positions.

When applied to a triacylglycerol fraction composed only of molecules with the same three component fatty acids, the Brockerhoff method allows determination of the molecular species present in the fraction with accuracy. An example is given for the palmitoyloleoyllinoleoylglycerol (16:0-18:1-18:2) isolated from cottonseed oil (19).

Triacylglycerols of unsaturation 012 were isolated by silver-ion TLC. They were separated according to carbon number by reverse-phase HPLC, and the fraction 16:0-18:1-18:2 was collected in sufficient quantities to be submitted to stereospecific analysis. This fraction was partially hydrolyzed by pancreatic lipase to obtain 2-monoacyl-sn-glycerols and deacylated by Grignard reagent to obtain a mixture of 1,2(2,3)-diacyl-sn-glycerols. Table 11.4 lists the fatty acid composition of the triacylglycerol, 16:0-18:1-18:2; the 2-monoacyl-sn-glycerols; and the 1,2(2,3)-diacyl-sn-glycerols obtained by GLC analysis and predicted from the triacylglycerol and monoacyl-sn-glycerol data.

Table 11.5 reports the fatty acid distributions of the three positions. Position sn-1 was determined from fatty acid composition of the formed 1-lyso-sn-phosphatide after snake venom phospholipase A reacted with 1,2-diacyl-sn-glycerophosphatide. Position sn-2 was determined from the fatty acid composition of the 2-monoacyl-sn-glycerol that had been formed by partial hydrolysis of 16:0-18:1-18:2 with pancreatic lipase. Position sn-3 was calculated from data for sn-1 and sn-2 and for the triacylglycerol 16:0-18:1-18:2. It is clear from the data that the fatty acid preference for each position is in the order:

$$sn\text{-}1: 16:0 > 18:1 > 18:2$$
$$sn\text{-}2: 18:2 > 18:1 \gg 16:0$$
$$sn\text{-}3: 16:0 \gg 18:1 = 18:2$$

TABLE 11.4
Fatty Acid Composition (mol%) of the Cottonseed-Oil Triacylglycerol 16:0-18:1-18:2

	Triacylglycerol	2 Monoacyl-sn-Glycerol[a]	1,2(2,3)Diacyl-sn-Glycerol Experimental[b]	Calculated[c]
16:0	34.6	2.3	27.5	26.5
18:1	33.0	41.9	34.7	35.2
18:2	32.4	55.8	37.8	38.3

[a]Obtained after hydrolysis by rat pancreatic lipase.
[b]Obtained after deacylation by Grignard reagent (bromoethylmagnesium).
[c]Calculated from data for triacylglycerol (1st column) and for 2-monoacyl-sn-glycerols (3rd column) according to the formula:
%X = (3 × %X in triacylglycerol)/4 + (%X in monoacyl-sn-glycerol)
Source: Bézard et al. (19).

TABLE 11.5
Cottonseed-Oil Palmitoyloleoyllinoleoylglycerol Fatty Acid Distribution (mol%) in the three Positions of the Glycerol Moiety

	Triacylglycerol	1-Lyso-sn-Phosphatide[a]	2-Monoacyl-sn-Glycerol[b]	Calculated[c]
Positions	1,2,3	1	2	3
16:0	34.6	46.1	2.3	55.4
18:1	33.0	34.8	41.9	22.3
18:2	32.4	19.1	55.8	22.3

[a]Obtained after deacylation by snake venom phospholipase A_2 of the phosphorylated derivatives of 1,2(2,3)diacyl-sn-glycerols issued from chemical deacylation of the triacylglycerol.
[b]Obtained after deacylation of the triacylglycerol by rat pancreatic lipase.
[c]Calculated from data according to the formula: %X = (3 × %X in triacylglycerol) – (%X in 1-lyso-sn-phosphatide) – (%X in 2-monoacyl-sn-glycerol).
Source: Bézard et al. (19).

TABLE 11.6
Determination of Molecular Species Composition (mol%) for Cottonseed-Oil Palmitoyoleoyllinoleoylglycerol

Triacyl-glycerol[a]	2-Triacyl-sn-Glycerols[b]	%	Triacylglycerol Molecular Species[c]		%	Equations[d]	Solutions[e]
			16:0-18:1-18:2	a		(1) $a + b = 41.9$	$a = 23.6$
	16:0-18:1-18:2	41.9					
16:0			18:2-18:1-16:0	b		(2) $a + e = 46.1$	$b = 18.3$
18:1	18:1-16:0-18:2	2.3	18:1-16:0-18:2	c		(3) $c + d = 2.3$	$c = 1.5$
18:2			18:2-16:0-18:1	d		(4) $b + d = 19.1$	$d = 0.8$
	16:0-18:2-18:1	55.8	16:0-18:2-18:1	e		(5) $e + f = 55.8$	$e = 22.5$
			18:1-18:2-16:0	f		(6) $c + f = 34.8$	$f = 33.3$

[a]Palmitoyloleoyllinoleoylglycerol amounting to 16.4% of cottonseed oil triacylglycerols.
[b]Triacylglycerols of which the fatty acid in position sn-2 is known.
[c]Molecules of triacyl-sn-glycerols, the percentage of which represents the six unknowns, a to f.
[d]Equation derived from the fatty acid composition of 2-monoacyl-sn-glycerols and of 1-lyso-sn-phosphatides reported in Table 11.5.
[e]Solutions found by solving the six equation system after slight approximations for c and d.
Source: Bézard et al. (19).

From the data for 2-monoacyl-*sn*-glycerols and for 1-monoacyl-*sn*-glycerophosphatide, the molecular composition of the 16:0-18:1-18:2 triacylglycerol can be calculated. Table 11.6 reports the mathematical method used and the results obtained. The four major molecular species are those with an unsaturated fatty acid at the *sn*-2 position.

Use of Chiral Compounds

Methods using the stereospecific action of phospholipases are time consuming, and require relatively high amounts of triacylglycerols for complete analyses to be achieved. An alternative approach is based on the separation of diastereomeric derivatives by adsorption liquid chromatography. This method was independently developed by Christie's group in the United Kingdom (21) and Verger's group in France (22). In the first step, triacylglycerols are partially deacylated by Grignard reagent to generate diacylglycerols. The second step involves reacting the *sn*-1,2(2,3)-diacyl-*sn*-glycerols with a chiral derivatizing reagent. The third step is to separate the diastereomers formed by HPLC on a silica-gel column.

A second approach has been adopted by Takagi et al. in Japan (23), Kuksis et al. in Canada (24) and the authors' laboratory in France (25). It is based on the separation of dinitrophenylurethane (DNPU) derivatives of mono- or diacyl-*sn*-glycerols by HPLC on columns containing a stationary phase with chiral moieties bonded to the silica support. The simplest method is to separate 1- and 3-monoacyl-*sn*-glycerols formed by partial chemical deacylation of triacylglycerols. However, the problem is to prepare representative monoacylglycerols by Grignard deacylation without significant acyl migration at any step of the analysis. For this reason the authors first preferred to use diacyl-*sn*-glycerols which are less sensitive to acyl migration (10,26).

Use of 1,2(2,3)-Diacyl-sn-Glycerols

To get a more accurate insight into the molecular composition of oil triacylglycerols, it is necessary to fractionate the highly complex triacylglycerol mixture to simpler mixtures characterized by unsaturation and/or chain length. This can be achieved by combined chromatographic methods.

In Semporé and Bézard (29) peanut-oil and cottonseed-oil triacylglycerols were first fractionated by silver-ion TLC according to degree of unsaturation. The triacylglycerol fractions produced were fractionated according to chain length by reverse-phase HPLC (6). Each fraction was submitted to Grignard degradation with ethylmagnesium bromide. The 1,2(2,3)-diacyl-*sn*-glycerols were isolated by TLC on borate-impregnated silica gel and converted to dinitrophenylurethane (DNPU) derivatives. Figure 11.3 shows the two chromatograms obtained in the analysis of the DNPU derivatives of 1,2(2,3)-diacyl-*sn*-glycerols issued from Grignard deacylation of the triacylglycerols 16:0-18:1-18:2 of (*a*) peanut oil and of (*b*) cottonseed oil (27). In the chromatograms, the three peaks correspond to the three enantiomer mixtures of 1,2(2,3)-diacyl-*sn*-glycerols: 18:1-18:2, 16:0-18:2, 16:0-18:1.

These three diacyglycerol fractions can be collected at the outlet of a nondestructive detector (differential refractometer) with no detectable cross-contamination between the peaks because of good separation. When analyzed by HPLC on a chiral column containing the chiral moiety [*N*-(*R*)-1-(α-naphthyl)ethylaminocarbonyl-(*S*)-valine] chemically bonded to γ-amino propyl silanized silica (Sumipax OA-4100, from Sumitomo, Japan), each mixture of enantiomers gave rise to two peaks (Figure 11.4) corresponding to the 1,2-diacyl-*sn*-glycerol with the lowest retention time and to the 2,3-isomer eluted later (28). In Figure 11.4, the chromatogram (*e*) corresponds to fraction 1 (18:1-18:2), (*d*) to fraction 2 (16:0-18:2), and (*b*) to fraction 3 (16:0-18:1).

Quantitatively, the relative proportion of the three fractions in Figure 11.3 can be accurately estimated from peak areas when detected by differential refractometry (27), and the relative proportion of the two diacylglycerol enantiomers in Figure 11.4 can also be accurately estimated from peak areas when detected by UV absorbance (28). From data for the composition of the mixtures of 1,2(2,3)-diacyl-*sn*-glycerols in Figure 11.3, and for composition of the two enantiomeric diacylglycerols in Figure 11.4, it is possible to calculate the molecular species composition of the major Diacid- and Triacid-Triacylglycerols of oil, as demonstrated for peanut oil and cottonseed oil. The mathematical method used is illustrated in Table 11.7 from Semporé and Bézard (29). To solve the set of six equations derived from experimental data, an approximation had to be made. Concerning the two molecular species comprising palmitic acid at the *sn*-2 position (*e* and *f* in Table 11.7)

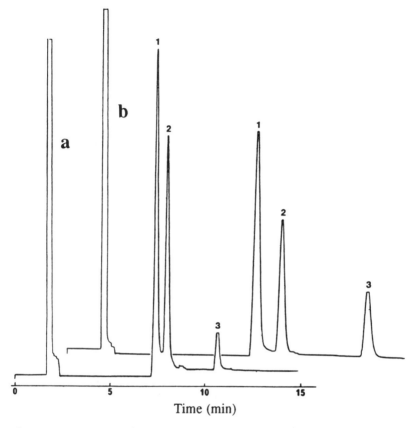

Figure 11.3. Reverse-phase HPLC separation of 3,5-dinitrophenyl isocyanate derivatives of 1,2(2,3)-diacylglycerols originating from Grignard degradation of (a) peanut-oil palmitoyloleoyllinoleoyl (16:0-18:1-18:2) at 19°C and (b) the same triacylglycerol of cottonseed oil at 10°C. Other analytical conditions: stainless steel column (250 mm × 4 mm i.d.) packed with 4 μm octadecylsilyl (C_{18}) reverse-phase material; eluent, acetonitrileacetone (60/40 v/v) at 1 mL/min; refractive index detection; isocratic analysis. Peaks: 1 = sn-1,2(2,3)-18:1 18:2; 2 = sn-1,2(2,3)-16:0 18:2; 3 = sn-1,2(2,3)-16:0 18:1. Source: Semporé and Bézard (27).

which are present in very low amounts. But for the diacid-triacylglycerols, comprising only two different fatty acids, no approximation was necessary [see Table 10.2 in Semporé and Bézard (29)]. As illustrated in Chapter 10 of this monograph, the determination of the fatty acid composition of the sn-2 position by pancreatic lipase is not necessary in this chromatographic method, while it was in the methods using stereospecific phospholipases (18–20).

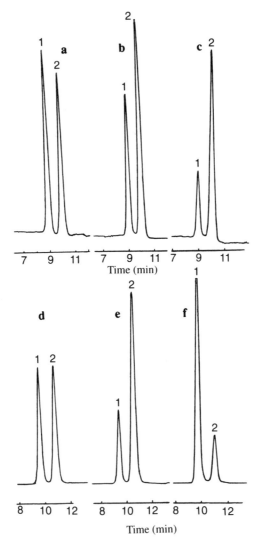

Figure 11.4. HPLC separation of diacylglycerol enantiomers as 3,5-dinitrophenyl isocyanate derivatives on the chiral OA-4100 column. Diacylglycerols derived from peanut-oil and cottonseed-oil triacylglycerols by chemical (b, c, e, and f) and enzymatic (a,d) partial deacylation. They were previously fractionated by reverse-phase HPLC. (a) 18:0-18:1; (b) 16:0-18:1; (c) 18:1-18:1; (d) 16:0-18:2; (e) 18:1-18:2; (f) 18:2-18:2. Peaks: 1 = sn-1,2-Enantiomers and 2 = sn-2,3-Enantiomers in each chromatogram. Detection, UV absorption (254 nm); mobile phase, hexane/ethylene dichloride/ethanol (80:20:1 v/v) flow rate 1 mL/min; analysis temperature, ambient (20°C). *Source:* Semporé and Bézard (28).

TABLE 11.7
Determination of the 1,2,3-Triacyl-sn-Glycerol Composition of Palmitoyloleoyllinoleoylglycerol Isolated from Peanut Oil

Triacyl-glycerol[a]	2-Triacyl-sn-Glycerols[a]	Molecular Species[a]	mol %	sn-1,2 DG[b]	mol %	sn-2,3-DG[b]	mol %
16:0 18:1 18:2	16:0-18:1-18:2	16:0-18:1-18:2	a	16:0-18:1	a/3	18:1-18:2	a/3
		18:2-18:1-16:0	b	18:2-18:1	b/3	18:1-16:0	b/3
		16:0-18:2-18:1	c	16:0-18:2	c/3	18:2-18:1	c/3
	16:0-18:2-18:1	18:1-18:2-16:0	d	18:1-18:2	d/3	18:2-16:0	d/3
		18:1-16:0-18:2	e	18:1-16:0	e/3	16:0-18:2	e/3
	18:1-16:0-18:2	18:2-16:0-18:1	f	18:2-16:0	f/3	16:0-18:1	f/3

Equations[c]

(1) $(a + b) + (c + d) = 48.65 \times 2 = 97.30$

(2) $(c + d) + (e + f) = 41.81 \times 2 = 83.62$

(3) $(a + b) + (e + f) = 9.54 \times 2 = 19.08$

(4) $\dfrac{b + d}{a + c} = \dfrac{38.68}{61.32}$

(5) $\dfrac{c + f}{d + e} = \dfrac{66.86}{33.14}$

(6) $\dfrac{a + e}{b + f} = \dfrac{36.08}{63.92}$

		Solutions	
	A[d]	B[d]	Mean
a	5.57	6.00	5.78
b	10.81	10.38	10.59
c	54.09	53.66	53.88
d	26.83	27.26	27.05
e	0.88	0.88	0.88
f	1.82	1.82	1.82

[a]The triacylglycerol comprises three 2-triacyl-sn-glycerols with known fatty acid at the sn-2 position, and six molcular species with known fatty acids at the 1, 2, and 3-positions (from left to right).
[b]1,2- and 2,3-diacyl-sn-glycerols generated by partial chemical deacylation.
[c]Equations (1), (2), and (3) are derived from the percentage of the individual 1,2(2,3)-diacyl-sn-glycerols and equations (4), (5), and (6) from the percentage of the enantiomers.
[d]Solution A is derived from equations (2), (5), and (1), (4), and solution B from equations (3), (6), and (1), (4).
Source: Semporé and Bézard (29).

Application to Peanut Oil. This calculation method was applied to peanut-oil triacylglycerols comprising one or two long-chain saturated fatty acids (16:0 to 24:0) to study the distribution of the acids between the external *sn*-1 and *sn*-3 positions of triacyl-*sn*-glycerols from an African peanut oil.

Methods. A pure triacylglycerol fraction was prepared from the lipid extract of a crude peanut oil from Burkina Faso (West Africa) by column silicic acid chromatography (30). The three triacylglycerol classes of unsaturation 001, 011, and 012, containing at least one saturated fatty acid, were isolated by silver-ion TLC (23), using hexane/diethyl ether/methanol (79:20:1 v/v) as the eluting solvent. In each class, triacylglycerols containing the same three component fatty acids were isolated by reverse-phase HPLC according to chain length using an octadecyl/silyl column (18) and acetonitrile/acetone (42:58 v/v) as the solvent (6).

The triacylglycerols of each fraction, were partially deacylated by ethylmagnesium bromide (26,31). The 1,2(2,3)-diacyl-*sn*-glycerols were separated by boric acid TLC, using petroleum ether/diethyl ether as the solvent (26).

The diacylglycerols were reacted with 3,5-dinitrophenol isocyanate (32) to form DNPU derivatives. The DNPU derivatives were fractionated according to chain length and unsaturation by reverse-phase HPLC (28), as illustrated in Figure 11.3. Each fraction corresponding to a mixture of two 1,2- and 2,3-diacyl-*sn*-glycerols were resolved into their two constituent enantiomers by chiral-phase HPLC (29), as illustrated in Figure 11.4, on a Sumipax OA-4100 column (Sumitomo, Japan) with hexane/ethylene dichloride/ethanol (80:20:1 v/v) as the solvent.

From data for 1,2(2,3)-diacyl-*sn*-glycerol composition (as in Figure 11.3) and for enantiomer composition (as in Figure 11.4), the molecular composition of the triacylglycerol fractions were calculated using the mathematical method described previously (29).

Results. Figure 11.5 illustrates the fatty acid composition of the peanut oil analyzed. Six saturated fatty acids were present, with only a trace amount of 26:0. Oleic acid and linoleic acid were the two major fatty acids, followed by palmitic acid.

Figure 11.6 shows the distribution of fatty acids between the three positions calculated from data obtained for the different triacylglycerol classes. It appears that saturated fatty acids were rarely encountered at the *sn*-2 position in contrast to linoleic acid, while oleic acid showed no clear-cut preference for any position.

Figure 11.7 shows the distribution of fatty acids between the two external positions, *sn*-1 and *sn*-3, expressed as percent for each fatty acid. For saturated fatty acids, the longer the chain length, the higher the preference for the *sn*-3 position. For instance, more than 85% of behenic acid (22:0) present in external positions was in *sn*-3. Oleic acid (18:1) occupied both the *sn*-1 and *sn*-3 positions

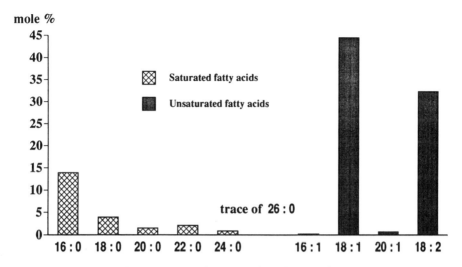

Figure 11.5. Fatty acid composition of peanut-oil total triacylglycerols.

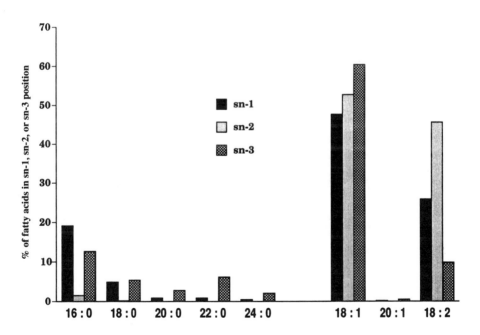

Figure 11.6. Distribution of fatty acids between the *sn*-1, *sn*-2, and *sn*-3 positions of the glycerol moiety in total peanut-oil triacylglycerols.

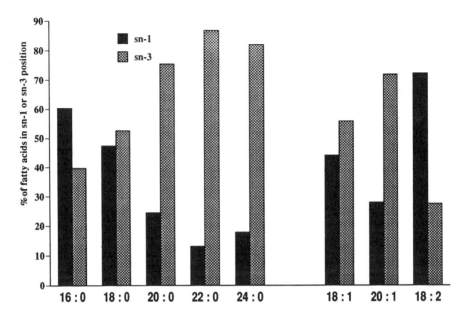

Figure 11.7. Distribution of fatty acids between the two external *sn*-1 and *sn*-3 positions in peanut-oil triacylglycerols.

equally, while eicosenoic acid (20:1), having a higher chain length, was found more often in the *sn*-3 position. On the other hand, the diunsaturated linoleic acid (18:2) was encountered more frequently in the *sn*-1 position.

Table 11.8 lists molecular species of triacylglycerols of the unsaturation class 010 having oleic acid (18:1) in the *sn*-2 position, with palmitic acid (16:0) and another saturated fatty acid in the external positions. The last column reports the

TABLE 11.8
Class 001 Triacyl-*sn*-Glycerols Containing Palmitic Acid (16:0) and Another Saturated Fatty Acid in External Positions and Oleic Acid (18:1) in the *sn*-2 Position

Triacylglycerol[a]	mol% in Oil[b]	Molecular Species[c]	mol% in Triacylglycerol	mol% in Oil
16:0 16:0 18:1	1.45	16:0-18:1-16:0	94.4	1.37
16:0 18:0 18:1	1.09	16:0-18:1-18:0	47.2	0.52
		18:0-18:1-16:0	46.3	0.51
16:0 20:0 18:1	0.50	16:0-18:1-20:0	35.0	0.17
		20:0-18:1-16:0	32.2	0.16
16:0 22:0 18:1	1.06	16:0-18:1-22:0	58.1	0.61
		22:0-18:1-16:0	36.5	0.38
16:0 24:0 18:1	0.41	16:0-18:1-24:0	59.0	0.24
		24:0-18:1-16:0	34.9	0.14

[a]Triacylglycerols with the same three component fatty acids but whose position is unknown.
[b]Triacylglycerols with the unsaturation 001 (0: saturated, 1: monounsaturated fatty acid).
[c]The position of fatty acids is *sn*-1, *sn*-2, *sn*-3 in that order from left to right.

percentage of the molecular species in total triacylglycerols of the oil. These percentages are very low, since the percentage of saturated fatty acids in peanut oil is low, except for 16:0 and to a lesser extent 22:0 (Figure 11.5).

Figure 11.8 illustrates the competition for the *sn*-3 position between 16:0 and the other saturated fatty acids. Data can be compared to those calculated assuming a 1-random-2-random-3-random distribution (10), that is a random distribution of the pool of fatty acids found in each position for total triacylglycerols (Figure 11.6). It can be seen that the competition was very low between 16:0 and the other saturated fatty acids and much lower than that predicted by the random hypothesis.

Table 11.9 reports the molecular species of the same class (010) of triacylglycerols, with 18:1 in the *sn*-2 position and stearic acid (18:0) and another saturated fatty acid in the external positions. The first two were already reported in Table 11.7. The others were in minor amounts in the oil. The triacyl-*sn*-glycerol containing 24:0 was not detected. Figure 11.9 illustrates the competition for the *sn*-3 position between 18:0 and the other saturated fatty acids. It shows that there was slight preference in favor of the long-chain fatty acids, 20:0 and 22:0, however it was not as high as the random distribution predicted.

Table 11.10 reports the triacylglycerol molecular species of the class 011, with 18:1 in both the internal and an external position, and a saturated fatty acid in the other external position. These triacyl-*sn*-glycerols were in relatively high concentrations in the oil since they totaled almost 35% of the triacyl-*sn*-glycerols. Those presenting a saturated fatty acid in an internal position were not reported.

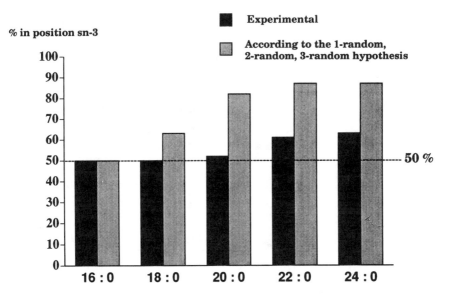

Figure 11.8. Percentage of saturated fatty acids in the *sn*-3 position in the peanut-oil class 001 triacylglycerols comprising oleic acid in the *sn*-2 position, palmitic acid and another saturated fatty acid in the external positions (see Table 11.8).

TABLE 11.9
Class 001 Triacyl-*sn*-Glycerols Containing Stearic Acid (18:0) and Another Saturated Fatty Acid in External Positions and Oleic Acid (18:1) in the *sn*-2 Position

Triacylglycerol[a]	mol% in Oil[b]	Molecular Species[c]	mol% in Triacylglycerol	mol% in Oil
16:0 18:0 18:1	1.09	16:0-18:1-18:0	47.2	0.52
		18:0-18:1-16:0	46.3	0.51
18:0 18:0 18:1	0.20	18:0-18:1-18:0	95.9	0.20
18:0 20:0 18:1	0.15	18:0-18:1-20:0	57.9	0.08
		20:0-18:1-18:0	39.7	0.06
18:0 22:0 18:1	0.09	18:0-18:1-22:0	61.0	0.13
		22:0-18:1-18:0	36.0	0.08

[a]Triacylglycerols with the same three component fatty acids but whose position is unknown.
[b]Triacylglycerols with the unsaturation 001 (0: saturated, 1: monounsaturated fatty acid).
[c]The position of fatty acids is *sn*-1, *sn*-2, *sn*-3 in that order from left to right.

Figure 11.10 illustrates the competition between 18:1 and the saturated fatty acids for the *sn*-3 position. When competing with 16:0, 18:1 was incorporated preferentially in the *sn*-3 position. But the opposite was true for the very long chain fatty acids 20:0, 22:0, and 24:0. Values found were close to those predicted by a 1-random-2-random-3-random distribution.

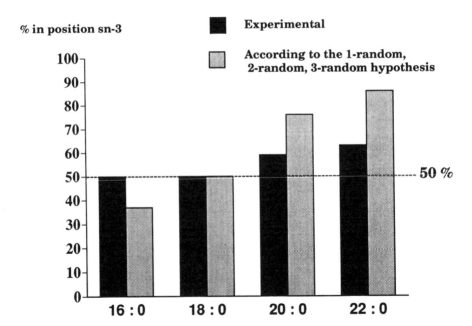

Figure 11.9. Percentage of saturated fatty acid in the *sn*-3 position in the peanut-oil class 001 triacylglycerols comprising oleic acid in the *sn*-2 position, stearic acid and another saturated fatty acid in the external positions (see Table 11.9).

TABLE 11.10
Class 011 Triacyl-sn-Glycerols Containing Oleic Acid (18:1) and a Saturated Fatty Acid in External Positions and 18:1 in the sn-2 Position

Triacylglycerol[a]	mol% in Oil[b]	Molecular Species[c]	mol% in Triacylglycerol	mol% in Oil
16:0 18:1 18:1	8.52	16:0-18:1-18:1	56.9	4.85
		18:1-18:1-16:0	36.2	3.08
18:0 18:1 18:1	2.96	18:0-18:1-18:1	40.3	1.19
		18:1-18:1-18:0	50.0	1.48
20:0 18:1 18:1	0.99	20:0-18:1-18:1	37.6	0.37
		18:1-18:1-20:0	58.3	0.58
22:0 18:1 18:1	1.43	22:0-18:1-18:1	2.2	0.03
		18:1-18:1-22:0	96.9	1.38
24:0 18:1 18:1	0.87	24:0-18:1-18:1	17.0	0.15
		18:1-18:1-24:0	81.0	0.70

[a]Triacylglycerols with the same three component fatty acids but whose position is unknown.
[b]Triacylglycerols with the unsaturation 001 (0: saturated, 1: monounsaturated fatty acid).
[c]The position of fatty acids is sn-1, sn-2, sn-3 in that order from left to right.

Table 11.11 shows the triacylglycerol molecular species of the class 012 with 18:1 in the internal position, and 18:2 and a saturated fatty acid in the external positions. These triacyl-sn-glycerols were present in very small amounts in the oil, especially when a very long chain saturated fatty acid (20:0 to 24:0) was present.

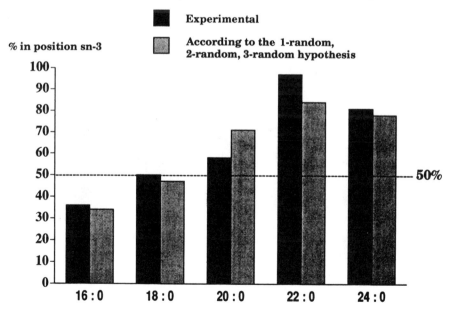

Figure 11.10. Percentage of saturated fatty acid in the sn-3 position in the peanut-oil class 011 triacylglycerols comprising oleic acid in the sn-2 position, oleic acid and a saturated fatty acid in the external positions (see Table 11.10).

TABLE 11.11
Class 012 Triacyl-sn-Glycerols Containing Oleic Acid (18:1) in sn-2 Position and Linoleic Acid (18:2) and a Saturated Fatty Acid in External Positions

Triacylglycerol[a]	mol% in Oil[b]	Molecular Species[c]	mol% in Triacylglycerol	mol% in Oil
16:0 18:1 18:2	11.80	16:0-18:1-18:2	5.8	0.68
		18:2-18:1-16:0	10.5	1.24
18:0 18:1 18:2	3.87	18:0-18:1-18:2	5.9	0.23
		18:2-18:1-18:0	19.8	0.77
20:0 18:1 18:2	1.37	20:0-18:1-18:2	0.5	trace
		18:2-18:1-20:0	36.0	0.49
22:0 18:1 18:2	2.22	22:0-18:1-18:2	0.4	trace
		18:2-18:1-22:0	38.6	0.86
24:0 18:1 18:2	0.88	24:0-18:1-18:2	4.0	0.04
		18:2-18:1-24:0	34.5	0.30

[a]Triacylglycerols with the same three component fatty acids but whose position is unknown.
[b]Triacylglycerols with the unsaturation 001 (0: saturated, 1: monounsaturated fatty acid).
[c]The position of fatty acids is sn-1, sn-2, sn-3 in that order from left to right.

Figure 11.11 illustrates the competition for the sn-3 position between 18:2 and a saturated fatty acid. The competition favored the saturated fatty acids, and the chain length increased with the amount in the sn-3 position. Experimental values were relatively close to those calculated assuming a random distribution. In Table

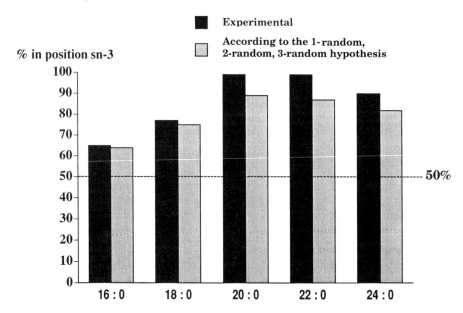

Figure 11.11. Percentage of saturated fatty acids in the sn-3 position in the peanut-oil class 012 triacylglycerols comprising oleic acid in the sn-2 position, linoleic acid and a saturated fatty acid in the external positions (see Table 11.11).

TABLE 11.12
Class 012 Triacyl-sn-Glycerols Containing Linoleic Acid (18:2) in the sn-2 Position and Oleic Acid (18:1) and a Saturated Fatty Acid in External Positions

Triacylglycerol[a]	mol% in Oil[b]	Molecular Species[c]	mol% in Triacylglycerol	mol% in Oil
16:0 18:1 18:2	11.80	16:0-18:2-18:1	53.7	6.34
		18:1-18:2-16:0	26.9	3.17
18:0 18:1 18:2	3.87	18:0-18:2-18:1	36.0	1.39
		18:1-18:2-18:0	30.4	1.18
20:0 18:1 18:2	1.37	20:0-18:2-18:1	2.9	0.04
		18:1-18:2-20:0	52.2	0.72
22:0 18:1 18:2	2.22	22:0-18:2-18:1	0.3	trace
		18:1-18:2-22:0	59.8	1.33
24:0 18:1 18:2	0.88	24:0-18:2-18:1	5.2	0.05
		18:1-18:2-24:0	53.1	0.47

[a]Triacylglycerols with the same three component fatty acids but whose position is unknown.
[b]Triacylglycerols with the unsaturation 001 (0: saturated, 1: monounsaturated fatty acid).
[c]The position of fatty acids is sn-1, sn-2, sn-3 in that order from left to right.

11.12 the triacylglycerol molecular species of the unsaturation class 021, with 18:2 in sn-2 position, are reported. The major triacyl-sn-glycerol was 1-palmitoyl-2-linoleoyl-3-oleoyl-sn-glycerol (16:0-18:2-18:1), which amounted to more than 6% of the total triacylglycerols. The quantity of its optical isomer, 1-oleoyl-2-linoleoyl-3-palmitoyl-sn-glycerol (18:1-18:2-16:0) was only one-half that value, clearly indicating a preference for 16:0 in the sn-1 position and for 18:1 in the sn-3 position. The amount of both isomers clearly indicates the preferential incorporation of 18:2 in the sn-2 position, when compared to the two isomers with 18:1 in the sn-2 position (Table 11.11 and Figure 11.11).

Figure 11.12 illustrates the competition between 18:1 and saturated fatty acids for the sn-3 position. It indicates the preference of very long chain saturated fatty acids, 20:0, 22:0, and 24:0, for the sn-3 position when compared with 18:1. For 20:0, 22:0, and 24:0, 90–100% were present in the sn-3 position. The experimental values were higher than those expected by the 1-random-2-random-3-random hypothesis.

The triacylglycerol 16:0-18:1-18:2 was interesting to analyze for molecular composition, since each of the three fatty acids were incorporated and positional preference could be expressed easily. The affinities are calculated from the percentages of each fatty acid esterified at each position for the six molecular species (Table 11.11 and 11.12) and Semporé and Bézard (29). Results are reported in Table 11.13. The relative affinities of each of the three component fatty acids for each position were:

sn-1: 16:0 > 18:1 > 18:2
sn-2: 18:2 > 18:1 >> 16:0
sn-3: 18:1 > 16:0 >> 18:2

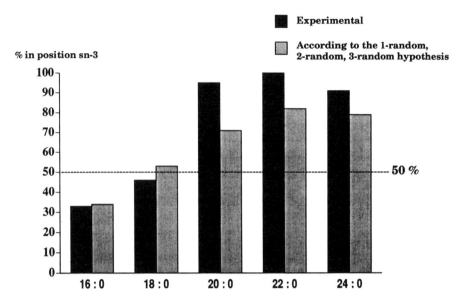

Figure 11.12. Percentage of saturated fatty acid in the *sn*-3 position in the peanut-oil class 012 triacylglycerols comprising linoleic acid in the *sn*-2 position; oleic acid and a saturated fatty acid in the external positions (see Table 11.12).

If the same triacylglycerol (16:0-18:1-18:2) from cottonseed oil and peanut oil are compared and analyzed according to the Brockerhoff method, the affinity for the *sn*-3 position is of the order:

$$16:0 > 18:1 = 18:2.$$

This confirms that the presence of fatty acids in the *sn*-3 position is influenced by the amount of the various acids in total triacylglycerols (10). The major molecular species of peanut-oil triacylglycerols, containing at least one saturated fatty acid are presented in Table 11.14. The experimentally determined percentages can be compared to those predicted by a 1-random-2-random-3-random hypothesis. Only 12 of those triacyl-*sn*-glycerols were found with percentages higher than 1%, and only

TABLE 11.13
Fatty Acid Distribution in the Glycerol Moieties of Peanut-Oil Palmitoyllinoleoylglycerol (Diacylglycerol method)

	sn-1	*sn*-2	*sn*-3
16:0	59.7	2.7	37.6
18:1	27.9	16.4	55.7
18:2	12.4	80.9	6.7

Source: Semporé and Bézard (29).

TABLE 11.14
Major Peanut-oil Triacylglycerol Molecular Species Comprising at Least One Saturated Fatty Acid

Molecular Species (percentage in the oil > 1)	mol% in Oil	
	Experimental[a]	Random[b]
16:0-18:2-18:1	6.34	5.27
16:0-18:1-18:1	4.85	6.10
18:1-18:2-16:0	3.17	2.74
18:1-18:1-16:0	3.08	3.17
18;1-18:1-18:0	1.48	1.36
18:0-18:2-18:1	1.39	1.35
18:1-18:1-22:0	1.38	1.56
16:0-18:1-16:0	1.37	1.15
18:1-18:2-22:0	1.33	1.35
18:2-18:1-16:0	1.24	1.72
18:0-18:1-18:1	1.19	1.56
18:1-18:2-18:0	1.18	1.17

[a]Percentages experimentally determined.
[b]Percentages calculated according to a 1-random, 2-random, 3-random distribution: % sn-XYZ = (mole% X at sn-1 position) × (mole% Y at sn-2 position) × (mole% Z at sn-3 position) × 10^{-4}.

four were more than 3%. For the first three molecular species, experimental values differed from random values: −17, +26, and −14%, respectively.

Use of 1-, 2-, 3-Monoacyl-sn-Glycerols. As stated previously, the simplest method to analyze the fatty acids present in the three positions of the glycerol moiety in triacyl-*sn*-glycerols is to analyze the fatty acid composition of representative 1-, 2-, and 3-monoacyl-sn-glycerols. The monoacyl-*sn*-glycerols were prepared from a simple mixture of triacyl-sn-glycerols by Grignard partial deacylation. The mixture of triacyl-*sn*-glycerols was the triacylglycerol 16:0-18:1-18:2 isolated from peanut oil by a double fractionation, first by silver-ion TLC (to isolate the class O12) and second by reverse-phase HPLC (to isolate the triacylglycerol 16:0-18:1-18:2). The mixture of monoacyl-*sn*-glycerols formed by Grignard degradation was isolated by boric acid TLC then analyzed and fractionated by reverse-phase HPLC (33).

Figure 11.13 shows the chromatogram obtained. Six fractions were separated in the order: sn-2-18:2, sn-1(3)-18:2, sn-2-18:1, sn-2-16:0, sn-1(3)-18:1, and sn-1(3)-16:0. The proportion of these six fractions was accurately determined from peak areas [detection by differential refractometry (33)]. The three fractions, 2, 5, and 6, corresponding to mixtures of the optical isomers 1- and 3-monoacyl-*sn*-glycerols were collected at the outlet of the detector, converted to DNPU derivatives (32) and analyzed by HPLC on a chiral phase (Sumipax OA-4100 from Sumitomo, Japan), using the same method as for diacyl-*sn*-glycerol DNPU derivatives (34).

Figure 11.14 shows a typical chromatogram obtained for racemic 1(3)-monopalmitoyl-*sn*-glycerols. The two optical isomers were well separated. Results

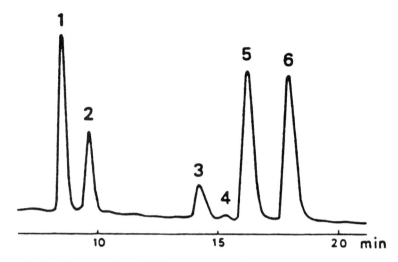

Figure 11.13. Reverse-phase HPLC separations of mixtures of underivatized monoacylglycerols obtained after Grignard degradation of the peanut-oil triacylglycerol 16:0-18:1-18:2. Mixtures of 1 = *sn*-2-18:2; 2 = *sn*-1(3)-18:2; 3 = *sn*-2-18:1; 4 = *sn*-2-16:0; 5 = *sn*-1(3)-18:1; 6 = *sn*-1(3)-16:0. Mobile phase, acetonitrile-water (85:15 v/v); flow-rate, 1.2 mL/min; analysis temperature 12°C. *Source:* Semporé and Bézard (33).

show that the proportion of the two isomers can be accurately determined from peak areas [detection by UV absorbance (34)]. The percentages of the three 2-monoacyl-*sn*-glycerols and the three 1(3)monoacyl-*sn*-glycerols analyzed by reverse-phase HPLC (Figure 11.13) are reported in Table 11.15A. The three 2-monoacyl-*sn*-glycerols represented 27.41% of the total. The remainder consisted of the three 1(3)-monoacyl-*sn*-glycerols. Differences between theoretical and experimental values could be due to a greater extent of acyl migration from position *sn*-2 to positions *sn*-1 and *sn*-3 than the reverse. When expressed as a percentage of fatty acids in position *sn*-2 (Table 11.13), values were not very different from those obtained using the 1,2(2,3)-diacyl-*sn*-glycerols (29). However, the percentage of linoleic acid (18:2) was lower when using monoacyl-*sn*-glycerols, and that of oleic acid (18:1) higher. Acyl migration extent could be greater for 18:2.

Results obtained in the analysis of 1(3)-monoacyl-*sn*-glycerols by HPLC on chiral phase for enantiomer separation are reported in Table 11.15. The presence of the 2-isomer means that acyl migration occurred after the 1(3)-monoacyl-*sn*-glycerols were collected, either before, during, or after derivatization, rather than during analysis by chiral-phase HPLC.

Assuming that acyl migration from the *sn*-1 or -3 position to position *sn*-2 was equal, the percentage of 2-monoacyl-*sn*-glycerol was discarded and the distribution of the three fatty acids in the *sn*-1 and 3 positions was recalculated from the new percentages of the 1 and 3 isomers (*sn*-1 + *sn*-3 = 100 for each fatty acid) and from the percentages of 1(3)-monoacyl-*sn*-glycerols (Table 11.15). Results are reported

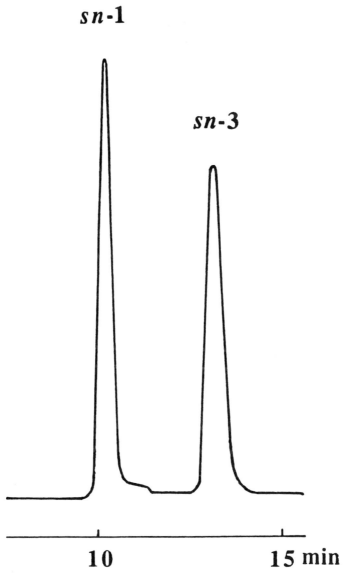

Figure 11.14. HPLC isocratic analysis of racemic monopalmitoylglycerol isomers as [di-]3,5-dinitrophenyl isocyanate (dinitrophenyl urethane = DNPU) derivatives at ambient temperature and with UV absorption detection (254 nm), on stainless steel (250 mm × 4 mm i.d.) chiral OA-4100 column. The column was packed with 5 μm of N-(R)-1-(α-naphthyl)ethylaminocarbonyl-(S)-valine chemically bonded to γ-aminopropyl silanized silica (Sumipax OA-4100); mobile phase, hexane-ethylene dichloride-ethanol (40:12:3 v/v) at 1 mL/min. *Source:* Semporé and Bézard (34).

TABLE 11.15
Fatty Acid Composition of Monoacyl-sn-Glycerols from Partial Chemical Deacylation of Peanut-Oil Palmitoyloleoyllinoleoyl (16:0-18:1-18:2)

	Part A	
	2-Monoacyl-sn-Glycerols[a]	1(3)-Monoacyl-sn-Glycerol[a]
16:0	0.56	32.46
18:1	5.59	29.07
18:2	21.26	11.06
	27.41	72.59

	Part B		
	1,3-Monoacyl-sn-Glycerols[b]		
	1(3)-16:0	1(3)-18:1	1(3)-18:2
sn-1	53.26	37.82	50.94
sn-2	7.34	7.12	6.37
sn-3	39.40	55.06	42.69

[a]Six fractions of 2- or 1(3)-monoacyl-sn-glycerols separated by reverse-phase HPLC (Figure 11.13). Results are expressed as percentage of the total.
[b]Monoacyl-sn-glycerols obtained after separation of the 1(3)monoacylglycerols by HPLC on chiral phase (Figure 11.14). Results are expressed as percentages of each isomer (mol%) found in the 1(3)-monoacyl-sn-glycerol.

in Table 11.16. Comparison between these data and those reported in Table 11.13 stereospecific analysis using the 1,2(2,3)-diacyl-sn-glycerols shows relatively high differences. If the difference between the sn-1 and -3 positions is calculated for each fatty acid, the values were 12.8, 15.3, and 2.5 in Table 11.16 and the corresponding values were 22.1, 27.8, and 5.7 in Table 11.13 for 16:0, 18:1, and 18:2, respectively. When using monoacyl-sn-glycerols, differences between the two external positions were almost halved, probably indicating a randomizing tendency between these two positions by extensive acyl migration. In the experimental conditions used, the method of stereospecific analysis using the monoacyl-sn-glycerols cannot be retained. Acyl migration should be minimized.

Apparently Takagi and Ando (35) have solved the problem of acyl migration, since the stereospecific analysis they carried out on synthetic triacyl-sn-glycerols showed little isomerization. However, the triacyl-sn-glycerols they used were symmetrical (18:1-18:0-18:1 and 16:0-18:3-16:0), so that possible acyl exchange between positions sn-1 and 3 could not be detected. It would be instructive to check

TABLE 11.16
Fatty Acid Distribution in the Glycerol Moieties of Peanut-Oil Palmitoyllinoleoylglycerol (Monoacylglycerol method)

	sn-1	sn-2	sn-3
16:0	51.1	2.0	38.3
18:1	32.4	20.4	47.7
18:2	16.5	77.6	14.0

isomerization with a triacid triacyl-*sn*-glycerol such as 16:0-18:2-18:1. If acyl migration could be effectively controlled at the different analytical steps, the method using monoacyl-*sn*-glycerols for stereospecific analysis of triacylglycerols would be preferred to that using diacyl-*sn*-glycerols, because of its greater simplicity. One way to limit acyl migration is to limit the number of analytical steps. In this view, the use of mass spectrometry coupled to GLC or HPLC as developed by Kuksis et al. (36,37) seems particularly suitable.

Acknowledgment

We are grateful to Joseph Gresti for his skillful assistance in the chromatographic analyses.

References

1. Daubert, B.F. (1949) *J. Am. Oil Chem. Soc. 26:*556.
2. Semporé, G., and Bézard, J. (1977) *Rev. Fr. Corps Gras 24:*611–21.
3. Barrett, C.B., Dallas, M.S.J., and Padley, F.B. (1963) *J. Am. Oil Chem. Soc. 40:*580.
4. Christie, W.W. (1988) *J. Chromatogr. 454:*273.
5. Plattner, R.D., Spencer, G.F., and Kleiman, R. (1977) *J. Am. Oil Chem. Soc. 54:*511.
6. Bézard, J.A., and Ouédraogo, M.A. (1980) *J. Chromatogr. 196:*279–93.
7. Bézard, J., Bugaut, M., and Clément, G. (1971) *J. Am. Oil Chem. Soc. 48:*134.
8. Bézard, J. (1971) *Lipids 6:*630.
9. Bailey, A.E. (1951) In *Industrial Oil and Fat Products,* 2nd Ed. Interscience, New York.
10. Litchfield, C. (1972) In *Analysis of Triglycerides.* Academic Press, New York and London. p. 248.
11. Mattson, F.H., and Beck, L.W. (1956) *J. Biol. Chem. 214:*115.
12. Savary, P., and Desnuelle, P. (1956) *Biochim. Biophys. Acta 50:*319.
13. Clément, G., Clément, J., and Bézard, J. (1962) *Arch. Sci. Physiol. 26:*213–25.
14. Semporé, G., and Bézard, J. (1982) *J. Am. Oil Chem. Soc. 59:*124–29.
15. Luddy, F.E., Barford, R.A., Herb, S.F., Magidman, P., and Riemenschneider, R.N. (1964) *J. Am. Oil Chem. Soc. 41:*693–96.
16. Bézard, J., Semporé, G., Descargues, G., and Sawadogo, A. (1981) *Fette Seifen Anstrich. 83:*17–23.
17. Ransac, S., Rogalska, E., Gargouri, Y., Dever, A.M.T.J., Paltauf, F., De Haas, G.H., and Verger, R. (1990) *J. Biol. Chem. 265:*20263.
18. Brockerhoff, H. (1967) *J. Lipid Res. 8:*167–69.
19. Bézard, J., Ouedraogo, M.A., and Semporé, G. (1990) *Rev. Fr. Corps Gras 37:*171–75.
20. Myher, J.J., and Kuksis, A. (1979) *Can. J. Biochem. 57:*117–24.
21. Laakso, P., and Christie, W.W. (1990) *Lipids 25:*349–53.
22. Rogalska, E., Ransac, S., and Verger, R. (1990) *J. Biol. Chem. 265:*2071.
23. Takagi, T., and Suzuki, T. (1990) *J. Chromatogr. 519:*237.
24. Itabashi, Y., Kuksis, A., Marai, L., and Takagi, T. (1990) *J. Lipid Res. 31:*1711.
25. Semporé, G., and Bézard, J. (1991) *J. Chromatogr. 557:*227.
26. Yurkowski, M., and Brocherhoff, H. (1966) *Biochim. Biophys. Acta 125:*55.
27. Semporé, G., and Bézard, J. (1991) *J. Chromatogr. 547:*89–103.

28. Semporé, B.G., and Bézard, J.A. (1991) *J. Chromatogr. 557*:227–40.
29. Semporé, G., and Bézard, J. (1991) *J. Am. Oil Chem. Soc. 68*:702–9.
30. Fillerup, D.L., and Mead, J.F. (1953) *Proc. Soc. Exp. Biol. Med. 83*:574.
31. Christie, W.W., and Hunter, M.L. (1980) *Biochem. J. 191*:637–43.
32. Itabashi, Y., and Takagi, T. (1987) *J. Chromatogr. 402*:257.
33. Semporé, B.G., and Bézard, J. (1992) *J. Chromatogr. 596*:185–95.
34. Semporé, B.G., and Bézard, J. (1994) J. *Liquid Chromatogr. 17*:1679–94.
35. Takagi, T., and Ando, Y. (1991) *Lipids 26*:542–47.
36. Itabashi, Y., Kuksis, A., and Myher, J.J. (1990) *J. Lipid Res. 31*:2119–26.
37. Marai, L., Kuksis, A., Myher, J.J., and Itabashi, Y. (1992) *Biomed. Mass Spectrom. 21*:541–47.

Chapter 12

Gas–Liquid Chromatography: Choice and Optimization of Operating Conditions

F.X. Mordret and J.L. Coustille

Institut des Corps Gras, Bordeaux-Pessac, France.

Introduction

Gas chromatography (GC) is widely used in the analysis of oils, fats, and their derivatives in order to determine the composition of lipid classes (i.e., fatty acids) or to provide the absolute amount of a component (i.e., cholesterol). Many papers have been published on the applications of GC in that field and many standardized methods are based on it. For example, important considerations concerning the analysis of fatty acids by GC were presented and discussed by Ackman (1). However, few books are specifically devoted to this subject, although Christie (2) is a useful practical guide.

Gas chromatography is a multiparameter technique, and the quality of the result depends on different factors: the column, operating conditions, injection system, preliminary treatment of the sample, and detection system. An important task of the analyst is to deal with all these parameters, the technical aspects of which are not always precise.

Basic Principles

It is necessary to remember the elementary process of GC (3), the partitioning of a solute between a mobile phase and a stationary phase, according to a distribution coefficient K_D that depends on the solute, the stationary phase, and the temperature. For a solute A:

$$K_{D,A} = \frac{C_{A(s)}}{C_{A(m)}}$$

Concentration of the solute A in the stationary phase S
Concentration of the solute A in the mobile phase m

A dynamic equilibrium is expressed by the coefficient K_D and two compounds A and B can be separated if the operating conditions are such that $K_{D,A} \neq K_{D,B}$.

Experimental measurements in GC concern the retention of the compounds (tr, tm) and the shape of the chromatographic peaks (σ, ω). Classical characteristics of the column such as the separation efficiency (n, h), the solute partition ratio (k), the selectivity factor (α) and the resolution (R) can be calculated from these data.

GC Columns

Although wall-coated open tubular (WCOT) capillary columns are now commonly used in lipid analysis, some laboratories are still working with the classical packed columns. A few years ago, starting with lab-made glass capillary columns, Prevot and Mordret (4) pointed out the capabilities of such columns in terms of fast analytical possibilities. As an example, in 1976 the separation of rapeseed oil fatty acid methyl esters in less than 3 min was described (Figure 12.1).

Modern capillary columns with fused silica tubes offer many advantages, such as high efficiency, large permeability, and high value of the phase ratio ß. The problem of adjustment of the sample size to the loading capacity of the capillary column has been resolved with the development of appropriate injection systems (5).

The characteristics of capillary columns are closely related to geometrical and chromatographic considerations. If the choice of the stationary phase is not optimized, it must be compensated for by increasing the column efficiency. High efficiency columns are required to analyze complex mixtures (i.e., fatty acids of marine oils).

The length L has direct implications on the efficiency (n). The average length of these columns is 25 m, which provides an efficiency of about 75,000 effective theoretical plates, suitable for most classical analysis (fatty acid methyl esters of vegetable oils and animal fats, sterols). Figure 12.2 shows the programmed temperature analysis achieved in the authors' laboratory (6) of the lipid extract from a French Hospital daily diet for nutritional purposes (determination of the polyunsaturated/saturated ratio). The chromatogram, with a large spectrum of fatty acids (chains from C_4–C_{20}) indicates the occurrence in the diet of dairy products, animal fats, and vegetable oils.

The internal diameter ($\phi_i = 2r$) is related to the column permeability B_o. The high value of permeability of the capillary columns allows one to work with important column lengths (50 or 100 m) without a noticeable modification of the average linear velocity $\vec{u}$ of the carrier gas and no loss of efficiency. In the Golay equation, if parameters are optimized, the minimal theoretical height of the plate h approaches the internal diameter value of the column (h → 2r).

The stationary phase plays an important part in the GC process by its composition and its thickness. The two main requirements of a stationary phase are a great thermal stability at the working temperature to prevent bleeding and loss of efficiency and a good physicochemical affinity to the compounds submitted for analysis; otherwise, column overloading and loss of resolution can occur.

An appropriate choice of the stationary phase allows one to obtain full resolution between peaks if the column is operating at minimal efficiency. Figure 12.3 shows the analysis of soybean oil fatty acids (as methyl esters) on three different 7-m capillary columns coated with apolar, medium polar, and polar stationary phases. Under the conditions of reduced efficiency, only the cyanopolysiloxane phase (CPSil 88) allowed the separation of all the compounds.

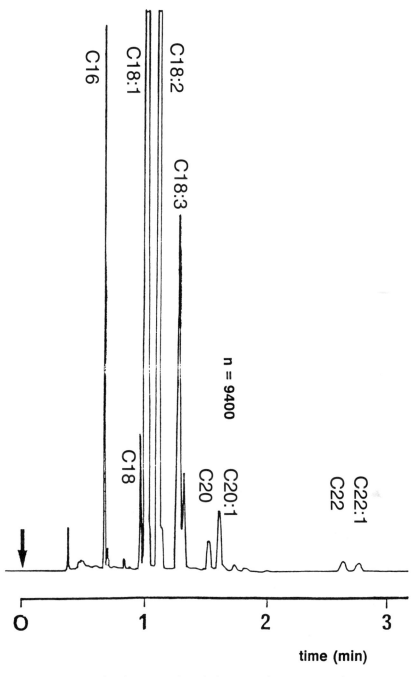

Figure 12.1. Rapeseed Oil Fatty Acid Methyl Esters. Glass WCOT column; $L = 25$ m; $d_f \pm 0.1$ µ; Oven temp.: 190°C; Carrier gas: H_2 1.2 b.

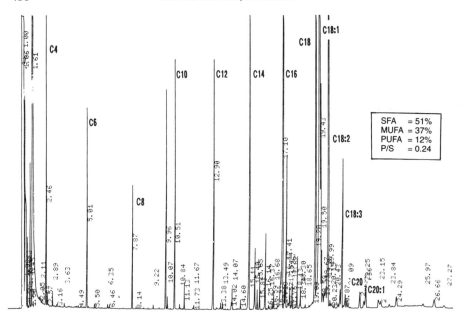

Figure 12.2. Lipid Extract Analysis of a French Hospital Daily Diet Methyl Esters of the Fatty Acids. WCOT capillary column DB Wax; L = 25 m; ϕ_i = 0.25 mm; d_f = 0.22; θ = 50°C—3 min to 200°C; 10°C/min; H_2 = 0.7b.

Increasing length and efficiency of the column and using an apolar stationary phase, such as SE 30 or OV1, resulted in fused peaks for linoleic and linolenic acids (Figure 12.4) because of a lack of polarity of the stationary phase. The stationary phase must be more polar as the difference in polarity of the compounds is reduced. The most typical application of this rule concerns the analysis of fatty acid geometrical isomers. A reference mixture of *cis* and *trans* isomers of linoleic and linolenic acids was analyzed (7) with two high-efficiency capillary columns (L = 50–60 m, ϕ_i = 0.25 mm) one coated with Carbowax 20M and the other with CPSil 88 phase (Figure 12.5). Using the polyethyleneglycol phase is not advisable for such a determination. An improvement may be obtained by using the strong polar phase CPSil 88, but all of the isomers are not clearly resolved.

Another example illustrating the importance of the stationary phase choice can be found in the field of sterol analysis. The application of such determination is well known for quality control of fats and oils. In past years, sterol compositions were achieved by GC from the total sterol fraction with nonpolar stationary phase (OV1, SE30, etc.) packed columns. In 1973, a Japanese team (8) using the medium polar phase OV17, demonstrated that the peak called "ß-sitosterol" until that time can be split into two peaks identified by GC/MS as ß-sitosterol (24-ethyl-Δ5-cholesten-3ß-ol) and 5-avenasterol (24-ethylidene-Δ5-cholesten-3ß-ol) which differ only by a double bond of a substituent (24 position) on the side chain. As a consequence of this study, all the data on sterol compositions had to be revised.

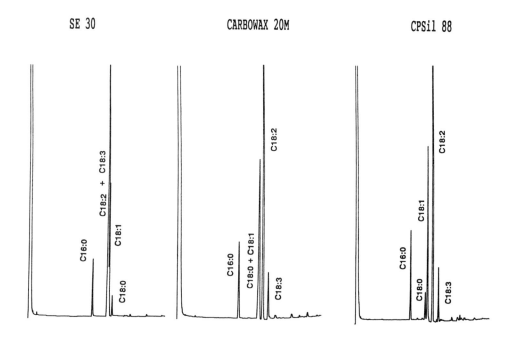

Figure 12.3. Effect of the Type of Stationary Phase on the Separation of Fatty Acid Methyl Esters. WCOT capillary columns; $L = 7$ m; $\phi_i = 0.32$ mm; $d_f = 0.24$ μ; H_2 = 0.4 b; on column injection; temperature programmed from 70°C to 200°C, 10°C/min.

The film thickness (d_f) of the stationary phase affects the ratio phase ß as a result of modifying the retention time and the resolution.

$$tr = tm + \frac{K2\, d_f}{r}\, tm$$

range of values
d_f: 0.1–5 μ
r: 0.1–0.35 mm

An increase of d_f improves resolution, but the analysis requires more time. The contribution of the film-thickness value to the separation is illustrated by the analysis of technical hexane which is a mixture of isomers (Figure 12.6). With a thick-film column, it is easier to characterize traces of volatile contaminants in the oil duct and improve separation of the oil volatile compounds and identification of individual isomers by GC/MS. On the other hand, a thin-film thickness and a large internal diameter of the column allows a rapid analysis.

The surface activity of the support must be reduced as much as possible by an appropriate deactivation technique to avoid adsorption and loss of compounds and can be checked by a specific test (i.e., Grob test). Figure 12.7 illustrates the deter-

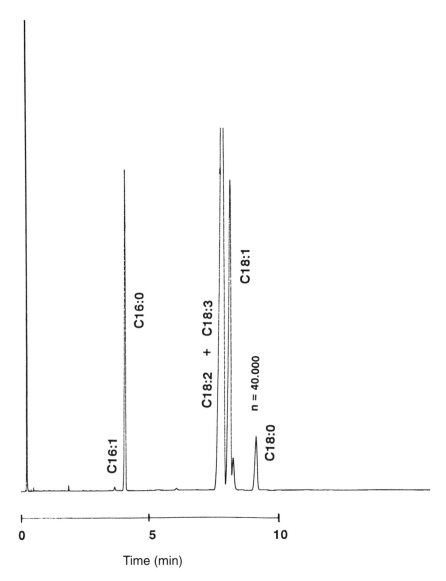

Figure 12.4. Soybean Oil Fatty Acid Methyl Esters. Column: $L = 25$ m; $\phi_i = 0.32$ mm; $d_f = 0.5$ µ; Temp. = 170°C; Stationary phase: OV1.

mination of acetic acid traces in fermentation products. This analysis, achieved by Mordret et al. (7), was difficult because of the high polarity and the low quantities of the solute. The column was selected with respect to its activity (ethylhexanoic acid retention) and the sample was introduced onto the column by "on column" injection.

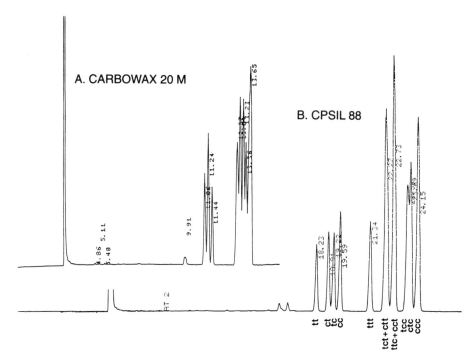

Figure 12.5. Separation of Geometrical Isomers for Linoleic and Linolenic Acids.

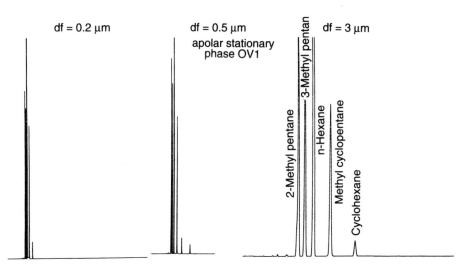

Figure 12.6.
Separation of Isomers in Technical Hexane. WCOT capillary column: L = 25 m; ϕ_i = 0.32 mm; Temp. = 40°C; Carrier H_2 = 60 cm/sec (CH_4).

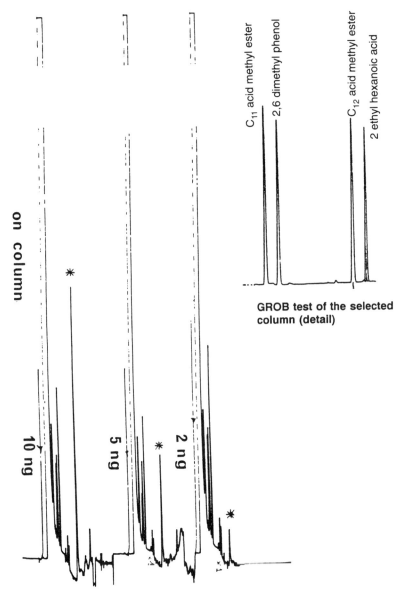

Figure 12.7. Determination of Acetic Acid in Fermentation Products.

Operating Conditions

Two parameters can be modified easily by the analyst and offer noticeable effects on resolution and retention (Figure 12.8). The column temperature must be adjusted, taking into account the compound volatility range. The column temperature is either isothermal or the temperature increase is controlled via a programmed temperature system. The retention of the solute in the stationary phase is correlated to the column temperature which must be high enough to avoid overloading but not too high, which causes loss of resolution.

The carrier gas velocity is very easy to modify. The relationship between the average linear velocity $\vec{u}$ and the height of an effective theoretical plate (HETP) h is well known (Van Deemter or Golay equation). The simplified representation presented here (Figure 12.8) allows one to understand how to adjust the carrier gas flow for high resolution on fast analytical conditions. Maximum resolution will be reached with the optimum value of $\vec{u}$. On either side of u_{opti}, increasing molecular diffusion or resistance to mass transfer contributes to the broadening of the peak and a loss of resolution. The left part of the graph can be considered a "prohibited area" while the right demonstrates that it is possible to achieve rapid analysis at the expense of a loss of resolution. Hydrogen is the optimal carrier gas, but it must be used with proper safety precautions.

The main problem an analyst faces is to find a compromise between two opposite goals: high resolution and fast analysis. The resolution R is a complex interplay of three chromatographic parameters:

$$R = \tfrac{1}{4} \left(\frac{\alpha - 1}{\alpha}\right) \cdot \left(\frac{k}{1+k}\right) \cdot (n)^{0.5}$$

Selectivity Capacity Efficiency

Stationary Phase Column Geometry

Appropriate Choice Adapted Film Thickness $(L; \phi_i)$

All the parameters must to be optimized. Sometimes, a maximum value of R is required and the general characteristics of the column will be a strong polar phase, high value of k', N between 150,000 and 200,000 plates. The retention time tr has to be reduced as much as possible:

$$tr = \frac{L}{\vec{u}} (1 + k')$$

In order to reduce retention times, the column characteristics are a short length (6–10 m), a high velocity of the carrier gas, and a small solute partition ratio. An application of these techniques was used to separate olive-oil triglycerides by GC (Figure 12.9 [9]). Usually this analysis is achieved on a short length column (5–7 m) coated with an apolar phase (SE 30 or similar stationary phase) with a high-velocity carrier gas. Under such conditions, triglycerides are separated according to their carbon numbers (C_{50}, C_{52}, C_{54} etc.), but triglycerides with different degrees of

COLUMN TEMPERATURE:

Below optimal value results in poor solubility of the solute in the stationary phase and column overloading.

Above optimal value results in too high volatility of the solute which remains in the mobile phase and a loss of resolution.

Short range of molecular weights requires isothermal conditions.

Large range of molecular weights requires programming temperature.

CARRIER GAS: Adjustment of the flow according to Van Dempter or Golay equations

$$h = \frac{B}{\vec{u}} + C\vec{u}$$

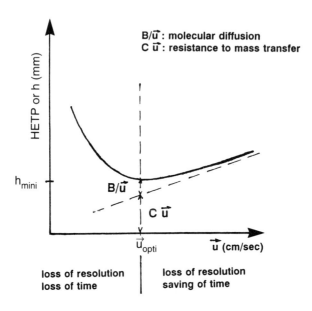

Figure 12.8. Choice of Operating Conditions.

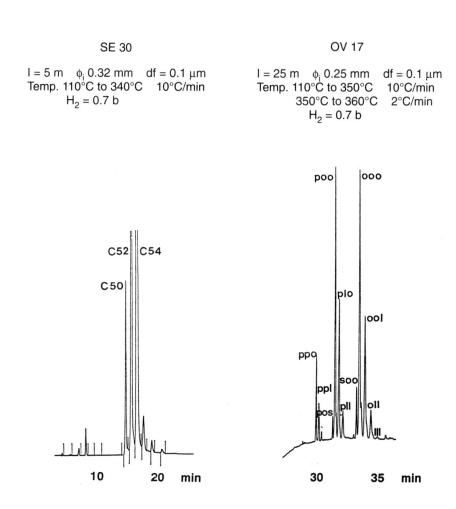

Figure 12.9. Analysis of Olive-Oil Triglycerides. WCOT Capillary Columns.

unsaturation and an identical carbon number are not separated. By increasing the column length (25 m), changing the stationary phase polarity (a medium polar phase, such as OV17), and optimizing the heating conditions of the oven, separations of individual triglycerides can be obtained, with results similar to those given by HPLC.

TABLE 12.1
Recommended Modes of Injection on Capillary Columns for Fats and Oils Analysis.

Compounds	Split	Splitless	On Column (or direct injection)
Triglycerides, partial glycerides, waxes			•
Fatty acids (as methyl esters)	•		•
Fatty acids (free, short chains)			•
Lipid traces (biological origin, environment, etc.)		•	
Hydrocarbons, alcohols, sterols[a]			•
Volatiles	•[b]		
Contaminants (HAP, PCB, pesticides)			•

[a]Other possibility: moving needle injector
[b]With static or dynamic headspace

Sample Injection Systems

The split injection was the first system described for capillary gas chromatography with the advantages of being simple, cheap, flexible, and universal. Its two major drawbacks are its lack of linearity (sample discrimination) for compounds with a broad range of molecular weights, and its unusability for trace analysis. This injector is widely used for the analysis of fatty acid methyl esters. The splitless injection with recondensation of the sample vapor on the first part of the column (solvent effect) represents an improvement for trace analysis. However, with the vaporizing system principle the same disadvantages remain: sample transfer, hot septum, and the activity of the injector walls. The "on column" injection is a nonvaporizing system, based on the solvent effect with the direct introduction of the sample solution onto the column. At the present time, it is considered the most appropriate injection technique for analysis of thermosensitive or polar substances, different volatilities, and high molecular weight compounds (free fatty acids, triglycerides, etc.). Figure 12.10 compares performances of split, splitless, and on column injection applied to a simple test mixture of methyl esters.

Additional possibilities can be considered, such as direct injection, programmed temperature vaporizing injection, and moving needle injection. Table 12.1 gives a list of the different compounds encountered in GLC of lipids and the recommended modes of injection on capillary columns. The on column injector appears to be the most appropriate system for analyzing fat and oil components or derivatives.

Pretreatment of the Sample

In some cases, before performing GC the sample has to be treated to transform it to a more suitable form, reducing the polarity, or increasing the volatility. Derivative formation is extensively used in lipid analysis; examples are preparation of fatty acid methyl esters or of trimethylsilylethers with hydroxy compounds. Many other possi-

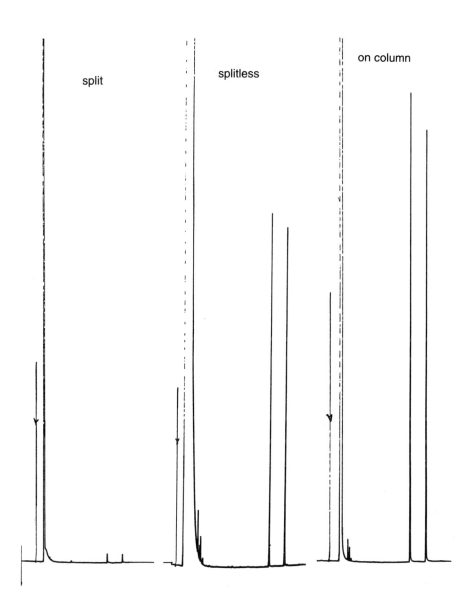

Figure 12.10. Performance of Three Injection Systems with $C_{10:0}$ and $C_{11:0}$ methyl esters; 25 ng of each in 1 μL of hexane.

bilities have been described. The reaction has to be quantitative, nondestructive, and not time consuming. The most useful methods are standardized (ISO, AOCS, IUPAC, AFNOR, etc.). The headspace techniques (static or dynamic) have been adopted in most of the laboratories for lipid analysis. The choice of the static or dynamic technique depends on the quantity level of the compound which is to be determined.

Detection Methods in GLC

The flame ionization detector (FID) is the most common in GC for lipid analysis. However, specific detectors, such as electron capture detector (ECD) find application in the determination of volatile compounds. To preserve the advantages of the capillary column, the interface between column and detector has to be optimized, which involves column-end positioning and flow-rate adjustment. However, coupling GC/MS offers many possible applications because of sensitivity and specificity.

References

1. Ackman, R.G. (1991) In *Analyses of Fats, Oils, and Lipoproteins.* Edited by E.G. Perkins, American Oil Chemists' Society, Champaign, Illinois.
2. Christie, W.W. (1989) *Gas Chromatography and Lipids. A Practical Guide.* The Oily Press. Ayr, Scotland.
3. Ettre, L.S. (1965) *Open Tubular Column in Gas Chromatography.* Plenum Press, New York.
4. Prevot, A.F., and Mordret, F.X. (1976) *Rev. Fse des Corps Gras 23:* 409–423.
5. Sandra, P. (1985) *Sample Introduction in Capillary Gas Chromatography.* Vol. 1. Huetig Verlag, Heidelberg.
6. Dabadie, H., Castera, A., Lacomere, R.P., Bernard, M., Mordret, F., Chazan, J.B., and Paccalin, J. (1991) *Cah. Nutr. Diét.* XXVI:(3)197.
7. Mordret, F., Prevot, A., Perrin, J.P., Coustille, J.L., and Morin, O. (1987) *Ann. Fals. Exp. Chim.* 80, No. 854, 9–24.
8. Itoh, T., Tamura, T., and Matsumuto, T. (1973) *J. Am. Oil Chem. Soc. 50:*122.
9. Coustille, J.L., Mordret, F., Morin, O., and Perrin, J.P. (1987) *Analysis of Triglycerides by Gas Chromatography. Application: Detection of Sunflower Oil in Olive Oil.* Eighth International Symposium on Capillary Chromatography. Riva del Garda, Italy. May 19–21.

Chapter 13
Recent Applications of Capillary Gas–Liquid Chromatography to Some Difficult Separations of Positional or Geometrical Isomers of Unsaturated Fatty Acids

Robert L. Wolff

ISTAB, Laboratoire de Lipochimie Alimentaire, Universite Bordeaux I, Allee des Facultes, 33405 Talence Cedex, France.

Introduction

Gas–liquid chromatography (GLC) on fused-silica capillary columns is now a widespread tool to study the composition of complex mixtures of fatty acids in detail. It has become a common practice to deal with chromatograms displaying 50–100 peaks. Examples are fish oils or dairy fat. However, some separations of structurally related fatty acids, such as positional or geometrical isomers, remain difficult to achieve through a single chromatographic run, and complementary techniques are needed to characterize these fatty acids.

One of the most powerful and economical means to fractionate and to isolate these fatty acids is silver-ion thin-layer chromatography (TLC). Following this first step, several techniques may be applied to characterize and quantitate those fatty acids that were not (or were only partly) resolved by GLC. It is beyond the scope of this chapter to present these techniques, as they are described and discussed at length in other chapters of this book.

Before employing these time-consuming procedures, it is advisable to carefully evaluate all parameters that can affect the resolution of the fatty acids that are poorly separated if at all. These parameters include the nature of the stationary phase in the column, the geometrical characteristics of the column, the inlet pressure of the carrier gas (and eventually its nature), the temperature of the oven, and the nature of the derivative form of the fatty acid. This last parameter is frequently ignored, and methyl esters are universally and almost exclusively used by lipid chemists.

In this chapter, some practical examples of separation of isomers with very similar chromatographic behavior or complex mixtures of isomers will be described. These examples are taken from recent work done by the author and his colleagues. Generally, 50-m long CP Sil 88 capillary columns (Chrompack, Middelburg, The Netherlands) were employed to analyze fatty acid esters. Successful resolution by GLC of otherwise overlapping isomers present in food or biological samples avoids any further use of other complementary fractionation techniques.

Separation of Petroselinic and Oleic Acids

It has been repeatedly stated (1–5) that petroselinic acid (*cis*-6 18:1 acid) and oleic acid (*cis*-9 18:1 acid) that occur together in most *Umbelliferae* (carrot, parsley, celery, etc.) seed oils (6) cannot be separated by GLC readily, at least when these acids are analyzed as methyl esters (1–5). However, other techniques, such as TLC on silver-impregnated alumina sheets (7), or reverse-phase high-performance liquid chromatography (HPLC) of the phenacyl ester derivatives (8), allow separation of the two acids. With the exception of HPLC separation, which does not seem to have been employed to fractionate natural mixtures extracted from *Umbelliferae* seeds, an accurate quantitation of petroselinic acid remains difficult to achieve. Time-consuming or complex analytical procedures that may need expensive equipment have to be used to complement GLC. One of the most popular procedures has been an oxidative cleavage of the double bonds in isolated monoenes followed by GLC of the resulting fragments (6,7,9–11). In the most extensive study on *Umbelliferae* seed oils (6), the petroselinic acid content was determined by combining data obtained by direct GLC analysis of methyl esters and data obtained by GLC analysis of aldehydes and aldesters prepared by ozonolysis of octadecenoic acids isolated first by silver-ion TLC and then by preparative GLC. More recently, the ^{13}C NMR chemical shifts of olefinic carbons of oleic and petroselinic acids have been used in conjunction with GLC of fatty acid methyl esters (FAME) to quantitate petroselinic acid (2,3).

Separation of the two octadecenoic acids by GLC, indicates that petroselinic and oleic acid methyl esters have equivalent chain lengths (ECL) on polar capillary columns that differ by 0.04–0.06 carbon units (1,12,13) (Table 13.1). Theoretically, this should be sufficient to achieve some kind of resolution between the two acids. Indeed, Wolff and Vandamme (14) have observed that this critical pair of fatty acids can be partly resolved when a mixture of almost equivalent amounts of methyl esters of the two acids is analyzed on a CP Sil 88 capillary column (100% cyanopropyl polysiloxane coating; 50 m × 0.25 mm i.d., 0.20 µm film; 141,000 theoretical plates) operated under optimum conditions (Figure 13.1). In this case, the

TABLE 13.1
Equivalent Chain Lengths of Fatty Acid Methyl and Isopropyl Esters on Capillary Columns Coated with Cyanoalkyl Polysiloxane Stationary Phases

Fatty acid	ECL				
	FAME				FAIPE
	Silar 10C[a] (12)[b]	SP-2340 (13)	CP Sil 84 (1)	CP Sil 88 (14)	CP Sil 88 (14)
cis-6 18:1	18.56	18.47	18.43	18.54	18.37
cis-9 18:1	18.61	18.53	18.47	18.60	18.45
Difference	0.05	0.06	0.04	0.06	0.08

Abbreviations: ECL, equivalent chainlength; FAME, fatty acid methyl ester; FAIPE, fatty acid isopropyl ester.
[a]Trademark of the column.
[b]References.

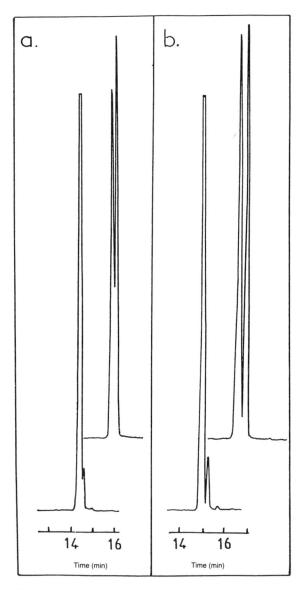

Figure 13.1. Chromatograms of mixtures of methyl esters (a) or isopropyl esters (b) of petroselinic (first peak eluted) and oleic acids. Upper chromatograms: artificial mixtures of almost equivalent quantities of authentic petroselinic and oleic acid derivatives. Lower chromatograms: petroselinic and oleic acid derivatives prepared from celery seed oil. Analyses on a CP Sil 88 fused silica capillary column (50 m × 0.25 mm i.d., 0.20 μm film) operated isothermally at 165°C with an inlet pressure of the carrier gas (helium) of 130 kPa. Adjusted retention times in min. *Source:* Wolff and Vandamme (14).

difference between ECL of the two acids is 0.06 carbon units (Table 13.1). However, when FAME prepared with an oil extract of *Umbelliferae* seeds are chromatographed under the same analytical conditions, oleic acid is almost reduced to a shoulder on the trailing edge of the main petroselinic acid peak (Figure 13.1). The two components cannot be individually quantitated with accuracy. To improve the separation, some authors have tried to use derivatives other than simple methyl esters. Ucciani et al. (4,5) have prepared the trimethylsilyloxy derivatives of methyl esters and have analyzed the resulting components on a 65-m long SE-30 capillary column (182,000 theoretical plates). The separation that was achieved is only partial and very similar to that obtained with simple methyl esters analyzed on the CP Sil 88 column (Figure 13.1, left chromatogram). Moreover, the quantitation of petroselinic and oleic acids needs the use of a mass spectrometer as a detector. It was claimed that a better resolution could be achieved when trimethylsilyloxy derivatives of methyl esters are analyzed on a Carbowax 20M capillary column (5). However, a severe drawback of this kind of analysis is that residual silylation reagents rapidly deteriorate the Carbowax 20M column (5). Other authors have used epoxy derivatives of methyl esters (15), but it would appear that the resolution was not improved greatly.

The use of isopropyl esters (16) instead of methyl esters increases the difference between the ECL of the two acids by 0.02 carbon units (from 0.06 to 0.08; Table 13.1 [14]). This is sufficient to obtain almost baseline resolution between the two fatty acids (Figure 13.1). This is true for either a mixture of almost equivalent quantities of the two acids, or for isopropyl esters prepared with an oil extract of *Umbelliferae* seeds. Consequently, the problem of petroselinic acid and oleic acid separation and quantitation is resolved by simply analyzing these components in the form of isopropyl esters by GLC on a CP Sil 88 capillary column. The preparation of isopropyl esters (16) is as simple as that of methyl esters, and the time of analysis by GLC is only a few minutes longer than that of methyl esters (14). So, this procedure can be applied routinely to the study of oils extracted from *Umbelliferae* seeds. There is no need for any other complementary techniques. The reason why isopropyl esters are better separated than methyl esters is as yet unknown. However, our observations indicate that when FAME are difficult to separate even under optimal analytical conditions, other derivatives should be assayed before changing one's column.

Separation of Polyunsaturated Fatty Acids from the *n*-7 Series

In the absence of dietary fat, the *n*-6 and *n*-3 fatty acid content of rat tissue lipids diminishes rapidly with time. Most of the fatty acids are then of endogenous origin (synthesized from acetate). Apart from saturated and monounsaturated acids of the *n*-9 and *n*-7 series (oleic, palmitoleic, and *cis*-vaccenic acids), some polyunsaturated acids of the same *n*-9 and *n*-7 series accumulate in tissue lipids. The main polyunsaturated acid that appears in lipids is 20:3*n*-9 acid, the end-product of the oleate (*n*-9) family. Generally, there is little difficulty in obtaining a clear-cut separation between this acid and other tissue fatty acids by GLC on either packed or capillary columns. On the other hand, several polyunsaturated acids of the n-7 (palmitoleate) series are incorporated in low amounts in rat lipids (17–27). These acids include 18:2n-7 (20,21,23,24), 18:3n-7 (18,24,27), 20:2n-7 (27), 20:3n-7 (17,20–24,27), and 20:4n-7 (19–25,27) acids. Fatty acids of the n-7 series are synthesized according to the metabolic pathway shown in Figure 13.2. To the best of our knowledge, there is no report indicating that these acids could be individualized by GLC on packed column. Thus, complex procedures were needed to detect and characterize these components (17–27). However, it would appear that highly efficient capillary columns are better suited to separate n-7 acids from other fatty acids (28,29).

In fact, Wolff et al. have observed that the use of a 50-m long CP Sil 88 capillary column allows the separation of several fatty acids associated with fat deficiency in the rat (30). The chromatogram shown in Figure 13.3 corresponds to an analysis of FAME prepared with liver mitochondria phospholipids of rats which were fed a fat-free diet for 66 d. About 10 supplementary peaks corresponding to unusual fatty acids have appeared in rat lipids during fat deficiency. These acids are normally absent from chromatograms obtained with fatty acids of rats fed an equilibrated diet providing sufficient essential fatty acids. Most of these components can be fractionated according to their number of ethylenic bonds by silver-ion TLC (Figure 13.4). Using this procedure, it is possible to isolate a tetraenoic acid that migrates along with 20:4n-6 acid. Due to the dietary conditions of the rats, it was highly probable that this minor fatty acid is the 20:4n-7 acid. This component is the only tetraenoic acid that has been shown to accumulate during fat deficiency (19–25,27). It should be noted that the major 20:4n-6 acid and the minor tetraenoic acid (supposed to be the 20:4n-7 acid) are not separated when analyzed on a CP Wax 52 CB capillary column (30). To ascertain the identity of the minor component, the tetraenoic acid fraction was submitted to partial hydrazine reduction. This treatment generates a complex mixture of trienes, dienes, and monoenes, which still contains some residual unreacted tetraenes.

The monoenes were separated from other partially hydrogenated components by silver-ion TLC and subjected to oxidative ozonolysis in BF_3/methanol. The resulting monomethyl and dimethyl esters (MME and DME, respectively) were then analyzed by GLC coupled with mass spectrometry. Besides major components originating from 20:4n-6 acid, it was possible to identify MME: C_7, C_{10}, C_{13}, C_{16}, and DME: C_4, C_7, C_{10}, and C_{13} (30). This pattern clearly shows that the initial tetraenoic acid fraction contained a fatty acid with its double bonds localized in positions 4, 7, 10, and 13. This corresponds to 20:4n-7 acid. Thus, chromatography on the CP Sil 88 capillary column is a simple and convenient means not only to detect the 20-4n-7 acid associated with fat deficiency, but also to quantitate this fatty acid which is separated from other fatty acids.

Some of the unusual fatty acids that appear during fat deficiency may correspond to other members of the n-7 series. However, due to the complexity of the fractions isolated by silver-ion TLC from the initial mixture of FAME prepared with rat liver mitochondria phospholipids (Figure 13.4), it was not easy to identify

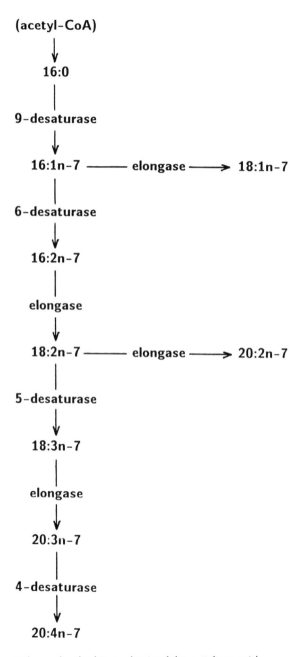

Figure 13.2. Pathway for the biosynthesis of the n-7 fatty acids.

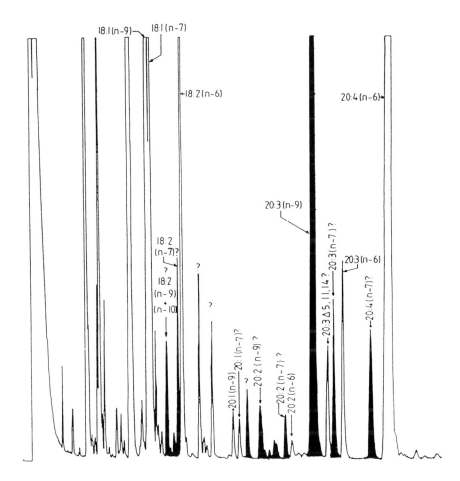

Figure 13.3. Partial gas–liquid chromatogram of FAME prepared with liver mitochondria phospholipids of rats fed a fat-free diet for 66 d. Analysis on a CP Sil 88 fused silica capillary column (50 m × 0.22 mm i.d., 0.22 μm film) operated isothermally at 185°C with an inlet pressure of the carrier gas (hydrogen) of 130 kPa. Black peaks correspond to fatty acids associated with fat deficiency. *Source:* Wolff et al. (30).

these components. However, the calculation method devised by Ackman et al. (31) was used to tentatively identify some of these acids. This calculation is based on the summation of the fractional chain lengths (FCL) of the individual monoenoic elements and a correction term which depends on the number and the relative position of ethylenic bonds (31).

Based on retention data of the C_{20} monoenes obtained after hydrazine treatment of the tetraenoic acids, on ECL of some standard C_{20} monoenes, and on inter-

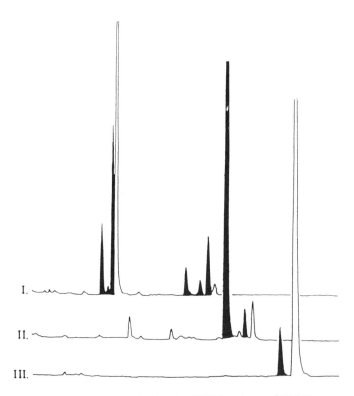

Figure 13.4. Chromatograms of silver-ion TLC-fractionated FAME prepared with liver mitochondria phospholipids of rats fed a fat-free diet for 66 d. Fraction I, dienes; fraction II, trienes; fraction III, tetraenes. Black peaks correspond to fatty acids associated with fat deficiency. Same chromatographic conditions as in Figure 13.3. *Source:* Wolff et al. (30).

polated values for some other C_{20} monoenes, it is possible to calculate the probable ECL of 20:2 and 20:3n-7 acids (Table 13.2). For example, 20:2n-7 acid has the structure *cis*-10,*cis*-13 20:2. Its ECL is thus the sum of the base value 20.00 and of the FCL of *cis*-10 20:1 and *cis*-13 20:1 acids, plus an adjustment for the dienoic structure. This adjustment can be deduced from a known dienoic C_{20} acid, for example the 20:2n-6 isomer. Adding the corresponding figures gives: 20.00 (base value) + 0.57 (FCL of *cis*-10 20:1 acid) + 0.69 (FCL of *cis*-13 20:1 acid) + 0.14 (diene adjustment) = 21.40 (Table 13.2). The calculated value is very close to the experimental value of 21.41 determined for one of the eicosadienoic acids isolated by silver-ion TLC (Figure 13.4).

TABLE 13.2
Retention Data for the Tentative Identification of Some Minor Polyenoic Acids Associated with Fat-Deficiency in the Rat

Fatty Acid[a]	FCL[c]	ECL[b] Exp[d]	ECL[b] Calc.	Diff[e]	Cor. ECL[f]
5-20:1	0.42	—	—	—	—
7-20:1	0.51	—	—	—	—
8-20:1	0.53	—	—	—	—
10-20:1	0.57	—	—	—	—
11-20:1	0.60	—	—	—	—
13-20:1	0.69	—	—	—	—
14-20:1	0.76	—	—	—	—
11,14-20:2	—	21.50	21.36	0.14	—
10,13-20:2	—	21.41	21.26	(+0.14)	21.40
8,11-20:2	—	21.27	21.13	(+0.14)	21.27
8,11,14-20:3	—	22.13	21.89	0.24	—
5,8,11-20:3	—	21.76	21.55	0.21	—
7,10,13-20:3	—	22.02	21.77	(+0.23)	22.00

[a]All ethylenic bonds are in the cis configuration.
[b]Equivalent chainlengths of fatty acid methyl esters determined with a CP Sil 88 capillary column (50 m × 0.22 mm i.d., 0.20 µm film) operated at 180°C; carrier gas (H_2), 130 kPa.
[c]Fractional chainlengths: FCL = ECL - 20.00.
[d]Exp., experimental values; Calc., calculated values (20.00 plus sum of FCL).
[e]Differences between experimental and calculated ECL values. Values in parentheses (ethylenic function adjustment) are used for the calculation of corrected ECL values.
[f]Corrected ECL (sum of FCL values, ethylenic function adjustment, and base value 20.00).

Similar reasoning allows the probable location of the 20:3n-7 acid peak (experimental ECL: 22.02; calculated ECL: 22.00). In this case, the adjustment for the three ethylenic bonds is deduced from data corresponding to 20:3n-6 and 20:3n-9 acids (Table 13.2). The tentative identification of 18:2n-7 acid is based on previously published biochemical studies (21), which have shown that this acid is the main isomer of 18:2n-6 acid that accumulates in rat-carcass lipids during essential fatty acid deficiency. 18:2n-7 acid is eluted just before 18:2n-6 acid on the CP Sil 88 capillary column. Thus, this column appears to be a simple and useful tool for the separation and the quantitation of several fatty acids of the palmitoleate series.

In the experimental conditions Wolff et al. (30) used newly weaned rats fed a fat-free diet for 66 d, and total fatty acids of the n-7 series account for only 3.3% of total fatty acids from rat liver mitochondria phospholipids. One may consider that this value is low. However, it should be emphasized that this low level is nevertheless higher than the total amount of n-3 acids remaining at this stage (2.7% [30]). Consequently, n-7 acids should not be considered as quantitatively unimportant compounds, and this should be taken into account when data for fatty acids from tissues of fat-deprived rats are tabulated.

Separation and Identification of α-Linolenic Acid Geometrical Isomers

Twenty years ago, Ackman et al. (32) demonstrated that steam-vacuum deodorization of Canbra oil (rapeseed oil low in erucic acid) induces the geometrical isomerization of up to 25% of α-linolenic acid (*cis*-9,*cis*-12,*cis*-15 18:3 acid, or all-*cis* 18:3*n*-3 acid). As a result of heat treatment, α-linolenic acid gives rise to isomers having the following structures: *cis*-9,*cis*-12,*trans*-15 18:3; *trans*-9,*cis*-12,*trans*-15 18:3; *trans*-9,*cis*-12,*cis*-15 18:3; and *cis*-9,*trans*-12,*cis*-15 18:3. At the time this experiment was performed, analysis by GLC of α-linolenic acid geometrical isomers (α-LAGI) on a Silar-5CP or on a BDS capillary column allowed the resolution of only three peaks, instead of five peaks corresponding to the five isomers (including the all-*cis* 18:3*n*-3 acid) present in the deodorized oils (32). Analyses of an NO_2-isomerized oil (containing all eight possible 18:3*n*-3 acid isomers) on BDS and Silar-5CP capillary columns allowed the separation of only four peaks (33). Thus, several overlaps between the different α-LAGI occurred under the analytical conditions.

Later, the separation of α-LAGI was improved by using Silar 10C (cyanopropyl polysiloxane) capillary columns (34–36). Five or six peaks could be fairly well separated with these columns. Similar results were obtained with an SP2340 capillary column, which was coated with a cyanopropyl polysiloxane stationary phase also (37). In another study (38), an uncommon Japanese capillary column (SS-4, cyanoethyl polysiloxane) allowed the characterization of eight peak apexes, but two pairs of isomers were poorly resolved. Moreover, peaks corresponding to mono- and di*trans* isomers of 18:3*n*-3 acid were not identified in this study (38). Recently, it has been shown that a 50-m long CP Sil 88 capillary column operated under optimal analytical conditions allowed an almost baseline resolution of seven peaks, with two isomers being eluted under the same peak (Figure 13.5a [39]). When isopropyl esters are used instead of methyl esters, eight peak apexes can be individually recognized (Figure 13.5b [39]). However, the resolution is not as good as with methyl esters.

Any heat treatment of α-linolenic acid containing oils (linseed, rapeseed, soybean, and walnut) can induce the formation of *trans* isomers of α-linolenic acid, provided the temperature is higher than 190–200°C. These heat treatments include vacuum deodorization in the presence of steam (32,40), but also simple heating in the presence of air (41) or nitrogen (36), or in ampoules sealed under vacuum (42,43). Some *trans* isomers of α-linolenic acid have also been found in partially hydrogenated soybean oil (44) and in soybean-oil margarines (45), but it is not clear whether geometrical isomers of 18:3*n*-3 acid in these products have been formed during partial hydrogenation or during deodorization. According to Ackman et al. (32), "slight hydrogenation. . .to reduce the 18:3 content is unlikely to generate methylene-interrupted geometrical isomers of these types in any high proportion of all-*cis* 18:3." Because the deodorization of oils is generally performed at 220–260°C, most soybean and rapeseed oils marketed in North America (32) and

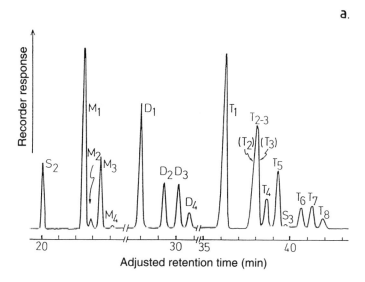

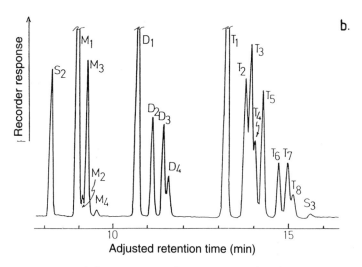

Figure 13.5. (a) Partial chromatogram of FAME prepared with NO$_2$-isomerized linseed oil. Analysis on a CP Sil 88 fused-silica capillary column (50 m × 0.33 mm i.d., 0.24 μm film) operated isothermally at 150°C with helium as carrier gas (inlet pressure, 80 kPa). (b) Partial chromatogram of fatty acid isopropyl esters prepared with NO$_2$-isomerized linseed oil. Same column as in (a), operated isothermally at 165°C with helium as carrier gas at an inlet pressure of 105 kPa. Abbreviations: S, saturated acids; M, monoenes; D, dienes; T, trienes. Identification of peaks in Table 13.3. *Source:* Wolff (39).

in some European countries (46,47) contain α-LAGI. Deodorized walnut oils (43) and foods containing deodorized rapeseed or soybean oils (46,48) may also contain α-LAGI. In some instances, the total amount of α-LAGI may account for as much as 3% of total fatty acids (46), that is for approximately 30% of the total 18:3n-3 acids. A survey of α-linolenic acid containing oils (soybean, rapeseed, and walnut) sold in several European countries has shown that about 9 samples out of 10 contain α-LAGI in noticeable, although variable, amounts (47). Thus, it is surprising that so few reports have dealt with these artifact fatty acids.

Because α-LAGI are widespread components of processed α-linolenic acid containing oils, there is a need to properly separate and identify these artifact components. As mentioned earlier, the CP Sil 88 capillary column appears to be a convenient tool to separate most of the eight theoretical α-LAGI (Figure 13.5). Fortunately, heating α-linolenic acid containing oils gives rise to only four α-LAGI (32,41,42), and the problem of separating five isomers (including the all-*cis* 18:3n-3 acid) is far less difficult than that of separating the eight isomers present in NO_2-isomerized oils. However, the resolution and the identification of all eight theoretical isomers are more than of academic interest. For example, it was claimed that the major isomer of 18:3n-3 acid that accumulates following partial hydrogenation of soybean oil is the *trans*-9,*trans*-12,*cis*-15 18:3 acid (44). This identification was later questioned (48), and it was suggested that this structure might have been erroneously established. If well-identified standards of α-LAGI (easily prepared by NO_2-isomerization of linseed oil) were available and properly analyzed to obtain the best chromatographic resolution, perhaps the identification of α-LAGI in partially hydrogenated soybean oil might have been easier. So, the problem is: Once α-LAGI are separated by GLC, how can they be identified?

The first step in the identification of α-LAGI is fractionation by silver-ion TLC. This procedure allows separation of fatty acids according to their number of *cis* and *trans* double bonds. Starting with FAME prepared with NO_2-isomerized linseed oil (39), this simple procedure allows separation of seven bands, including four fractions containing trienoic acids (Figure 13.6). Following the two saturated and *trans*-monoenoic acid bands (not shown in Figure 13.6), a third band containing the *cis*-monoenoic acids moves along with the all-*trans* linoleic acid. Just beneath this band, a fast-moving trienoic acid containing band is a mixture of the all-*trans* 18:3n-3 acid and of the two mono*trans* isomers of linoleic acid. The next band includes the di*trans* isomers of 18:3n-3 acid and the all-*cis* linoleic acid. The mono*trans* isomers of 18:3n-3 acid are present in the following band, and the slowest moving band is made of the all-*cis* isomer exclusively. From this isomer distribution pattern following silver-ion TLC fractionation, it appears that one *cis* double bond has the same effect on the migration rate of FAME as two *trans* ethylenic bonds.

The second step in the identification of α-LAGI is to determine the order of elution of components inside each of the di- and mono*trans* isomer fractions. Coherent results concerning the elution order of di*trans* isomers (*cis*-9,*trans*-12,*trans*-15 18:3 < *trans*-9,*cis*-12,*trans*-15 18:3 < *trans*-9,*trans*-12,-*cis*-15 18:3)

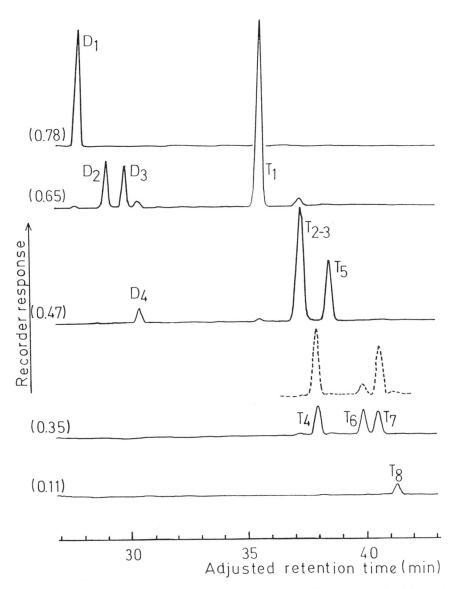

Figure 13.6. Partial chromatograms of FAME prepared with NO_2-isomerized linseed oil and fractionated by silver-ion TLC. The dotted insert corresponds to monotrans isomers of α-linolenic acid isolated from FAME prepared with a sample of deodorized rapeseed oil. Values between parentheses on the left of chromatograms are the Rf values of fractions isolated by silver-ion TLC. Chromatographic conditions and lettering of peaks are as in Figure 13.5(a). *Source:* Wolff (39).

have been obtained by either deduction or experiment in at least three different laboratories (12,13,35). In spite of a small difference in their ECL values, two of these isomers (the *cis*-9,*trans*-12,*trans*-15 and *trans*-9,*cis*-12,*trans*-15 18:3 acids; peak T_{2-3} in Figures 13.5 and 13.6) were not separated, at least when analyzed as methyl ester derivatives (13,34–36,39). However, if isopropyl esters are used instead of methyl esters, a fairly good separation between the two acids is obtained (39).

The location of the *trans*-9,*cis*-12,*trans*-15 18:3 acid is simplified by the fact that this isomer is the main di*trans* isomer that accumulates in deodorized or heated α-linolenic acid containing oils (32,41). It can be isolated from other mono*trans* isomers by silver-ion TLC fractionation of either methyl or isopropyl esters (32,41,48). Its ECL (19.87) is slightly but significantly different from the ECL of peak T_{2-3} (19.84 in Table 13.3). This suggests that the leading edge of peak T_{2-3} is the front edge of the *cis*-9,*trans*-12,*trans*-15 18:3 acid peak and that its trailing edge is the back edge of the *trans*-9,*cis*-12,*trans*-15 18:3 acid peak.

According to different literature data, the elution order of mono*trans* isomers of linolenic acid is apparently variable. With only one exception (12), all authors (13,34–36,41) indicate that the first mono*trans* isomer that elutes from cyanopropyl polysiloxane coated capillary columns is the *cis*-9,*cis*-12,*trans*-15 18:3 isomer. The exception to this rule, Scholfield (12), was in fact a tentative identification only. A detailed comparison between experimental ECL published in this study (12) and calculated ECL has shown that this identification was probably erroneous (39).

The situation is more complex for the two other mono*trans* isomers (12,13,34,35,41,48). In some studies (13,34,35), it was claimed that the *cis*-9,*trans*-12,*cis*-15 18:3 isomer elutes slightly after the *trans*-9,*cis*-12,*cis*-15 18:3 isomer. From other reports (12,41,48), the *trans*-9,*cis*-12,*cis*-15 18:3 isomer would elute after the *cis*-9,*trans*-12,*cis*-15 18:3 isomer. The discrepancies between these observations have been discussed elsewhere (39). There too, the typical profile of mono*trans* isomers that accumulate in deodorized (32) or heated (41) oils provides further information on the elution order of the *trans*-9,*cis*-12,*cis*-15 18:3 and *cis*-9,*trans*-12,*cis*-15 18:3 isomers. In α-linolenic acid containing oils submitted to heat treatments, the least abundant mono*trans* isomer is the *cis*-9,*trans*-12,*cis*-15 18:3 acid (32,41). This compound elutes between the two other mono*trans* isomers (Figure 13.6 [39,41,48]). Consequently, the order of elution of the mono*trans* isomers of linolenic acid should be *cis*-9,*cis*-12,*trans*-15 18:3 < *cis*-9,*trans*-12,*cis*-15 18:3 < *trans*-9,*cis*-12,*cis*-15 8:3. Indeed, this elution order was established experimentally by Grandgirard et al. (41) who analyzed methyl esters prepared with heated rapeseed oil on a Silar 10C capillary column. The two groups of mono- and di*trans* isomers of 18:3*n*-3 acid are not separated as integral entities during GLC. The *cis*- 9,*cis*-12,*trans*-15 18:3 isomer emerges from the column before the *trans*-9,*trans*-12,*cis*-15 18:3 acid. Using the CP Sil 88 capillary column, these two isomers are baseline resolved (peaks T_4 and T_5 in Figure 13.5a [39]). This was not achieved with the Silar 10C columns (13,34–36) for which the two isomers were mixed under a single peak.

TABLE 13.3
Chromatographic Characteristics of Fatty Acid Methyl and Isopropyl Esters Prepared with NO_2-Isomerized Linseed Oil and Analyzed on a CP Sil 88 Capillary Column

Peak Number[b]	Rf (silver-ion TLC)[c]	ECL[a] FAME	FAIPE	Fatty Acid Structure
S_2	0.95	18.00	18.00	18:0
S_3		20.00	20.00	20:0
M_1	0.90	18.32	18.24	trans-9 18:1
M_2		18.39	18.31	trans-11 18:1
M_3	0.78	18.45	18.37	cis-9 18:1
M_4		18.55	18.46	cis-11 18:1
D_1		18.95	18.80	trans-9,trans-12 18:2
D_2	0.65	19.10	18.93	cis-9,trans-12 18:2
D_3		19.18	19.02	trans-9,cis-12 18:2
T_1		19.71	19.45	trans-9,trans-12,trans-15 18:3
D_4	0.47	19.24	19.06	cis-9,cis-12 18:2
T_2		19.84	19.59	cis-9,trans-12,trans-15 18:3
T_3		19.87	19.62	trans-9,cis-12 trans-15 18:3
T_5		19.96	19.70	trans-9,trans-12,cis-15 18:3
T_4	0.35	19.91	19.65	cis-9,cis-12,trans-15 18:3
T_6		20.06	19.81	cis-9,trans-12,cis-15 18:3
T_7		20.11	19.85	trans-9,cis-12,cis-15 18:3
T_8	0.11	20.15	19.89	cis-9,cis-12,cis-15 18:3

Abbreviations: S, saturates; M, monoenes; D, dienes; T, trienes; FAME, fatty acid methyl esters; FAIPE, fatty acid isopropyl esters.
[a]Equivalent chainlengths established under chromatographic conditions described in Figure 13.5.
[b]Peak numbers correspond to Figures 13.5 and 13.6.
[c]Rf values of fractions separated by silver-ion TLC.

Another way to establish the elution order of all eight α-LAGI is based on the following observations. The elution order of linoleic acid mono*trans* isomers by GLC on cyanoalkyl polysiloxane stationary phases is well established. The *cis*-9,*trans*-12 precedes the *trans*-9,*cis*-12 isomer, and both isomers emerge from the column after the di*trans* isomer but before the natural di*cis* isomer (1,12,13,49). The elution order of 9,12-18:2 geometrical isomers (*trans*-9,*trans*-12 < *cis*-9,*trans*-12 < *trans*-9,*cis*-12 < *cis*-9,*cis*-12) may be used to predict the elution order of α-LAGI if one assumes that adding a third double bond (either of the *trans* or *cis* configuration) to the dienes will not modify the elution order of the corresponding trienes. This assumption is justified by the fact that the ECL of the trienoic acids will be equal to the sum of the FCL of the dienoic isomers, of the FCL of the 15-ethylenic bond, and of an adjustment for the supplementary diethylenic system (plus the base value 18.00). Provided the adjustment does not depend too much on the geometry of double bonds in the supplementary diethylenic system, the elution order of the resulting trienoic acids will be the same as that of the initial dienoic acids. If one adds a *trans*-15 double bond to the four dienes, the elution order of the resulting trienoic acids will then be *trans*-9,*trans*-12,*trans*-15 < *cis*-9,*trans*-12,*trans*-15 < *trans*-9,*cis*-12,*trans*-15 < *cis*-9,*cis*-12,*trans*-15. Adding a *cis*-15 double bond will result in the following elution order: *trans*-9,*trans*-12,*cis*-15 < *cis*-9,*trans*-12,*cis*-15 < *trans*-9,*cis*-12,*cis*-15 < *cis*-9,*cis*-12,*cis*-15.

Geometrical isomers of other octadecadienoic acids related to 18:3*n*-3 acid (12,15-18:2 and 9,15-18:2 acids) are also eluted off the column in the order *trans*,*trans* < *cis*,*trans* < *trans*,*cis* < *cis*,*cis* (12,13). Adding the appropriate third ethylenic bond will give geometrical isomers of 18:3*n*-3 acid which should elute in an order compatible with the sequences: *trans*-9,*trans*-12,*trans*-15 < *cis*-9,*trans*-12,*trans*-15 < *trans*-9,*cis*-12,*trans*-15 < *cis*-9,*cis*-12,*trans*-15 and *trans*-9,*trans*-12,*cis*-15 < *cis*-9,*trans*-12,*cis*-15 < *trans*-9,*cis*-12,*cis*-15 < *cis*-9,*cis*-12,*cis*-15.

Although one cannot predict how the two series will interact (it depends on the FCL value of the 15-ethylenic bond, which is linked to the nature of the column and to the operating conditions), experimental data obtained with the CP Sil 88 column indicate that all isomers containing a *trans*-15 double bond elute before those containing a *cis*-15 double bond (39). This is not always the case for other cyanopropyl polysiloxane coated capillary columns, on which the last eluting component with a *trans*-15 double bond (*cis*-9,*cis*-12,*trans*-15 18:3) can elute with, or after, the first eluting component with a *cis*-15 double bond (*trans*-9,*trans*-12,*cis*-15 18:3 [12,13,34–36]). Taking into account all data presented here, the elution order of the eight α-LAGI on the CP Sil 88 capillary column is finally: *trans*-9,*trans*-12,*trans*-15 < (*cis*-9,*trans*-12,*trans*-15 < *trans*-9,*cis*-12,*trans*-15) < *cis*-9,*cis*-12,*trans*-15 < *trans*-9,*trans*-12,*cis*-15 < *cis*-9,*trans*-12,*cis*-15 < *trans*-9,*cis*-12,*cis*-15 < *cis*-9,*cis*-12,*cis*-15. Two isomers (*cis*-9,*trans*-12,*trans*-15 and *trans*-9,*cis*-12,*trans*-15 18:3 acids) have ECL that are not sufficiently different to allow their resolution, at least when they are analyzed as methyl esters (Table 13.3). When isopropyl esters are analyzed under conditions that allow the best resolution between all eight α-LAGI, the difference in ECL values of the two preceding isomers is not increased (Table 13.3). However, their elution time is shortened considerably (from 38 to approximately 14 min) and the two components are partly resolved (Figure 13.5 [39]).

As mentioned previously, industrially deodorized oils contain at most four *trans* isomers of 18:3*n*-3 acid that give very typical chromatographic profiles (Figure 13.7). These isomers have the structures *trans*-9,*cis*-12,*trans*-15 (minor), *cis*-9,*cis*-12,*trans*-15 (major), *cis*-9,*trans*-12,*cis*-15 (minor), and *trans*-9,*cis*-12,*cis*-15 (major) and are eluted off the column in this order (Figure 13.7). All of these isomers precede the all-*cis* 18:3*n*-3 acid.

Initially, it was thought that the individual proportions of these isomers relative to total *trans*-octadecatrienoic acids were constant (46). Based on the analysis of twelve French α-linolenic acid containing oils, these proportions were *trans*-9,*cis*-12,*trans*-15 18:3 acid, 4.9 ± 1.5%; *cis*-9,*cis*-12,*trans*-15 18:3 acid, 47.8 ± 1.7%; *cis*-9,*trans*-12,*cis*-15 18:3 acid, 6.5 ± 0.7%; and *trans*-9,*cis*-12,*cis*-15 18:3 acid, 41.1 ± 1.0% (46).

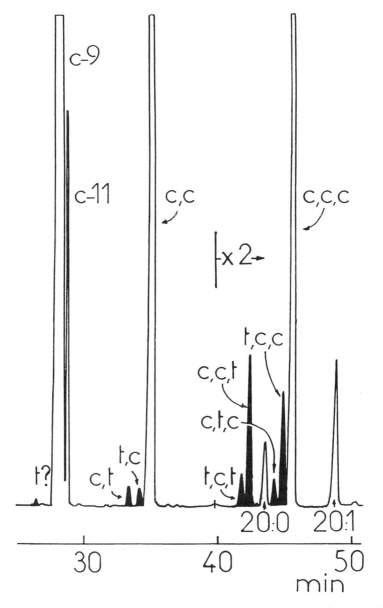

Figure 13.7. Partial chromatogram of FAME prepared with a sample of commercial deodorized rapeseed oil. Chromatographic conditions as in Figure 13.5(a). The configurations of double bonds are given in the order 9 and 12 for 18:2n-6 acid geometrical isomers, and in the order 9, 12, and 15 for 18:3n-3 acid geometrical isomers. Abbreviations: c, cis; t, trans. Source: Wolff (47).

Later, as the number of samples was increased, this conclusion was revised (47), and it was shown that the relative proportions of individual α-LAGI depend on the degree of isomerization (DI) of α-linolenic acid. The DI is defined as the percentage of total *trans* isomers relative to total octadecatrienoic acids. As the DI increases from 10 to 30%, the relative proportions of the two main mono*trans* isomers decrease slightly in a linear fashion, while the proportion of the minor mono*trans* isomer practically remains constant. The proportion of the di*trans* isomer increases linearly with the DI. These tendencies were further confirmed using linseed oil as a model (42). In this study, aliquots of linseed oil (containing approximately 45% α-linolenic acid) were heated to temperatures ranging from 190 to 260°C for 2–16 h in ampoules sealed under vacuum. Degrees of isomerization as high as approximately 70% could be obtained under such conditions.

It has been shown that the disappearance of the all-*cis* 18:3n-3 acid follows first-order kinetics (42,50) and that the formation of the *trans*-9,*cis*-12,*trans*-15 18:3 isomer follows a two-step reaction (42). The di*trans* isomer is issued from both the *cis*-9,*cis*-12,*trans*-15 and *trans*-9,*cis*-12,*cis*-15 18:3 isomers. The probabilities of isomerization of double bonds in positions 9, 12, and 15 at the very beginning of the reaction are approximately 0.42, 0.05, and 0.53, respectively (42,47 [Figure 13.8]). At that moment, the probability of a simultaneous *cis-trans* isomerization of double bonds in positions 9 and 15 is 0. As the reaction continues, the *cis*-9,*cis*-12,*trans*-15 and *trans*-9,*cis*-12,*cis*-15 18:3 isomers are progressively transformed into the *trans*-9,*cis*-12,*trans*-15 isomer through a second geometrical isomerization, while the *cis*-9,*trans*-12,*cis*-15 isomer remains mostly unchanged. Further modifications rarely occur, and the *trans*-9,*trans*-12,*cis*-15, *cis*-9,*trans*-12,*trans*-15, and *trans*-9,*trans*-12,*trans*-15 isomers are formed in trace amounts only, even under the harsher conditions of heating (Figure 13.9).

Commercial deodorized oils generally have DI that range from a few percent to almost 30%; the *cis*-9,*cis*-12,*trans*-15 18:3 isomer represents 46–53% of the total α-LAGI, the *trans*-9,*cis*-12,*cis*-15 18:3 isomer between 38 and 42%, and the *cis*-9,*trans*-12,*cis*-14 18:3 isomer from 5.4–7.8% (42,43,46). The *trans*-9,*cis*-12,*trans*-15 isomer varies from trace amounts to almost 10% of total α-LAGI. These values apply to any α-linolenic acid containing oil presenting a DI less than 30% (soybean, rapeseed, walnut, and linseed oils) and are thus independent of the initial concentration of α-linolenic acid in the oils (from approximately 6% [soybean oil] to 45% [linseed oil] [42,43,46,47]). Consequently, any commercial deodorized α-linolenic acid containing oil may be used as a source of standards for the identification of four (five, if one includes the all-*cis* 18:3n-3 acid) of the eight α-LAGI. Alpha-linoleic acid geometrical isomers can be identified either by their order of elution or by their relative proportions. However, care should be taken not to confuse these isomers with the 20.0 acid (Figure 13.7). The *trans*-9,*cis*-12,*trans*-15 18:3 isomer can be easily located because its relative proportion increases with temperature and heating time (42). Standard α-LAGI may also be prepared conveniently in the laboratory simply by heating a sample of α-linolenic acid containing oil in a glass ampoule (sealed under vacuum) at 220–260°C for a few hours (42).

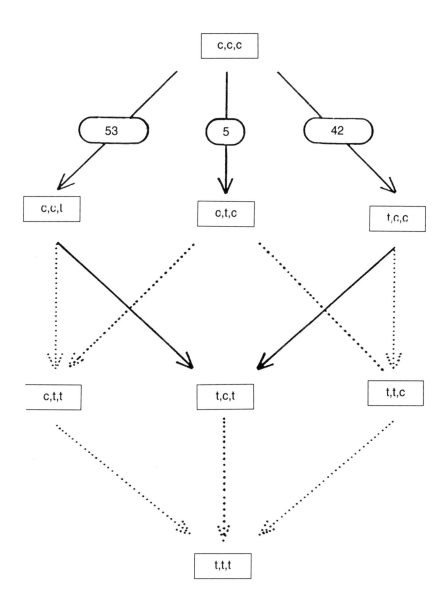

Figure 13.8. Schematic representation of the proposed reactional sequences that lead to the formation of α-linolenic acid geometrical isomers when the all-*cis* 18:3 *n*-3 acid is heated under vacuum. Values are the probabilities of isomerization of the ethylenic bonds at the very beginning of the reaction. Dotted arrows correspond to possible reactions that occur at a very low rate.

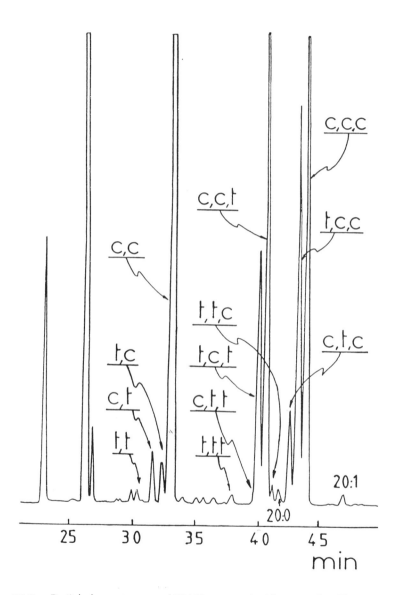

Figure 13.9. Partial chromatogram of FAME prepared with a sample of linseed oil that was heated under vacuum at 245°C for 16 h. The degree of isomerization of 18:3n-3 acid was approximately 69%. Chromatographic conditions as in Figure 13.5(a). The configurations of double bonds are given in the order 9 and 12 for 18:2n-6 acid geometrical isomers, and in the order 9, 12, and 15 for 18:3n-3 acid geometrical isomers. Abbreviations: *c, cis; t, trans. Source:* Wolff (42).

The use of the CP Sil 88 capillary column for the identification and the quantitation of α-LAGI is not restricted to the study of heated (deodorized) oils. This column is also useful when fatty acids from biological samples are analyzed (Figure 13.10). As the essential all-*cis* 18:3*n*-3 acid present in commercial α-linolenic acid containing oils is almost always accompanied by some of its geometrical isomers, the metabolic fate of these isomers in animals and humans is of interest. Piconneaux et al. (51) have shown that α-LAGI present in heated linseed oil fed to rats can be incorporated in total lipids of the liver. Moreover, one α-LAGI, most probably the *cis*-9,*cis*-12,*trans*-15 18:3 acid, is further elongated and desaturated to *cis*-5,*cis*-8,*cis*-11,*cis*-14,*trans*-17 20:5 and *cis*-4,*cis*-7,*cis*-10,*cis*-13,*cis*-16,*trans*-19 22:6 acids (*trans*-17 20:5*n*-3 and *trans*-19 22:6*n*-3 acids, respectively [51,52]). These very long chain polyunsaturated fatty acids are also incorporated into rat lipids (51), and particularly into phospholipids (53). Although its structure has not been for-

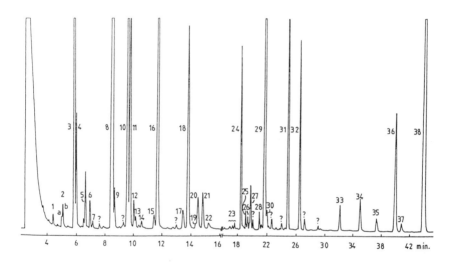

Figure 13.10. Chromatogram of FAME and free aldehydes prepared with heart mitochondria phospholipids (other than cardiolipin) of rats fed a concentrate of α-LAGI following a fat-free diet. Identification of peaks: 1. 14:0; 2. 15:0 and 15:0 aldehyde; 3. 16:0; 4. 16:0 aldehyde; 5. 16:1*n*-7; 6. 17:0; 7. 17:0 aldehyde; 8. 18:0; 9. 18:0 aldehyde; 10. 18:1*n*-9; 11. 18:1*n*-7; 12. 18:1*n*-9 aldehyde; 13. 18:1*n*-7 aldehyde; 14. 18:2*n*-9 (tentative identification); 15. 18:2*n*-7; 16. 18:2*n*-6; 17. *trans*-9,*cis*-12,*trans*-15 18:3*n*-3; 18. *cis*-9,*cis*-12,*trans*-15 18:3*n*-3; 19. *cis*-9,*trans*-12,*cis*-15 18:3*n*-3; 20. *trans*-9,*cis*-12,*cis*-15 18:3; 22. 20:1; 23. 20:2 (several isomers); 24. 20:3*n*-9; 25. 5,11,14-20:3 (tentative identification); 26. 20:3*n*-7; 27. 20:3*n*-6; 28. 20:4*n*-7; 29. 20:4*n*-6; 30. 20:3*n*-3; 31. *trans*-17 20:5*n*-3; 32. 20:5*n*-3; 33. 22:4*n*-6; 34. 22:5*n*-6; 35. *trans*-19 22:5*n*-3; 36. 22:5*n*-3; 37. *trans*-19 22:6*n*-3; 38. 22:6*n*-3. If no precision is given, ethylenic bonds in unsaturated fatty acids or aldehydes have the *cis* configuration. Analysis on a CP Sil 88 capillary column operated isothermally at 185°C using hydrogen as a carrier gas (inlet pressure, 130 kPa). *Source:* Wolff et al. (53).

mally established, a third component (probably the *cis*-7,*cis*-10,*cis*-13,*cis*-16,*trans*-19 22:5 acid [*trans*-19 22:5*n*-3 acid], the intermediary compound in the formation of *trans*-19 22:6*n*-3 acid) is also found in rat lipids (51,52).

As shown in Figure 13.10, the three mono*trans n*-3 very long chain polyunsaturated acids are well separated from other fatty acids during GLC, despite the fact that they differ from their all-*cis* counterparts by a single *trans* ethylenic bond. Examination of the incorporation of α-LAGI into rat phospholipids has shown that there is an important selectivity in favor of the *cis*-9,*cis*-12,*trans*-15 18:3 isomer, which is preferentially esterified to positions 1(1″) of cardiolipins (53). This isomer resembles the essential all-*cis* 18:2*n*-6 acid by several aspects, which would indicate that the *trans*-15 double bond is perceived by some enzymatic systems as a single bond. On the other hand, there is no discrimination against any of the isomers between the initial step of intestinal absorption and their final deposition into adipose-tissue triglycerides (53).

Apparently, α-LAGI have not yet been detected in human tissues (54–56). This indicates that these isomers do not accumulate in any noticeable amount in humans. However, in an unpublished experiment, the author has observed that dietary α-LAGI are indeed incorporated in human serum triglycerides after a single meal containing approximately 40 g of a commercial deodorized soybean oil in which α-LAGI accounted for about 3% of total fatty acids (Figure 13.11). This means that α-LAGI are effectively absorbed by the human digestive tract and that they readily cross the intestinal barrier. The fact that α-LAGI have not been characterized in human tissues is probably linked to their low dietary intake which results in a considerable dilution of these isomers in body tissues. It is the author's opinion that α-LAGI are probably present in tissue lipids, but that their presence cannot be easily recognized without a preliminary concentration of these compounds prior to GLC analysis.

Separation and Partial Identification of γ-Linolenic Acid Geometrical Isomers

Oils containing γ-linolenic acid (*cis*-6,*cis*-9,*cis*-12 18:3 acid or all-*cis* 18:3*n*-6 acid) are sometimes deodorized to improve their quality (57). Among components removed by deodorization are oxidation products, such as aldehydes; hydrolysis products, such as free fatty acids; residual extraction solvents; and potential pesticide residues. Moreover, peroxides are destroyed by this heat treatment. The higher the temperature, the better the quality in this respect. However, it has been shown recently that heating of borage oil at temperatures equal to or higher than 200°C induces the appearance of some artifacts derived from γ-linolenic acid (58). Using adapted analytical techniques (silver-ion TLC; comparison of the ECL with those of isomers present in NO_2-isomerized borage oil on two different capillary columns, CP Sil 88 and DB Wax; partial hydrazine reduction; oxidative ozonolysis; GLC coupled with mass spectrometry; and GLC coupled with Fourier-transform infrared spectroscopy), Wolff and Sebedio (58) have demonstrated that these artifacts are γ-

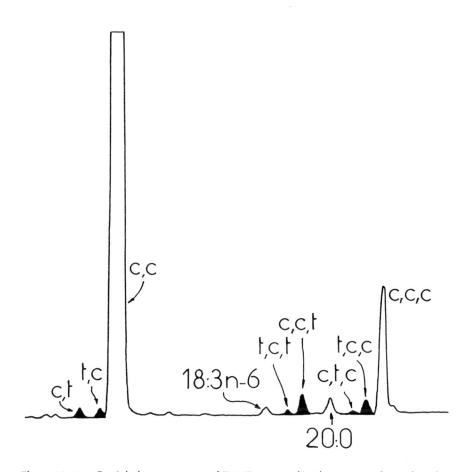

Figure 13.11. Partial chromatogram of FAME prepared with serum triglycerides of a human volunteer who ate a single meal containing approximately 40 g of commercial deodorized rapeseed oil. Three percent of total fatty acids were α-LAGI. Blood was withdrawn 4 h after the meal. The configurations of double bonds are given in the order 9 and 12 for 18:2n-6 acid geometrical isomers, and in the order 9, 12, and 15 for 18:3n-3 acid geometrical isomers. Analysis on a CP Sil 88 capillary column. Abbreviations: c, cis; t, trans.

linolenic acid geometrical isomers (γ-LAGI). They are formed either upon heating under vacuum or after steam-vacuum deodorization.

Heat-induced isomerization of γ-linolenic acid is similar to that of α-linolenic acid in several aspects. The number and relative isomer structures are the same and the DI are almost identical when the two fatty acids are heated under identical conditions. The two main isomers that accumulate upon heat treatment are the *cis*-6,*cis*-9,*trans*-12 and the *trans*-6,*cis*-9,*cis*-12 18:3 acids. The two minor isomers that are also present in the heated oil include the third mono*trans* isomer (*cis*-6,*trans*-9,*cis*-12 18:3 acid) and one di*trans* isomer, the *trans*-6,*cis*-9,*trans*-12 18:3 acid (58). The distribution pattern of these isomers indicates that the less reactive double bond regarding *cis*–*trans* isomerization is that located in position 9. In α-linolenic acid, it is the double bond in position 12 which is the most resistant to geometrical isomerization. In both cases, it is the internal double bond which isomerizes at the lowest rate.

The presence of γ-LAGI is an excellent method to detect whether a γ-linolenic acid containing oil has been subjected to deodorization at temperatures higher than 200°C. On the other hand, the appearance of γ-LAGI may be used as a means to control the deodorization process. Due to the high price of these oils, this quality aspect should not be neglected. There is thus a need to separate γ-LAGI from the all-*cis* 18:3*n*-6 acid, and eventually to identify these artifacts.

Unfortunately, separation and identification of γ-LAGI by GLC appear to be much more difficult than that of α-LAGI. Starting with NO_2-isomerized borage oil (containing the eight γ-LAGI), only five peaks could be separated using the same 50-m long CP Sil 88 capillary column that allowed baseline resolution of seven peaks when α-LAGI were analyzed. Such a separation is obtained when either methyl or isopropyl esters are analyzed (58). However, one drawback of analyzing methyl esters of γ-LAGI on the CP Sil 88 column is that the all-*cis* 18:3*n*-6 acid coelutes with one of the two main and the minor mono*trans* isomers formed during heat treatment. This precludes any direct estimation of the true amount of total *trans* isomers that appear during heat treatment.

When fatty acid isopropyl esters prepared with a heated oil are analyzed, the all-*cis* 18:3*n*-6 acid is fairly well resolved from the *trans* isomers on the CP Sil 88 column (Figure 13.12). With these derivatives, the total amount of *trans* isomers can be determined easily. Unfortunately, one of the main mono*trans* isomer is mixed with the minor di*trans* isomer (peak Y in Figure 13.12), and the other main mono*trans* isomer is mixed with the minor *cis*-6,*trans*-9,*cis*-12 18:3 isomer (peak Z in Figure 13.12).

A 30-m long DB Wax capillary column allows the resolution of only four peaks when γ-LAGI present in NO_2-isomerized borage oil are analyzed; the number of peaks is independent of the derivative form of the fatty acids (58). However, when heated borage oil is the source of γ-LAGI, three peaks corresponding to *trans* isomers are separated from the parent all-*cis* 18:3*n*-6 acid, and are eluted off the column after this last fatty acid (Figure 13.12). Two main peaks, each corresponding to one of the two major mono*trans* isomers, are eluted first (peaks A and B in Figure 13.12). A third peak containing the minor mono*trans* and one di*trans* isomers is partly separated from the second mono*trans* isomer eluted (peak C in Figure 13.12).

Consequently, analyses of FAME on the DB Wax capillary column or of fatty acid isopropyl esters on the CP Sil 88 capillary column appear to be the best means to quantitate γ-LAGI in deodorized α-linolenic acid containing oils. Analysis of

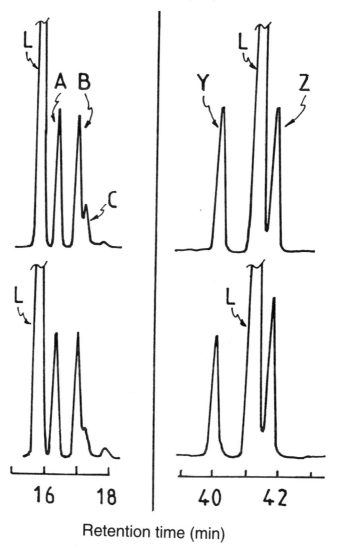

Figure 13.12. Comparison of partial chromatograms of fatty acids from borage oil heated under vacuum at 240°C for 6 h (upper tracings) and from borage oil deodorized at 240°C for 2 h (lower tracings). Injections were made at different loads. Left chromatogram, analyses of FAME on a DB Wax capillary column (30 m × 0.32 mm i.d., 0.5 μm film) operated isothermally at 180°C with an inlet pressure of the carrier gas (hydrogen) of 90 kPa. Right chromatograms, analyses of fatty acid isopropyl esters on a CP Sil 88 capillary column (50 m × 0.33 mm i.d., 0.24 μm film) operated isothermally at 160°C with an inlet pressure of the carrier gas (helium) of 100 kPa. L, all-*cis* α-linolenic acid; A, B, C, Y, and Z, *trans* isomers of α-linolenic acid (see text for their partial identification).

methyl esters on the DB Wax column yields more insight into the detailed composition of γ-LAGI than analysis of isopropyl esters on the CP Sil 88 column. Moreover, the time of analysis of γ-LAGI on the DB Wax column operated under optimal conditions is less than one-half that on the CP Sil 88 column (approximately 18 min instead of 42 min [Figure 13.12]). So, it is more beneficial to identify γ-LAGI using the DB Wax column.

An indirect approach based on the hypothesis that γ-linolenic isomerizes like α-linolenic acid was used. Thus, each individual double bond should isomerize at a given rate according to its position relative to the carboxylic group. The double bonds in positions 6, 9, and 12 in γ-linolenic acid should isomerize at the same rates as the double bonds in positions 9, 12, and 15, respectively, in α-linolenic acid. Therefore, the distribution pattern of γ-LAGI and α-LAGI in borage and linseed oil samples that were heated under vacuum at two temperatures were compared (Table 13.4). Under both conditions, the main α-LAGI that accumulates in linseed oil is the *cis*-9,*cis*-12,*trans*-15 18:3 acid. From this observation, it is likely that the main γ-LAGI that accumulates in borage oil (peak B in Figure 13.12) is the *cis*-6,*cis*-9,*trans*-12 18:3 acid. The second major mono*trans* isomer in linseed oil is the *trans*-9,*cis*-12,*cis*-15 18:3 acid. Therefore, the second major mono*trans* isomer in borage oil (peak A in Figure 13.12) should be the *trans*-6,*cis*-9,*cis*-12 18:3 acid. The proportions of the mixture of the γ-LAGI (*cis*-6,*trans*-9,*cis*-12 and *trans*-6,*cis*-9,*trans*-12 18:3 acids) in the third peak eluted on the DB Wax column (peak C in Figure 13.12) are the same as those of the sum of the *cis*-9,*trans*-12,*cis*-15 and *trans*-9,*cis*-12,*trans*-15 18:3 acids in the heated linseed oil samples. Thus, the elution order of methyl esters of γ-LAGI on the DB Wax capillary column should be *cis*-6,*cis*-9,*cis*-12 18:3 < *trans*-6,*cis*-9,*cis*-12 18:3 < *cis*-6,*cis*-9,*trans*-12 18:3 < (*cis* 6,*trans*-9,*cis*-12 18:3 + *trans*-6,*cis*-9,*trans*-12 18:3).

TABLE 13.4
Comparison of the Relative Proportions of Individual α- and γ-Linolenic Acid Geometrical Isomers in Heated Oils

γ-LAGI[a]	Proportions[b] in Heated Borage Oil		α-LAGI[c]	Proportions in Heated Linseed Oil	
	240°C, 6 h	260°C, 5 h		240°C, 6 h	260°C, 5 h
A	42.7	32.5	t,c,c[d]	40.3	30.6
B	45.1	35.2	c,c,t	46.5	37.6
C	12.2	32.2	c,t,c + t,c,t	13.2	31.9

[a]Peak lettering of γ-linolenic acid geometrical isomers as in Figure 13.12 (analyses on the DB Wax capillary column).
[b]Proportions of individual *trans* isomers are relative to their total and are expressed in weight percentages.
[c]α-linolenic acid geometrical isomers analyzed on a CP Sil 88 capillary column. Chromatographic conditions are as described in the legend of Figure 13.5a.
[d]c, *cis*; t, *trans*.

Separation and Identification of Pinolenic Acid Geometrical Isomers

Pinolenic acid (*cis*-5,*cis*-9,*cis*-12 18:3 acid) is an uncommon C_{18} nonmethylene-interrupted trienoic acid that is relatively abundant (15% of the total fatty acids in the oil analyzed) in pine seed oil (59). This fatty acid shares the presence of three *cis* ethylenic bonds, two of these being located in positions 9 and 12, with α- and γ-linolenic acids. Thus, pinolenic acid can be of some help to understand the mechanism of the *cis–trans* isomerization that occurs upon heating α- or γ-linolenic acids. Consequently, the separation and identification of pinolenic acid geometrical isomers (PAGI) are of interest.

To identify PAGI that may appear upon heat treatment of pine seed oil, the NO_2-isomerized pine seed oil was first studied. Isomerization was achieved as described for linseed (39) and borage (58) oils. A chromatogram of FAME prepared with the modified oil and analyzed on a 50-m long CP Sil 88 capillary column is given in Figure 13.13. Note the excellent resolution that is obtained inside each family of isomers: monoenes, nonmethylene-interrupted dienes (5,9-18:2 acids), methylene-interrupted dienes (9,12-18:2 acids), nonmethylene-interrupted trienes (5,9,12-18:3 acids), peaks M, D, D', and T, respectively, in Figure 13.13. For the last group, seven peaks are baseline resolved. Enlargement of one of the peaks (peak T_{6-7} in Figure 13.13) indicates that it contains two isomers that are on the verge of separating. Another peak (peak T_1 in Figure 13.13) is eluted as a shoulder on the leading edge of one mono*trans* isomer of 18:2*n*-6 acid (peak D'3).

The fact that PAGI are easier to separate than γ-LAGI on the CP Sil 88 column is probably linked to the ECL contribution of the double bond nearest the carboxylic group. For example, the difference on cyanoalkyl polysiloxane stationary phases between ECL of *cis*-6 and *cis*-9 18:1 acids is only approximately 0.05 carbon units (1,12). On the other hand, the difference between ECL of *cis*-5 and *cis*-9 18:1 acids is approximately 0.18 carbon units (1,12).

The complex mixture was fractionated by silver-ion TLC following the same procedure used for linseed (39) and borage (58) oils. With this technique, it is possible to separate the tri*trans*, di*trans*, and mono*trans* isomers of the all-*cis* pinolenic acid (Figure 13.14). As important is the fact that it is possible to separate geometrical isomers of the minor C_{18} nonmethylene-interrupted dienoic acid, *cis*-5,*cis*-9 18:2 acid, which is initially present in low amounts (approximately 2% of total fatty acids) in pine seed oil. This uncommon fatty acid is easily located in comparison with fatty acids prepared with *Taxus baccata* (yew) seed oil, a relatively rich source of *cis*-5,*cis*-9 18:2 acid (60).

Geometrical isomers of this nonmethylene-interrupted dienoic acid are separated according to their number of *cis* and *trans* double bonds by silver-ion TLC. The fast moving diene is the all-*trans* isomer (peak D_1 in Figure 13.14). Following this band is a band that contains a mixture of the *trans*-5,*cis*-9 and *cis*-5,*trans*-9 18:2 acids (peaks D_2 and D_3 in Figure 13.14) among other components. Note that the

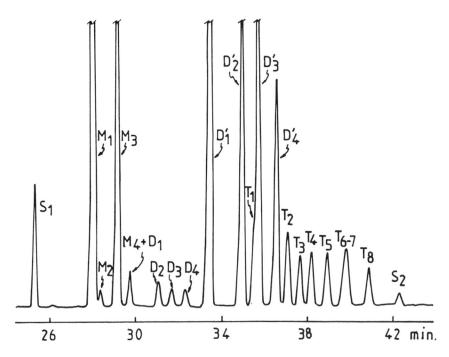

Figure 13.13. Partial chromatogram of FAME prepared with NO_2-isomerized pine seed oil. Identification of peaks as in Table 13.5. Analyses on a CP Sil 88 capillary column (50 m × 0.33 mm i.d., 0.24 μm film) operated isothermally at 160°C with an inlet pressure of the carrier gas (helium) of 100 kPa. Abbreviations: S, saturates; M, monoenes; D, nonmethylene-interrupted dienes (5,9-18:2 acid isomers); D', methylene-interrupted dienes (9,12-18:2 acid isomers); T, nonmethylene-interrupted trienes (5,9,12-18:3 acid isomers).

geometrical isomers of 5,9-18:2 acid move at a slightly lower rate during silver-ion TLC than the corresponding isomers of 9,12-18:2 acid. However, when isomers of 5,9-18:2 acid are compared to isomers of the all-*cis* 5,9,12-18:3 acid, the rule established for 18:2*n*-6 and 18:3*n*-3 isomers still holds true: one *cis* double bond has the same effect on the migration rate of nonmethylene-interrupted acids as two *trans* double bonds.

The identification of the *trans*-5,*cis*-9 18:2 acid is made possible by comparison with the same compound present in small amounts in *Aquilegia vulgaris* (columbine) seed oil (61). Consequently, all four isomers of 5,9-18:2 acid are identified. Their order of elution on the CP Sil 88 capillary column is *trans*-5,*trans*-9 18:2 < *cis*-5,*trans*-9 18:2 < *trans*-5,*cis*-9 18:2 < *cis*-5,*cis*-9 18:2 (Table 13.5). As mentioned previously, the elution order of 18:2*n*-6 acid geometrical isomers is well established on cyanoalkyl polysiloxane coated capillary columns: *trans*-9,*trans*-12

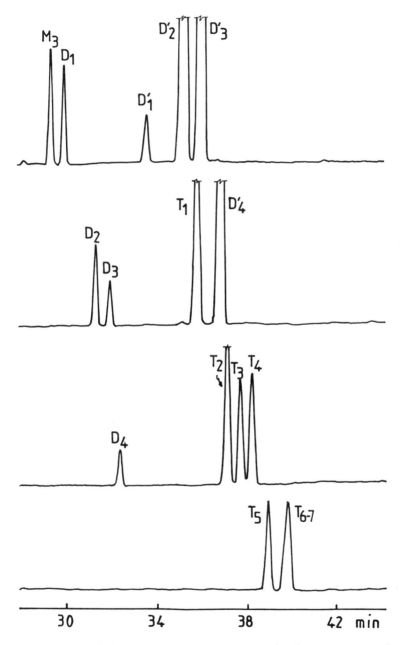

Figure 13.14. Partial chromatograms of FAME prepared with NO_2-isomerized pine seed oil and fractionated by silver-ion TLC. Chromatographic conditions and peak lettering as in Figure 13.13.

TABLE 13.5
Experimental and Calculated Chromatographic Retention Data for Fatty Acid Methyl Esters Prepared with NO_2-isomerized Pine Seed Oil

Peak Number[a]	ECL[b] Exp.	Calc.	Diff[c]	Fatty Acid Structure
S_1	18.00	—	—	18:0
S_2	20.00	—	—	20:0
M_1	18.40	—	—	trans-9 18:1
M_2	18.47	—	—	trans-11 18:1
M_3	18.57	—	—	cis-9 18:1
M_4	18.66	—	—	cis-11 18:1
D_1	18.66	—	—	trans-5,trans-9 18:2
D_2	18.82	—	—	cis-5,trans-9 18:2
D_3	18.89	—	—	trans-5,cis-9 18:2
D_4	18.98	—	—	cis-5,cis-9 18:2
D'_1	19.09	—	—	trans-9,trans-12 18:2
D'_2	19.27	—	—	cis-9,trans-12 18:2
D'_3	19.36	—	—	trans-9,cis-12 18:2
D'_4	19.45	—	—	cis-9,cis-12 18:2
T_1	19.34	19.35	0.01	trans-5,trans-9,trans-12 18:3
T_2	19.51	19.51	0.00	cis-5,trans-9,trans-12 18:3
T_3	19.57	19.60	0.03	trans-5,cis-9,trans-12 18:3
T_4	19.62	19.62	0.00	trans-5,trans-9,cis-12 18:3
T_5	19.69	19.69	0.00	cis-5,cis-9,trans-12 18:3
T_{6-7}	19.76	19.78	0.02	cis-5,trans-9,cis-12 18:3
		19.77	0.01	trans-5,cis-9,cis-12 18:3
T_8	19.87	19.87	0.00	cis-5,cis-9,cis-12 18:3

Abbreviations: Exp., experimental values; Calc., values calculated as described in the text.
[a]Peak numbers refer to Figure 13.13, 13.14 and 13.15.
[b]Equivalent chainlengths determined on a CP Sil 88 capillary column under conditions described in the legend of Figure 13.13.
[c]Differences between experimental and calculated ECL values.

18:2 < cis-9,trans-12 18:2 < trans-9,cis-12 < cis-9,cis-12 18:2. Note that the elution order of geometrical isomers of the two dienes is the same: trans,trans < cis,trans < trans,cis < cis,cis.

Using the ECL of 5,9-18:2 acid and 9,12-18:2 acid geometrical isomers together with the ECL of trans-9 and cis-9 18:1 acids (both present in the NO_2-isomerized pine seed oil), it is possible to calculate the ECL of all PAGI. For example, the calculated ECL for trans-5,trans-9,trans-12 18:3 acid will be the sum of the base value 18.00 plus the FCL of the trans-5,trans-9 18:2 acid, plus the FCL of the trans-9,trans-12 18:2 acid, minus the FCL of the trans-9 18:1 acid (which is counted twice). The result of this calculation is equivalent to summing the FCL of each of the three trans-5, trans-9, and trans-12 18:1 acids and the two dienoic adjustments (plus the base value 18.00). For example, the calculated figure for the ECL of trans-5,trans-9,trans-12 18:3 acid is: 18.00 + 0.66 + 1.09 - 0.40 = 19.35. The experimental ECL value for this acid is 19.34 (Table 13.5). When all combinations are taken into account, calculated values for ECL of all eight PAGI differ from experimental values by only 0.00–0.03 carbon units (Table 13.5). This excellent agreement between calculated and experimental figures allows identification of all eight PAGI.

The number of *trans* ethylenic bonds is also supported by the migration rate during silver-ion TLC. For at least one isomer, *trans*-5,*cis*-9,*cis*-12 18:3 acid (columbinic acid), the identification could be confirmed by comparison with the same acid present in great abundance (more than 50% of total fatty acids) in *Aquilegia vulgaris* seed oil (61,62). The elution order of PAGI on the CP Sil 88 capillary column is thus: *trans*-5,*trans*-9,*trans*-12 18:3 < *cis*-5,*trans*-9,*trans*-12 18:3 < *trans*-5,*cis*-9,*trans*-12 18:3 < *trans*-5,*trans*-9,*cis*-12 18:3 < *cis*-5,*cis*-9,*trans*-12 18:3 < (*cis*-5,*trans*-9,*cis*-12 18:3 + *trans*-5,*cis*-9,*cis*-12 18:3) < *cis*-5,*cis*-9,*cis*-12 18:3 (Table 13.5).

It should be noted that the principle developed for the elution order of α-LAGI can be applied to PAGI. Isomers containing a *trans* ethylenic bond in position 12 are eluted according to the elution order of the 5,9-18:2 isomers. The same holds true for those isomers containing a *cis* double bond in position 12. However, the two series are not fully separated as in the case for α-LAGI: the *trans*-5, *trans*-9,*cis*-12 18:3 acid, the first element of the family with a *cis*-12 double bond, elutes before the *cis*-5,*cis*-9,*trans*-12 18:3 acid, the last element of the family with a *trans*-12 double bond.

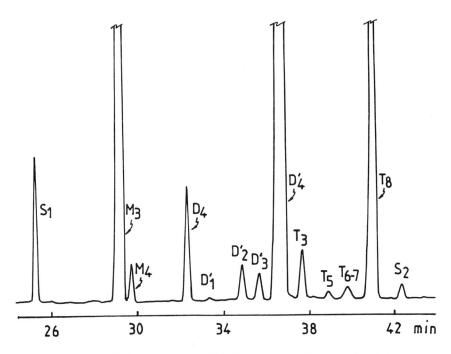

Figure 13.15. Partial chromatogram of FAME prepared with a sample of pine seed oil that was heated under vacuum in a sealed ampoule at 240°C for 6 h. Chromatographic conditions and peak lettering as in Figure 13.13.

When a sample of pine seed oil is heated under vacuum in a sealed ampoule (42,58) at 240°C for 6 h, one can note the appearance of several small artifacts (Figure 13.15) in the chromatographic zone where PAGI should elute. Their level is relatively low when compared to the content of α-LAGI and γ-LAGI in samples of linseed and borage oils heated under the same conditions (42,58). The DI obtained in pine seed oil is only one-fourth of the DI obtained in linseed and borage oils (8.3% instead of 31.8% and 33.1%, respectively). Moreover, the main artifact that accumulates is the *trans*-5,*cis*-9,*trans*-12 18:3 acid (approximately 64% of total *trans* isomers). This identification is based on the ECL of the peak and on its characteristic migration during silver-ion TLC. Three other isomers, identified by ECL and by their migration rate during silver-ion TLC, are also present. These compounds are the three mono*trans* isomers of pinolenic acid, present in almost equivalent quantities.

These observations are at variance with those made with α- and γ-linolenic acids. Upon heating, these two acids primarily produce to two mono*trans* isomers. One of the main differences between pinolenic acid and both α- and γ-linolenic acids is the distance between the two extreme ethylenic bonds: seven carbon atoms in pinolenic acid instead of six in the two other trienoic acids. Apparently, this is sufficient to modify the reaction mechanism of *cis–trans* isomerization. However, it should be emphasized that in all three cases, the internal double bond is the most resistant to geometrical isomerization. Our observations made with α-LAGI, γ-LAGI, and PAGI generated by heat treatments indicate that the mechanisms of geometrical isomerization of the internal and external ethylenic bonds are different.

References

1. Christie, W.W. (1989). *Gas Chromatography and Lipids*. The Oily Press, Ayr, Scotland, pp. 85–128.
2. Charvet, A.-S., Comeau, L.C., and Gaydou, EM. (1991). *J. Am. Oil Chem. Soc.* 68:604–607.
3. Mallet, J.F., Gaydou, E.M., and Archavlis, A. (1990). *J. Am. Oil Chem. Soc.* 67:607–610.
4. Ucciani, E., Mallet, G., and Chevolleau, S. (1991). *Rev. Fr. Corps Gras* 38:109–115.
5. Ucciani, E., Chevolleau, S., Mallet, G., and Morin, O. (1989). *Rev. Fr. Corps Gras* 36:433–436.
6. Kleiman, R., and Spencer, G.F. (1982). *J. Am. Oil Chem. Soc.* 59:29–38.
7. Breuer, B., Stuhlfauth, and Fock, H.P. (1987). *J. Chromatogr. Sci.* 25:302–306.
8. Borch, R.F. (1975) *Anal. Chem.* 47:2437–2439.
9. Privett, O.S., Nadenick, J.D., Weber, R.P., and Pusch, F.J. (1963). *J. Am. Oil Chem. Soc.* 40:28–30.
10. Seher, H., and Gundlach, U. (1982). *Fette Seifen Anstrichm.* 84:342–349.
11. Mbayoudel, K., and Comeau, L.C. (1989). *Rev. Fr. Corps Gras* 36:427–431.
12. Scholfield, C.R. (1981). *J. Am. Oil Chem. Soc.* 58:662–663.
13. Ratnayake, W.M.N., and Beare-Rogers, J.L. (1990). *J. Chromatogr. Sci.* 28:633–639.
14. Wolff, R.L., and Vandamme, F.F. (1992). *J. Am. Oil Chem. Soc.* 69:1228–1231.

15. Lognay, G., Charlier, M., Severin, M., and Wathelet, J.-P. (1987). *Rev. Fr. Corps Gras 34*:407–411.
16. Wolff, R.L., and Fabien, R.J. (1989). *Lait 69*:33–46.
17. Fulco, A.J., and Mead, J.F. (1959). *J. Biol. Chem. 234*:1411–1416.
18. Fulco, A.J., and Mead, J.F. (1960). *J. Biol. Chem. 235*:3379–3384.
19. Klenk, E., and Oette, K. (1960). *J. Physiol. Chem. 318*:86–99.
20. Privett, O.S., Blank, M.L., and Romanus, C. (1963). *J. Lipid Res. 4*:260–265.
21. Sand, D., Sen, N., and Schlenk, H. (1965). *J. Am. Oil Chem. Soc. 42*:511–516.
22. Mohrhauer, H., and Holman, R.T. (1965). *J. Am. Oil Chem. Soc. 42*:639–643.
23. Klenk, E., and Tschöpe, G. (1963). *Hoppe-Seyler's Z. Physiol. Chem. 334*:193–200.
24. Lemarchal, P., and Munsch, N. (1965). *C. R. Acad. Sci. 260*:714–716.
25. Sprecher, H. (1968). *Biochim. Biophys. Acta 152*:519–530.
26. Spence, M.W. (1971). *Lipids 6*:831–835.
27. Schmitz, B., Murawski, U., Pflüger, M., and Egge, H. (1977). *Lipids 12*:307–313.
28. Pullarkat, R.K., and Reha, H. (1976). *J. Chromatogr. Sci. 14*:25–28.
29. Karmiol, S., and Bcttgcr, W.J. (1990). *Lipids 25*:73 77.
30. Wolff, R.L., Sebedio, J.-L., and Grandgirard, A. (1990). *Lipids 25*:859–862.
31. Ackman, R.G., Manzer, A., and Joseph, J. (1974). *Chromalographia 7*:107–114.
32. Ackman, R.G., Hooper, S.N., and Hooper, D.L. (1974). *J. Am. Oil Chem. Soc. 51*:42–49.
33. Ackman, R.G., and Hooper, S.N. (1974). *J. Chromatogr. Sci. 12*:131–138.
34. Snyder, J.M., and Scholfield, C.R. (1982). *J. Am. Oil Chem. Soc. 59*:469–470.
35. Rakoff, H., and Emken, E.A. (1982). *Chem. Phys. Lipids 31*:215–225.
36. Grandgirard, A., Julliard, F., Prevost, J., and Sebedio, J.-L. (1987). *J. Am. Oil Chem. Soc. 64*:1434–1440.
37. Heckers, H., Melcher, F.W., and Schloeder, U. (1977). *J. Chromatogr. 136*:311–317.
38. Kobayashi, T. (1980). *J. Chromatogr. 194*:404–409.
39. Wolff, R.L. (1992). *J. Chromatogr. Sci. 30*:17–22.
40. Devinat, G., Scamaroni, L., and Naudet, M. (1980). *Rev. Fr. Corps Gras 27*:283–287.
41. Grandgirard, A., Sebedio, J.-L., and Fleury, J. (1984). *J. Am. Oil Chem. Soc. 61*:1563–1568.
42. Wolff, R.L. (1993). *J. Am. Oil Chem. Soc. 70*:425–430.
43. Wolff, R.L. (1993). *Sci. Alim. 13*:155–163.
44. Perkins, E.G., and Smick, C. (1987). *J. Am. Oil Chem. Soc. 64*:1150–1155.
45. Ratnayake, W.M.N., Hollywood, R., O'Grady, E., and Beare-Rogers, J.L. (1990). *J. Am. Oil Chem. Soc. 67*:804–810.
46. Wolff, R.L. (1992). *J. Am. Oil Chem. Soc. 69*:106–110.
47. Wolff, R.L. (1993). *J. Am. Oil Chem. Soc. 70*:219–224.
48. Wolff, R.L., and Sebedio, J.-L. (1991). *J. Am. Oil Chem. Soc. 68*:719–725.
49. Sebedio, J.-L., Grandgirard, A., and Prevost, J. (1988). *J. Am. Oil Chem. Soc. 5*:362–368.
50. O'Keefe, S.F., Wiley, V.A., and Wright, D. (1993). *J. Am. Oil Chem. Soc. 70*:915–917.
51. Piconneaux, A., Grandgirard, A., and Sebedio, J.-L. (1985). *C. R. Acad. Sci. 300*:353–358.
52. Grandgirard, A., Piconneaux, A., Sebedio, J.-L., Semon, E., and Le Quere, J.-L. (1989). *Lipids 24*:799–804.

53. Wolff, R.L., Combe, N.A., Entressangles, B., Sebedio, J.-L., and Grandgirard, A. (1993). *Biochim. Biophys. Acta 1168:*285–291.
54. Enig, M.G., Budowski, P., and Blondheim, H. (1983). *Hum. Nutr. Clin. Nutr. 38C:*223–230.
55. Adlof, R.O., and Emken, E.A. (1986). *Lipids 21:*543–547.
56. Hudgins, L.C., Hirsch, J., and Emken, E.A. (1991). *Am. J. Clin. Nutr. 53:*474–482.
57. Uzzan, A., Helme, J.-P., and Klein, J.-M. (1992). *Rev. Fr. Corps Gras 39:*339–343.
58. Wolff, R.L., and Sebedio, J.-L. (1993). *J. Am. Oil Chem. Soc.* in press.
59. Takagi, T., and Itabashi, Y. (1982). *Lipids 17:*716–722.
60. Madrigal, R.V., and Smith, C.R., Jr. (1975). *Lipids 10:*502–504.
61. Demirbüker, M., Blomberg, L.G., Olsson, N.U., Bergqvist, M., Herslöf, B.G., and Jacobs, F.A. (1992). *Lipids 27:*436–441.
62. Kaufman, H.P., and Barve, J. (1965). *Fette Seifen Anstrichm. 67:*14–16.

Chapter 14

Determination of *trans* Fatty Acids in Dietary Fats

W.M.N. Ratnayake

Nutrition Research Division, Food Directorate, Health Protection Branch, Health Canada, Ottawa, Ontario, K1A OL2, Canada.

Introduction

Several unusual *cis* and *trans* isomers of naturally occurring unsaturated fatty acids are found in many dietary fats. In partially hydrogenated vegetable oils (PHVO), *cis* and *trans* isomers of oleic acid are the main components with double-bond positions located from $\Delta 5-\Delta 16$ (1). In addition, PHVO contain various positional and geometric isomers of linoleic acid with *trans* or nonmethylene-interrupted double bonds (2). The isomers of linoleic acid are generally prevalent in mildly hydrogenated vegetable oils. Levels up to 7% have been found in some margarines (2,3). Hydrogenated fish oils contain numerous *cis* and *trans* isomers of mono- and polyunsaturated fatty acids with a wider range of chain lengths (4). *trans* fatty acids also occur naturally in dairy products, especially those from ruminant animals. Rumen microorganisms biohydrogenate dietary polyunsaturated fatty acids to *trans* fatty acids with $18:1\Delta 11t$ being the most prevalent isomer.

Recent reports indicate that *trans* fatty acids are hypercholesterolemic as that of saturated fatty acids and adversely affect the low-density lipoprotein (LDL)/high-density lipoprotein (HDL) cholesterol ratio (5–7). Furthermore, *trans* fatty acids have been implicated with increased serum levels of lipoprotein (a) (Lp[a]) when compared to oleic and linoleic acids (8). High Lp(a) is independent of and a greater risk factor than high serum cholesterol for coronary heart disease (9). These widely publicized adverse health effects of dietary *trans* fatty acids and the widespread use of PHVO create a need for accurate determination of total *trans* unsaturation and the detailed fatty acid composition, including the levels of *cis*- and *trans*-monounsaturates, in PHVO and dietary fats made from PHVO. In Canada (10) and the United States (11), voluntary nutritional labeling regulations of food require that monounsaturates only of the *cis* configuration be declared on the label and polyunsaturates are restricted to all-*cis* methylene-interrupted structures. These labeling regulations also require accurate determination of fatty acid composition of dietary fats of PHVO origin.

This short review examines the methods available, particularly the official methods, for the routine determination of total *trans* content and levels of *cis*- and *trans*-octadecenoates in dietary fats of PHVO origin. A recent review by Firestone and Sheppard (12) is highly recommended to anyone who wants to have more details of *trans* fatty acid analysis.

Determination of Total Trans Content

A number of methods are described in the literature for the determination of total *trans* content, including infrared (IR), Raman and nuclear magnetic spectroscopy (NMR), gas–liquid chromatography (GC), GC coupled to Fourier Transform (FT) IR Spectroscopy (GC/FTIR), reversed-phase and silver-ion high-performance liquid chromatography (HPLC), and silver nitrate thin-layer chromatography ($AgNO_3$-TLC) in conjunction with GC (12). Of the various methods, IR spectroscopy has been the method of choice for routine determination of total *trans* content in dietary fats. The methods based on TLC and HPLC are generally used for isolation of *trans* fatty acids for subsequent structural identification, while Raman and NMR spectroscopy, and GC/FTIR are more suited for structural elucidation of pure *trans* fatty acids.

An isolated *trans* double bond absorbs in the IR region at a wave number of approximately 967 cm^{-1}, equivalent to a wavelength of 10.3 µm, as a result of the deformation of the C-H bonds adjacent to the *trans* double bond. The measurement of the intensity of this absorption under controlled conditions is the basis of the official methods of AOCS (13), AOAC (14), and IUPAC (15) for the determination of total *trans* unsaturation in fats. The IR method is intended for use on all IR spectrophotometers so that the absorptivity values used in the calculation of percent isolated *trans* must be determined for each instrument.

The AOCS method is similar in many respects to the AOAC method and can be used for either triglycerides, methyl esters, or unesterified fatty acids. The absorbance or transmittance is recorded by scanning a CS_2 solution of the fat sample (20 mg/mL CS_2) from 1110 (9 µm) to 910 cm^{-1} (11 µm) against a CS_2 blank. A baseline is drawn from 990 (10.10 µm) to 939 cm^{-1} (10.65 µm) for unesterified acids, from 998 (10.02 µm) to 944 cm^{-1} (10.59 µm) for methyl esters, or from 995 (10.05 µm) to 937 cm^{-1} (10.67 µm) for triglycerides and the absorptivity is calculated. The *trans* content is calculated by comparing this absorptivity to that of a standard solution of either elaidic acid, methyl elaidate, or trielaidin in CS_2.

The AOCS method, despite the use of the baseline technique to correct for any background absorption, suffers from a few drawbacks (12). The spectra of long-chain fatty acids exhibit a band of medium intensity at about 943 cm^{-1} (10.6 µm), produced by a vibration of the carboxyl group. Correction for the contribution of this band to the 10.3 µm band of isolated double bonds, can be made by the baseline technique. However, if the isolated *trans* content is sufficiently small, the correction will become the major factor in the measured absorption at 10.3 µm and quantitative accuracy is not attainable. Therefore, long-chain fatty acids containing less than 15% isolated *trans* isomers must be converted to methyl esters prior to IR analysis (13). Another limitation of the IR method is that it is not applicable to samples containing more than 5% conjugated unsaturation, since conjugated double bonds absorb near the 10.3 µm band of isolated *trans* double bonds. Further, it was reported that the AOCS method gives higher values when fat samples are analyzed as triglycerides, particularly for samples containing less than 15% (16), whereas samples analyzed as methyl esters produce *trans* levels which are 1.5–3% low for *trans* values from 1–15% (12). The official method adopted by AOAC (965.34) prescribes incorporation of appropriate correction factors to compensate for the higher absorption of triglycerides and the lower absorption of methyl esters (14).

The IUPAC (15) method specifies the use of methyl esters and measures the absorption against a blank containing methyl stearate at the same concentration as the sample. The *trans* content is calculated using a calibration curve of absorption versus % isolated *trans* unsaturation developed using a series of CS_2 solutions containing different ratios of methyl elaidate and methyl stearate. The use of methyl stearate removes the interference from methyl ester absorption and gives greater accuracy to 1% *trans* content.

Madison et al. (17) proposed a two-component calibration procedure similar to that of IUPAC (15). However, they suggested standard mixtures of methyl elaidate and methyl linoleate for the development of the calibration curve. Calibration and test solutions are scanned from 900 to 1050 cm^{-1} against a CS_2 blank. A baseline is drawn between peak minima at about 935 and 1020 cm^{-1}, and the baseline-corrected absorbance of the *trans* peak (967 cm^{-1}) is obtained. The baseline for the test-sample spectrum is drawn exactly as the baseline in the standard spectrum, by overlaying the two spectra. This method allows analysis of *trans* contents in the 0.5–36% range with increased accuracy. This two-component calibration procedure compensates for the low bias of the AOCS method (13) in methyl esters and eliminates the need to calculate correction factors in the AOAC official method (14). A recent collaborative study coordinated by Ratnayake (18) tested a slightly modified procedure of Madison et al. (17). Methyl oleate was used instead of methyl linoleate for the development of the calibration curve. A good agreement among the participating laboratories (reproducibility relative standard deviation, RSD_R, ranged between 8.8 to 11.7%) was obtained for samples containing moderate to high content of *trans* unsaturation (15–34% trans) (see Table 14.1). However, for sample A

TABLE 14.1
Statistical Evaluation of GC-IR Collaborative Study of PHVO Samples

PHVO Sample	IR *trans* Mean[a]	S_R	RSD_R	18:1*t* Mean[a]	S_R	RSD_R	18:1*c* Mean[a]	S_R	RSD_R
A	5.17	1.79	34.56	4.88	1.77	36.39	24.93	0.95	3.79
B	15.54	1.76	11.31	14.92	1.41	9.48	24.70	1.75	7.08
C1[b]	18.92	2.21	11.69	17.37	2.18	12.53	28.11	1.94	6.89
C2[b]	19.09	1.97	10.32	17.53	1.81	10.34	28.17	2.01	7.14
D	30.06	2.69	8.94	26.64	2.55	9.58	34.38	2.11	6.14
E	34.48	3.90	11.31	32.60	2.53	7.78	34.28	3.61	10.52
R	21.63	1.90	8.79	19.37	1.87	9.65	32.16	2.10	6.53

Abbreviations: S_R, Reproducibility Standard Deviation; RSD_R, Reproducibility Relative Standard Deviation; PHVO, Partially hydrogenated vegetable oil (blend of partially hydrogenated soybean oil and cottonseed oil).
[a]Mean for 12 laboratories.
[b]C1 and C2 are blind duplicates.

(Table 14.1), the sample with the lowest *trans* content (5.2%), the agreement among the laboratories was less satisfactory (RSD_R = 34.5%). This suggests that an accurate measure of low levels of *trans* unsaturation (<5%) by IR is difficult.

Determination of Trans by FTIR

The new technique of FTIR offers several advantages over the conventional dispersive IR, including the high signal to noise ratio (S/N) obtained by averaging multiple spectral scans, rapid and comprehensive data collection allowing simple integration of peaks, and digital background substraction (19). Use of computerized FTIR eliminates time-consuming tasks encountered with conventional procedures, such as manually drawing the baseline and measuring the peak heights.

Lanser and Emken (20) developed a computer-assisted procedure for the estimation of isolated *trans* unsaturation in fats, using the peak area of the *trans* absorbance band at 966 cm^{-1} from FTIR spectra of FAME in CS_2. The area under a peak depended on the baseline chosen. They observed that absorbance minima, more specifically the minimum at the higher wave number, varied with *trans* unsaturation. This required adjustment of the baseline according to the *trans* content. Samples with more than 10% *trans* produced an absorption band with minima at 944 and 988 cm^{-1}; whereas at less than 10% *trans*, the peak minima are at 944 and 985 cm^{-1}, and below 5% *trans*, the peak minima are at 944 and 973 cm^{-1}. The calculation of *trans* content in hydrogenated oils containing less than 5% was improved by the use of appropriately selected integration limits.

Use of thin cells and neat methyl esters is the basis of the FTIR method proposed by Sleeter and Matlock (19) for determination of *trans* unsaturation in fats. Fourier Transform Infrared uses a Michelson Interferometer, which allows all wavelengths of light to pass through the sample simultaneously, whereas conventional dispersive spectrophotometers, which use diffraction gratings, only limited amounts of light pass through the sample. Due to the increased amount of light at all wavelengths, FTIR allows analysis of neat products using thin cells with pathlengths of ≈0.1 mm, eliminating possible errors due to sample weight and CS_2 dilution.

Use of CS_2 in dispersive instruments frequently leads to stratification due to vapor- and air-bubble formation within the cell. Sleeter and Matlock (19) use neat mixtures of methyl elaidate and methyl linoleate for calibration as proposed by Madison et al. (17). The area of the *trans* peak was integrated from 945–990 cm^{-1}. Quantitation was obtained by fitting measured *trans* areas of the calibration mixtures with a second-order polynomial. This provides a correlation coefficient of 0.9998 and standard error of 0.11% over a range of 0–50%.

Trans content can also be determined by measuring peak heights, which give a slightly increased error.

Determination of Fatty Acid Composition

Gas chromatography of FAME is undoubtedly the most convenient and widely used analytical method for determining the fatty acid composition (21). Slightly polar stationary phases, such as polyglycol Carbowax-20M, are normally employed for the analysis of fatty acids of natural origin, in which the double bonds of unsaturated fatty acids are almost exclusively of *cis* configuration. However, with these stationary phases, the separation of *cis/trans* isomers is not possible. With highly polar cyanosilicone stationary phases, such as SP-2560, SP-2340, OV-275, or CP-SIL-88, *cis* and *trans* isomers could be separated to a far greater extent than with polar stationary phases.

Based on an interlaboratory study, a 6.1 m × 2 mm i.d. column packed with OV-275 was recommended by both AOCS (22) and AOAC (23) for determination of *trans* unsaturation in partially hydrogenated oils. However, a complete resolution is not possible with the OV-275 column, since some of the *trans*-monoenes are hidden under the larger *cis* isomer peak (24).

In many lipid laboratories, capillary columns coated with cyanosilicone stationary phases appeared to gain acceptance for *cis/trans* isomer separation (25,26). The American Oil Chemists' Society (27) and AOAC (28) recently recommended the use of a 60 m × 0.25 mm i.d. flexible fused capillary column coated with SP-2340 to determine the general fatty acid composition, including the levels of *cis*- and *trans*-octadecenoates of partially hydrogenated oils. This same method is recommended for determination of total *trans* unsaturation. The direct capillary GC procedure was based on the assumption that 18:1c and 18:1t isomers are completely separable on the SP-2340 column. However, because of the numerous isomers present in PHVO, a satisfactory separation of 18:1t as a group from the *cis* isomers is not feasible on SP-2340 (29) or any other currently available capillary column (2,30). On SP-2340 and other polar columns, the 18:1t isomers with Δ values lower than 11 are well separated from the 18:1c isomers, but the 18:1t isomers with high Δ values (i.e., Δ12–Δ16) overlap with the *cis* isomers. Because of this overlap, the direct GC method greatly underestimates the total 18:1t in favor of the *cis* isomers (31). In some margarines, the underestimation in determining the total 18:1t can be as high as 32% (30). The levels of 18:1t isomers of high Δ values may depend on the hydrogenation conditions and the source oil, and this will result in variation in the extent of overlaps of the isomers from one PHVO to another. The concentration of the methyl esters applied to the GC could also influence the *cis* and *trans* resolution.

Sampugna et al. (25) proposed the use of appropriate correction factors to compensate for the *cis* and *trans* overlaps. They found a linear relationship between the correction factors by comparing the results obtained by silver nitrate TLC/GC for 18:1t and 18:1c with the proportion of total 18:1 isomers in the sample. Gas chromatography combined with other chromatographic techniques (particularly silver-ion chromatography) has been suggested (24,29,31–33), but these procedures are lengthy and are not suitable for routine analysis of dietary fats.

Combined GC-IR

Ratnayake et al. (30) proposed use of a combined capillary GC and IR method for accurate determination of 18:1*t* and 18:1*c* and the general fatty acid composition of PHVO. The total *trans* unsaturation determined by IR was related to the capillary GC weight percentages of the component *trans* FAME by the mathematical formula:

IR *trans* = %18:1*t* + 0.84 × %18:2*t* + 1.74 × %18:2*tt* + 0.84 × %18:3*t*

where 0.84, 1.74, and 0.84 are the correction factors relating GC weight percentages to the IR *trans*-equivalents for mono*trans* octadecadienoic (18:2*t*), *trans,trans*-octadecadienoic (18:2*tt*) and mono*trans*-octadecatrienoic (18:3*t*) acids, respectively. This formula forms the basis for the determination of 18:1*t* and 18:1*c* in PHVO. In capillary columns coated with polar cyanoalkylsiloxane stationary phases such as SP-2340 and SP-2560, 18:2*t*, 18:2*tt*, and 18:3*t* are separated as distinct groups without any serious interferences or overlaps (29,30, see also Figure 14.1) and therefore, levels of these *trans* polyunsaturates are obtained directly by GC analysis. Infrared spectroscopy provides the total *trans* unsaturation and therefore total 18:1*t* is calculated from the mathematical formula. 18:1*c* is obtained as the difference between total 18:1, which is the sum of all the 18:1 isomer peaks in GC, and 18:1*t*.

This combined GC-IR procedure was studied collaboratively in 12 laboratories using seven PHVO samples (18). The isolated *trans* unsaturation was determined by an IR procedure similar to that described by Madison et al. (17). Use of either FTIR or conventional dispersive IR instrumentation and either SP-2340, SP-2380

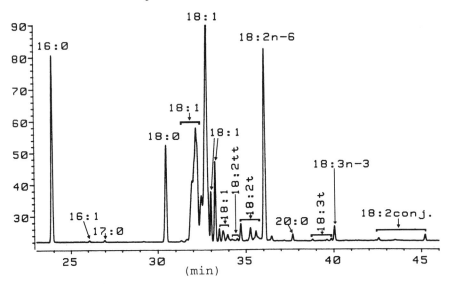

Figure 14.1. Chromatogram of methyl esters from a partially hydrogenated canola oil sample, using 100 m × 0.25 mm fused silica capillary column coated with SP-2560. The column temperature was programmed at a rate of 1°C/min from 165 to 210°C and held at that temperature for 35 min. The detector and the injection port temperatures were 235 and 225°C, respectively, and the H_2 carrier gas pressure was 20 psig.

TABLE 14.2
Statistical Evaluation of GC-IR Collaborative Study of PHVO for Blind Duplicates: Samples C1 and C2.

	Mean[a]	S_r	S_R	RSD_r	RSD_R
IR trans	19.04	1.22	2.04	6.42	10.72
18:1t	17.45	1.26	1.92	7.24	11.00
18:1c	28.16	1.12	1.87	3.97	6.05

Abbreviations: S_r, Repeatability Standard Deviation; S_R, Reproducibility Standard Deviation; RSD_r, Repeatability Relative Standard Deviation; RSD_R, Reproducibility Relative Standard Deviation.
[a] n = 12 laboratories.

and CP-Sil-88, or SP-2560 cyanoalkylsiloxane stationary phase had no effect on the fatty acid composition data. Reproducibility relative standard deviations for 15–35% *trans* content determined by IR were in the range of 8–12%, whereas RSD_R for 5% *trans* was 35%, suggesting that accurate measurement of *trans* unsaturation by IR of oils containing 5% or less *trans* fatty acids is difficult. The same pattern as those of total *trans* by IR was found for RSD_R values for 18:1t by GC-IR; 36% for the test samples with 5% *trans* (sample A in Table 14.1) and 8–13% for test samples with 15–31% *trans*. Reproducibility relative standard deviation values for 18:1c ranged from 4–10% (Table 14.1), and RSD_R values for 16:0, 18:0, total saturates, and 18:2n-6 (major components) were less than 5% (18). That the GC-IR method is capable of good precision is demonstrated by the excellent agreement for a pair of duplicate samples (Table 14.2).

Ratnayake (18) compared the values for 18:1t and 18:1c obtained from the collaborative study with those determined by two other methods; combined $AgNO_3$-TLC/GC and AOCS Official Method Ce 1c-89 (27), a one-step direct GC method, also recommended by AOAC (28) for the determination of fatty acid composition and total *trans* content in PHVO. The mean values obtained by the GC-IR study were in agreement with the values determined by the tedious, combined procedure of $AgNO_3$-TLC/GC (Table 14.3), which confirms the accuracy and reliability of the GC–IR method. The AOCS direct GC method gave substantially lower values for 18:1t and higher values for 18:1c than those of the other two methods (Table 14.3), a consequent of ignoring the overlap of *cis* and *trans* isomers. The error in determining the 18:1t and 18:1c by the AOCS direct GC method was large for samples containing high amounts of *trans* unsaturation. This is because the proportion of high Δ value 18:1 isomers, which coelute with 18:1c isomers, was high in samples with high *trans* content and consequently, the extent of overlap of 18:1c and 18:1t isomers in GC cyanosilicone capillary columns becomes substantial.

Conclusions

With any of the current official methods a fairly good quantitative estimate of *trans* unsaturation can be obtained by IR spectrophotometry. Whether low levels of *trans* unsaturation would be determined by IR is doubtful. For these, use of direct GC is

TABLE 14.3
Comparison of 18:1t and 18:1c Levels of the Collaborative Samples Determined by $AgNO_3$-TLC/GC, GC-IR, and Direct GC.

Sample Code	% 18:1t			% 18:1c		
	$AgNO_3$-TLC/GC[a]	GC-IR[b]	Direct GC[c]	$AgNO_3$-TLC/GC[a]	GC-IR[b]	DIRECT GC[c]
A	5.1	4.9	4.4	24.7	24.9	25.9
B	15.2	14.9	12.3	24.1	24.7	26.0
C_1	18.9	17.4	14.7	27.2	28.1	30.4
C_2	18.9	17.5	14.7	27.2	28.2	30.4
D	26.1	26.6	19.6	35.0	34.4	41.8
E	31.9	32.6	23.4	33.0	34.3	41.6
R	19.9	19.4	16.8	31.0	32.2	36.0

[a]Values ($n = 1$) determined in the author's laboratory. 18:1t and 18:1c isolated by $AgNO_3$-TLC were quantitatively analyzed by GC in the presence of 17:0 internal standard.
[b]Mean values ($n = 12$) from GC-IR collaborative study.
[c]Values ($n = 1$) determined in the author's laboratory using AOCS Official Method Ce 1c-89.

recommended (e.g., AOCS Official Method Ce 1c-89 [27]). Alternatively, accuracy at lower *trans* levels could be improved with the use of FTIR by analyzing neat methyl esters in 0.1 mm IR cells.

The combined GC-IR method should be useful for routine analysis of *cis*- and *trans*-octadecenoates and the general fatty acid composition in dietary fats made from PHVO and animal fats, provided that the *trans* content is more than 5%. For samples containing less than 5% *trans*, detailed fatty acid composition and the total *trans* unsaturation can be obtained through GC analysis alone (direct GC); combining the IR and GC data is unnecessary. This is because the proportion of high Δ value 18:1t isomers ($\geq \Delta 12$) is low in samples with low *trans* content, and hence the overlap of 18:1c and 18:1t isomers in GC is almost negligible. The GC-IR method, as is, cannot be applied to hydrogenated fish oils because these fats contain a complex mixture of *cis/trans* isomers of polyunsaturated fatty acids with a wide range of chain lengths. Their isomers are not easily resolvable by GC, impeding the development of a simple mathematical relationship between the total *trans* content determined by IR and the GC weight percentages of the component *trans* fatty acids.

References

1. Dutton, H.J. (1979) in *Geometrical and Positional Fatty Acid Isomers,* Emken, E.A., and Dutton, H.J., The American Oil Chemists' Society, Champaign, IL, pp. 1–16.
2. Ratnayake, W.M.N., and Pelletier, G. (1992) *J. Am. Oil Chem. Soc.* 69:95–105.
3. Ratnayake, W.M.N., Hollywood, R., and O'Grady, E. (1991) *Can. Inst. Food Sci. Technol. J. 24:*81–86.
4. Ackman, R.G. (1982) in *Nutritional Evaluation of Long-Chain Fatty Acids in Fish Oils,* Barlow, S.M. and Stansby, M.E., Academic Press, London, England, pp. 25–88.
5. Mensink, R.P., and Katan, M.B. (1990) *N. Engl. J. Med. 323:*439–445.

6. Zock, P.L., and Katan, M.B. (1992) *J. Lipid Res. 33:*399–410.
7. Troisi, R., Willett, W.C., and Weiss, S.T. (1992) *Am. J. Clin. Nutr. 56:*1019–1024.
8. Mensink, R.P., Zock, P.L., Katan, M.B., and Hornstra, G. (1992) *J. Lipid Res. 33:*1493–1501.
9. Sandkamp, M., Funke, H., Schute, H., Kohler, E., and Assmann, G. (1990) *Clin. Chem. 36:*20–23.
10. *Guide for Food Manufacturers and Advertisers.* Revised Edition 1988. Consumer and Products Branch, Bureau of Consumer Affairs, The Ministry of Consumer and Corporate Affairs, Canada.
11. *Federal Register.* January 6, 1993, Volume 58(3), Book II. Department of Health and Human Services, Food and Drug Administration 21 CFR.
12. Firestone, D., and Sheppard, A. (1992) in *Advances in Lipid Methodology,* Christie, W.W., The Oily Press, Ayr, Scotland, pp. 273–322.
13. *Official Methods and Recommended Practices of the American Oil Chemists' Society* (1989) 4th edn. Firestone, D., Method 14-61, American Oil Chemists' Society, Champaign, IL.
14. *Official Methods of Analysis of the Association of Official Analytical Chemists* (1990) 15th edn. Helrich, K., Method 965.34, Association of Official Analytical Chemists, Arlington, VA.
15. International Union of Pure and Applied Chemistry, Commission on Oils, Fats and Derivatives. (1987) *Standard Methods for the Analysis of Oils, Fats, and Derivatives,* 7th edn., Paquot, C., and Hautfenee, A., Method 2.207, Blackwell Scientific Publications, London, England.
16. Firestone, D., and LaBouliere, P. (1965) *J. Assoc. Off. Anal. Chem. 48:* 437–443.
17. Madison, B.L., Depalma, R.A., and D'Alonzo, R.P. (1982) *J. Am. Oil Chem. Soc. 59:* 178–181.
18. Ratnayake, W.M.N. *J. Assoc. Off. Anal. Chem.* Submitted for publication.
19. Sleeter, R.T., and Matlock, M.G. (1989) *J. Am. Oil Chem. Soc. 66:*121–129.
20. Lanser, A.C., and Emken, E.A. (1988) *J. Am. Oil Chem. Soc. 65:*1483–1487.
21. Ackman, R.G., and Ratnayake, W.M.N. (1989) in *The Role of Fats in Human Nutrition,* Vergroesen, A.J., and Crawford, M., Academic Press, London, pp. 441–514.
22. *Official Methods and Recommended Practices of the American Oil Chemists' Society* (1989) 4th edn. Firestone, D., Method Cd 17-85. American Oil Chemists' Society, Champaign, IL.
23. *Official Methods of Analysis of the Association of Official Analytical Chemists* (1990) 15th edn., Helrich, K., Method 985.21, Association of Official Analytical Chemists, Arlington, VA.
24. Smith, L.M., Dunkley, W.L., Franke, A., and Dairiki, T. (1978) *J. Am. Oil Chem. Soc. 55:*257–261.
25. Sampugna, J., Pallansch, L.A., Enig, M.G., and Keeney, M. (1982) *J. Chromatogr. 249:*245–255.
26. Heckers, H., Melcher, F.W., and Schloeder (1977) *J. Chromatogr. 136:*311–317.
27. *Official Methods and Recommended Practices of the American Oil Chemists' Society* (1990) 4th edn. Firestone, D., Method Ce 1c-89. American Oil Chemists' Society, Champaign, IL.
28. General Referee Report on Oils and Fats (1990) *J. Assoc. Off. Anal. Chem. 73:*105.

29. Ratnayake, W.M.N., and Beare-Rogers, J.L. (1990) *J. Chromatog. Sci. 28:*633–639.
30. Ratnayake, W.M.N., Hollywood, R., O'Grady, E., and Beare-Rogers, J.L. (1990) *J. Am. Oil Chem. Soc. 67:*804–810.
31. Conacher, H.B.S. (1976) *J. Chromatog. Sci. 14:*405–411.
32. Sebedio, J.-L., Farquharson, T.E., and Ackman, R.G. (1982) *Lipids 17:*469–475.
33. International Union of Pure and Applied Chemistry, Commission on Oils, Fats and Derivatives. (1987) *Standard Methods for the Analysis of Oils, Fats, and Derivatives,* 7th edn., Paquot, C., and Hautfenne, A., Method 2.208. Blackwell Scientific Publications, London, England.

Chapter 15

Gas Chromatography–Mass Spectrometry and Tandem Mass Spectrometry in the Analysis of Fatty Acids

Jean-Luc Le Quéré

INRA. Laboratoire de Recherches sur les Arômes, 17, rue Sully, 21034 Dijon, France.

Introduction

Most lipids from natural sources contain complex mixtures of polyunsaturated fatty acids. These fatty acids are adequately separated, generally as their methyl esters, using high-resolution gas chromatography (GC) and polar capillary columns. Their structural differences are based on certain parameters: chain length, chain type (straight, branched, cyclic, etc.); and number, position, and geometry of the double bonds. The double bond configuration can be determined with infrared spectroscopy, using a hyphenated technique, such as gas chromatography coupled to Fourier transform infrared spectrometry (GC-FTIR), for complex mixtures.

Determination of chain length is achieved routinely by catalytic perhydrogenation combined with the equivalent chain length (ECL) method in GC. Structural modification of the carbon chain, such as branching or the presence of rings, can be determined by mass spectrometry only in some favorable cases for saturated species.

The location of double-bond positions cannot be directly achieved by gas chromatography–mass spectrometry (GC–MS), because positional isomers show almost identical mass spectra, due to double bond migration under electron impact conditions.

The mass spectra of saturated species often are interpreted more readily than those of related unsaturated compounds. However, positional and geometrical isomers cannot be distinguished after hydrogenation, as they give the same hydrogenated products (same retention time and same mass spectrum).

GC–MS in the Determination of Structural Modifications of Fatty Acids

Catalytic Hydrogenation

In order to acquire better GC and MS correlations between unsaturated species occurring in complex mixtures and their hydrogenated counterparts, Le Quéré et al.

have developed a simple on-line hydrogenation method (1). This technique allows selective hydrogenation of all the unsaturated species after chromatographic separation (Figure 15.1). Using hydrogen as the carrier gas, hydrogenation takes place in a capillary reactor connected to the outlet of the analytical column in the oven of the gas chromatograph and before the mass spectrometer. The reactor is a fused silica tube coated with palladium acetylacetonate. Selective hydrogenation of olefinic bonds is achieved after a normal chromatographic run. Structural information, such as carbon skeleton and double-bond equivalents, can be deduced, and structural correlations between the saturated and unsaturated components can be made.

Structures of cyclic fatty acid monomers (CFAM) isolated from heated linseed oil have been discerned using this method in a GC–MS instrument (1). The total ion currents of the CFAM mixture obtained with and without the Pd reactor in the GC oven could be superimposed (Figure 15.2). Mass spectra of the pure hydrogenated components were acquired for each of the 10 major peaks. The carbon skeleton of peak 1 was identified as a propylcyclopentenyl decenoate because the mass spectrum of its hydrogenated counterpart 1H clearly displayed the characteristic fragment ions of methyl (2-propylcyclopentyl)-decanoate (Figure 15.3 [1]). The ionized molecule was increased by four mass units from 292 to 296 and the peaks for diagnostic ions B (m/z 185), C (m/z 111), D (m/z 253), D-32 (m/z 221), and D-32-18 (m/z 203) were identified. This on-line hydrogenation method presents the advantage of counting the number of double bonds and giving structural features, such as chain length and structural modifications in a single GC–MS run.

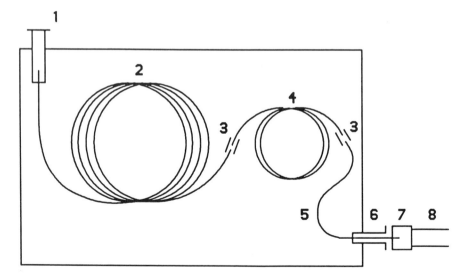

Figure 15.1. Capillary reactor installation for on-line hydrogenation in a GC/MS system: 1, injector; 2, analytical column; 3, zero dead-volume connector; 4, Pd capillary reactor; 5, capillary tubing; 6, heated interface; 7, ionization chamber; and 8, mass analyzer.

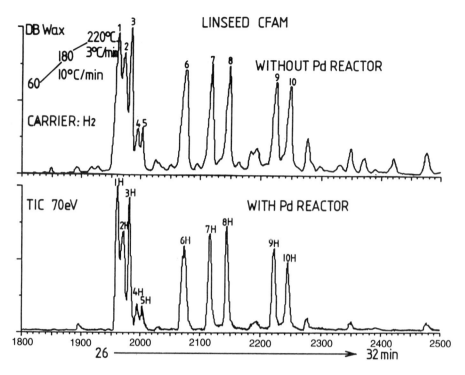

Figure 15.2. Total ion currents of cyclic fatty acid monomers isolated from heated linseed oil with and without postcolumn on-line hydrogenation. Pd capillary reactor: 60 cm × 0.32 mm i.d.

Localization of Carbon-Carbon Double Bonds

Determining the position of the double bonds is still a challenge, however, and has been the subject of comprehensive reviews (2,3). Among the available methods, chemical degradation by means such as ozonolysis is not adapted to complex mixtures or to polyenoic acids. Chemical modification of the double bonds helps overcome this problem to a certain extent. The first method cited is the conversion of unsaturated fatty acid methyl esters (FAME) via stereospecific oxidation of the double bonds with OsO_4 followed by derivatization of the resulting polyhydroxy compounds into polytrimethylsilyl ethers. These ethers give rise to ionized fragments from which double-bond positions can be definitively placed. A typical spectrum is shown in Figure 15.4. Fragmentation of the *bis*-trimethylsilyl ether derivative of methyl oleate gives rise to two intense fragment ions (m/z 215 and m/z 259, respectively) as a result of favored cleavage of the site of the initial double bond, the bond located between the two ether-bearing carbons. Locating up to five double bonds in pentaenoic acids is possible by observing diagnostically important fragments (2). Methyl ethers may be used also, but only the spectra of all

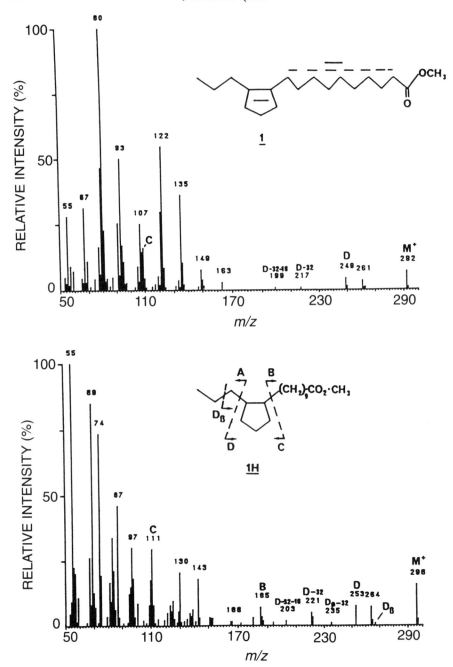

Figure 15.3. Mass spectra of peak 1 from Figure 15.2 before (1) and after on-line hydrogenation (1H). Peak 1H was unambiguously identified as methyl (2-propylcyclopentyl)-decanoate. Le Quéré, J.-L. et al. (1).

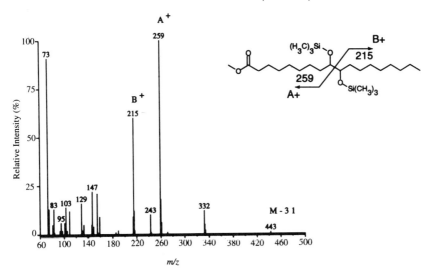

Figure 15.4. Mass spectrum of the *bis*-trimethylsilyl-derivative of methyl oleate with characteristic fragments.

trimethylsilyl ether derivatives are characterized by a clear, distinctive fragmentation pattern according to Dommes et al. (2).

Alkylthiolation of the double bonds, via iodine-catalyzed addition of dimethyldisulfide, also produces ionized fragments characteristic of the position of the double bonds (Figure 15.5). This derivation procedure is particularly suited for trace components and has been used extensively for analysis of unsaturated fatty acids of biological origin where only a limited amount of material is available (4–6). For practical reasons, however, its use is limited to mono- and, to a lesser extent, diunsaturated fatty acids.

Localization of Structural Features Including Unsaturations

Pyridine-containing derivatives, such as picolinyl esters, have been shown to be suitable for direct mass spectrometric structural analysis of acids containing straight, branched, unsaturated, cyclic, or oxygenated chains. The subject has been reviewed recently (7–9). In electron impact conditions, these fatty acid derivatives stabilize the charge on the nitrogen atom far from the site of structural interest during ionization, and radical-induced cleavage of the hydrocarbon chain predominates. Distinctive fragmentations, according to the position of structural features, produce ionized fragments of diagnostic value. As an example, the mass spectrum of the picolinyl ester of 10-octadecenoic acid is shown in Figure 15.6 (10). Distinctive features are the doublet of abundant ions 14 amu apart at *m/z* 288 and 302, representing cleavage of the allylic and subsequent bonds in the hydrocarbon moiety, and the gap of 26 amu between the ions at *m/z* 248 and 274, representing

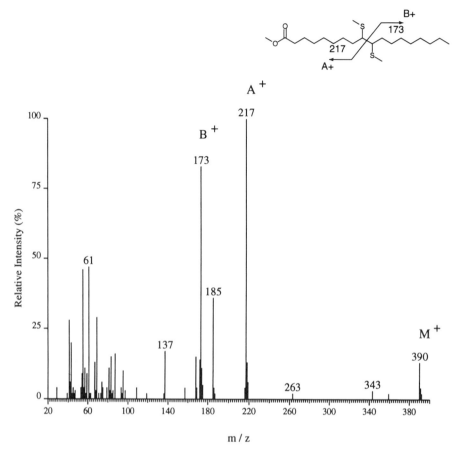

Figure 15.5. Mass spectrum of the *bis*-methylthio-derivative obtained by iodine-catalyzed addition of dimethyldisulfide to methyl oleate, showing characteristic ionized molecular fragments for double-bond position.

cleavage on either side of the double bond. The technique has been successfully applied in the structural analysis of cyclic monoenoic acids isolated from frying oils (11).

The 4,4-dimethyloxazoline (DMOX) derivatives, first introduced by Zhang et al. (12), are prepared by reacting either fatty acids or methyl esters with 2-amino-2-methylpropanol and lead to fragmentation patterns suitable for structural investigations. The location of double bonds can be determined by observing a 12 amu gap between the most intense ions in consecutive clusters of fragments, rather than the usual 14. The 12 mass rule is equivalent to the rule applied to pyrrolidide derivatives (13), but the diagnostic ions of the DMOX derivatives are more intense, leading to clearer and more distinctive patterns. For example, the method has been

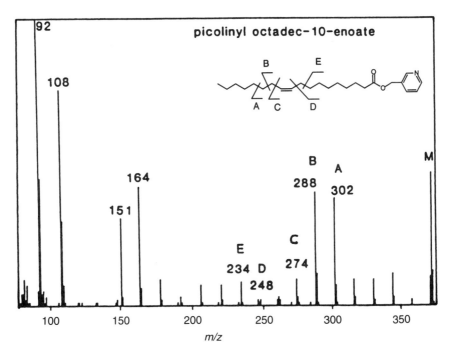

Figure 15.6. Mass spectrum of picolinyl 10-octadecenoate. Reproduced from Christie et al. (10).

applied successfully to the structural determination of polyunsaturated fatty acids (Figure 15.7 [14]) and of cyclopropenoid fatty acids (15).

MS/MS and GC–MS/MS in the Analysis of Fatty Acids

A very interesting approach for structural studies of fatty acids was recently introduced using tandem mass spectrometry, and especially Fast Atom Bombardment MS/MS (FAB-MS/MS). The topic has been the subject of comprehensive reviews (16–18). The mass-analyzed ion kinetic energy (MIKE) spectra obtained after collisional activation of the carboxylate anion of monounsaturated fatty acids desorbed by fast atom bombardment (FAB) show interesting features (Figure 15.8 [19]): the observed pattern always shows three low intensity peaks between two enhanced peaks. The two enhanced peaks correspond to allylic cleavages, hence they allow the determination of the position of the double bond (here elaidic acid:*trans*-$C_{18:1}\Delta 9$). This fragmentation pattern, discovered by Gross and coworkers at the University of Nebraska in 1983, is known as charge remote fragmentation (17).

In this fragmentation, the cleavage reactions are not charge mediated, bond cleavage occurs at a site in the ion that is remote from the charge site and the mech-

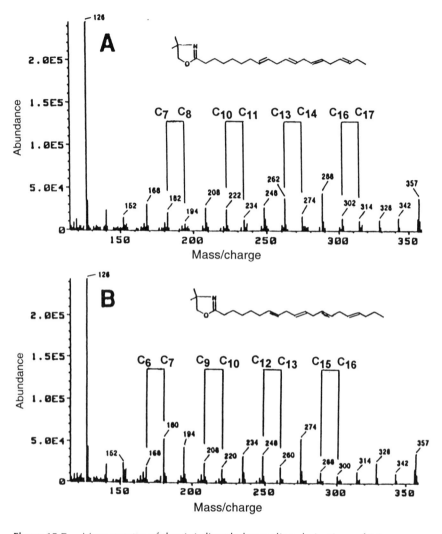

Figure 15.7. Mass spectra of the 4,4-dimethyloxazoline derivatives of (a) 8,11,14,17-eicosatetraenoic acid. (b) 7,10,13,16-eicosatetraenoic acid. Reproduced from Luthria and Sprecher (14).

anism of fragmentation does not involve any significant intervention of the charge. Charge-remote fragmentations of ions are mechanistic analogies to gas-phase thermal decompositions of neutral molecules; bonds distant from the charge-site are cleaved via simple rearrangements and homolytic bond dissociations governed by classical thermochemistry (20). The main mechanism is a 1,4-elimination of H_2 and of a neutral alkene, while the charge remains stable at the carboxylate end. The only structural requirement for inducing charge-remote fragmentations is that the charge is stable and localized at a specific site.

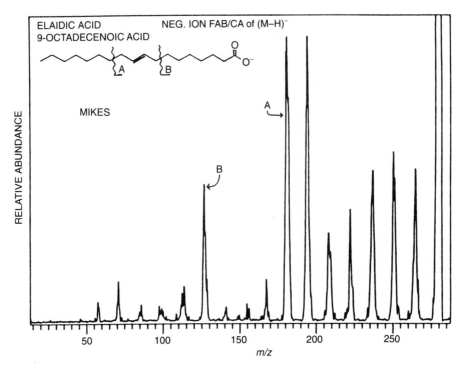

Figure 15.8. MIKE spectrum of elaidic acid obtained by collisional activation of the carboxylate anion desorbed by FAB. Reproduced from Tomer et al. (19) with permission of the American Chemical Society.

Charge-remote fragmentations may be obtained with any double-focusing mass spectrometer, of either reverse or forward geometry. In the first case, MIKE spectra are obtained, and in the second case B/E linked scans give daughter-ion spectra, with greater resolution.

Taking advantage of charge-remote fragmentation, other structural features of fatty acids can be also identified and located. These include branching (21), epoxides, rings, and hydroxyl groups (22). This capability is illustrated by the spectrum of the 10-(2'-propyl-4'-cyclopentenyl)-9-decenoate, a synthetic model of cyclic fatty acids isolated from heated linseed oil, shown in Figure 15.9. The main characteristic fragment *a* corresponds to cleavage of the alkyl substituent α to the cyclopentene moiety. The second important fragment *b* is formed by an allylic cleavage on the carboxylate chain and the third *c* helps to locate the double-bond position and defines the ring dimension.

As previously outlined, a requirement for inducing charge-remote fragmentation is that the charge is stable, and localized at a specific functional group.

In this respect, it may be advantageous to substitute positive ions for carboxylate anions. Positive ions could be metal cationic species, for example, with lithium (23), or ions formed by protonation at sites of high proton affinity, for example

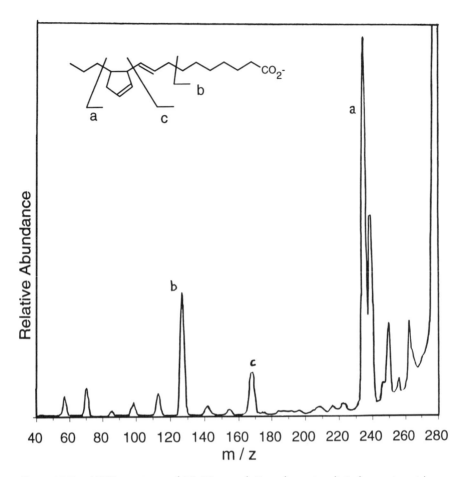

Figure 15.9. MIKE spectrum of 10-(2'-propyl-4'-cyclopentenyl)-9-decenoic acid obtained by collisional activation of the carboxylate anion desorbed by FAB.

nitrogen in picolinyl esters (24). Cationization is a prerequisite for the analysis of polyunsaturated fatty acids containing four or more double bonds (25).

When complex mixtures of isomeric forms of fatty acids are being analyzed, a chromatographic separation is necessary. Promé and co-workers (26) developed an elegant GC–MS/MS method to generate high yields of gas-phase carboxylate anions from electron capture ionization of pentafluorobenzyl fatty acid esters. They applied this method to the analysis of mycobacterial polyunsaturated fatty acids (27). A typical spectrum is shown in Figure 15.10 (MS/MS spectrum of a fatty acid with 38 carbon atoms and 6 double bonds) allowing the determination of the positions of the double bonds, which are ethylene-interrupted (27).

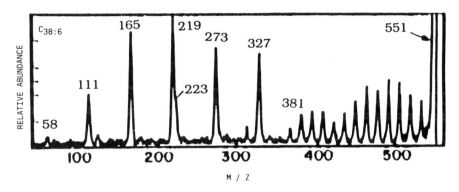

Figure 15.10. MIKE spectrum of an acid having 38 carbon atoms and 6 double bonds, obtained in GC–MS/MS, by collisional activation of the carboxylate anion produced by dissociative electron capture of its pentafluorobenzyl ester. Reproduced from Aurelle et al. (27) by permission of the authors and of John Wiley & Sons, Ltd.

This method was applied to characterize the structures of the cyclic fatty acid monomers (CFAM) isolated from heated linseed and sunflower oils (28). Typical spectra with important characteristic fragments defining structural features are shown in Figures 15.11 and 15.12 for hydrogenated cyclopentyl and cyclohexyl acids. It should be emphasized that picolinyl esters should be amenable to GC–MS/MS also and that methyl esters themselves give an abundant carboxylate anion in negative chemical ionization conditions (29).

Conclusion

As yet, a universal method for determining structural modifications of polyunsaturated fatty acids does not exist. Mass spectrometry of intact or derivatized fatty acids or esters has been extensively used for this purpose with some success. The

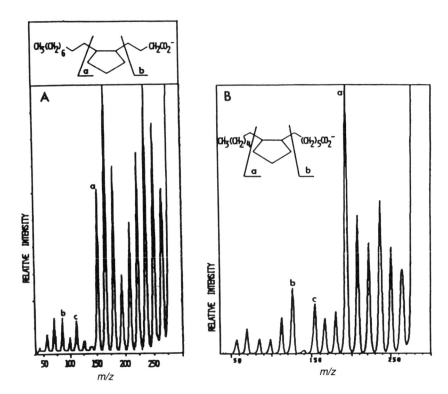

Figure 15.11. MIKE spectra of hydrogenated disubstituted cyclopentyl acids isolated from heated sunflower oil, obtained in a GC–MS/MS analysis of their pentafluorobenzyl esters.

number of real structural problems solved with tandem mass spectrometry is now significant, and the number of examples in the area of fatty acid analysis is growing rapidly.

Acknowledgments

The author is indebted to Robert Henry, INSA Lyon, for the synthesis of the cyclopentenyl fatty acid methyl ester, and to François Couderc, CNRS Toulouse, for its MS/MS spectrum.

References

1. Le Quéré, J.-L., Sémon, E., Lanher, B., and Sébédio, J.-L. (1989) *Lipids 24:*347–350.
2. Dommes, V., Wirtz-Peitz, F., and Kunau, W.H. (1976) *J. Chromatogr. Sci. 14:*360–366.
3. Jensen, N.J., and Gross, M.L. (1987) *Mass Spectrom. Rev. 6:*497–536.
4. Dunkelblum, E., Tan, S.H., and Silk, P.J. (1985) *J. Chem. Ecol. 11:*265–277.
5. Scribe, P., Guezennec, J., Dagaut, J., Pepe, C., and Saliot, A. (1988) *Anal. Chem. 60:*928–931.

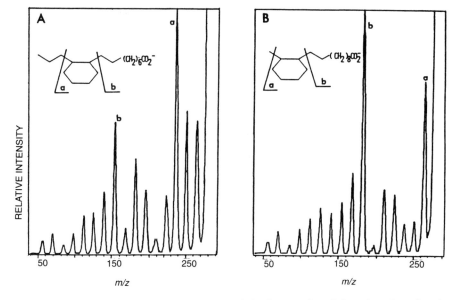

Figure 15.12. MIKE spectra of hydrogenated disubstituted cyclohexyl acids isolated from heated sunflower oil, obtained in a GC–MS/MS analysis of their pentafluorobenzyl esters.

6. Carballeira, N.M., and Sepulveda, J.A. (1992) *Lipids* 27:72–74.
7. Harvey, D.J. (1990) *Spectroscopy (Ottawa)* 8:211–244.
8. Harvey, D.J. (1992) in *Advances in Lipid Methodology.* Vol. 1. Edited by Christie, W.W., The Oily Press, Ayr, pp. 19–80.
9. Christie, W.W. (1993) *INFORM* 4:85–91.
10. Christie, W.W., Brechany, E.Y., and Holman, R.T. (1987) *Lipids* 22:224–228.
11. Christie, W.W., Brechany, E.Y., Sébédio, J.-L., and Le Quéré, J.-L. (1993) *Chem. Phys. Lipids* 66:143–153.
12. Zhang, J.Y., Yu, Q.T., Liu, B.N., and Huang, Z.H. (1988) *Biomed. Environ. Mass Spectrom.* 15:33–44.
13. Andersson, B.A., and Holman, R.T. (1974) *Lipids* 9:185–190.
14. Luthria, D.L., and Sprecher, H. (1993) *Lipids* 28:561–564.
15. Spitzer, V. (1991) *J. Am. Oil Chem. Soc.* 68:963–969.
16. Gross, M.L. (1989) In *Advances in Mass Spectrometry.* Edited by Longevialle, P., Heyden and Son Ltd, London, Vol. 11A, pp. 792–811.
17. Gross, M.L. (1992) *Int. J. Mass Spectrom. Ion Proc.* 118/119:137–165.
18. Le Quéré, J.-L. (1993) In *Advances in Lipid Methodology,* Vol. 2. Edited by W.W. Christie, The Oily Press, Dundee, pp. 215–245.
19. Tomer, K.B., Crow, F.W., and Gross, M.L. (1983) *J. Amer. Chem. Soc.* 105:5487–5488.
20. Adams, J. (1990) *Mass Spectrom. Rev.* 9:141–186.
21. Jensen, N.J., and Gross, M.L. (1986) *Lipids* 21:362–365.
22. Tomer, K.B., Jensen, N.J., and Gross, M.L. (1986) *Anal. Chem.* 58:2429–2433.

23. Davoli, E., and Gross, M.L. (1990) *J. Am. Soc. Mass Spectrom.* 1:320–324.
24. Deterding, L.J., and Gross, M.L. (1988) *Org. Mass Spectrom.* 23:169–177.
25. Adams, J., and Gross, M.L. (1987) *Anal. Chem.* 59:1576–1582.
26. Promé, J.-C., Aurelle, H., Couderc, F., and Savagnac, A. (1987) *Rapid Commun. Mass Spectrom.* 1:50–52.
27. Aurelle, H., Treilhou, M., Promé, D., Savagnac, A., and Promé, J.-C. (1987) *Rapid Commun. Mass Spectrom.* 1:65–66.
28. Le Quéré, J.-L., Sébédio, J.-L., Henry, R., Couderc, F., Demont, N., and Promé, J.C. (1991) *J. Chromatogr.* 562:659–672.
29. Bambagiotti, A.M., Coran, S.A., Vincieri, F.F., Petrucciani, T., and Traldi, P. (1986) *Org. Mass Spectrom.* 21:485–488.

Chapter 16
Mechanism for Separation of Triacylglycerols in Oils by Liquid Chromatography: Identification by Mass Spectrometry

S. Héron, J. Bleton, and A. Tchapla

Laboratoire d'Etudes des Techniques et Instruments d'Analyse Moléculaire, Institut Universitaire de Technologie d'Orsay, Plateau du Moulon, BP 127, 91403 Orsay Cedex, France.

Introduction

Oils and fats are a very important class of compounds with applications in a number of industries, such as the cosmetic and food industries, and fundamental research, for example, nutrition, archeological, biomedical, or pharmaceutical investigations (1). This paper will only report recent chromatographic results about the most abundant components: the triacylglycerols.

The structural differences of fatty acid residues are based on the following parameters: chain length, chain type (straight, branched, or hydroxylated), and the number and position of the double bond. If each glycerol position is esterified by long-chain fatty acids, seven different classes of triacylglycerols can be described. From simplest to the most difficult to separate they are

1. Homogeneous saturated triacylglycerols with a different total number of carbon atoms (tripalmitin, PPP; tristearin, SSS);

2. Homogeneous triacylglycerols with the same total number of carbon atoms but differing by degree of unsaturation (tristearin, SSS; triolein, OOO; trilinolein, LLL; trilinolenin, LnLnLn);

3. Mixed saturated triacylglycerols with the same total number of carbon atoms but differing in the chain length distribution (myristol-palmito2-stearin3, MPS; tripalmitn, PPP);

4. Mixed triacylglycerols containing the same number of carbon atoms and the same total number of unsaturations but differing in the distribution of these unsaturations within the three chains (trilinolein, LLL; oleol-linoleo2-linolenin3, OLLn);

5. Mixed triacylglycerols containing the same number of carbon atoms and the same total number of unsaturations but differing in the location of the unsaturations or the stereochemistry within the chain (tri α-eleostearin, EEE; trilinolenin, LnLnLn);

6. Mixed triacylglycerols containing the same fatty acid residues but differing in

the distribution within the glyceryl chain (dilinoleol-3 olein2, LOL; dilinoleol-2 olein3, LLO); and

7. Mixed asymmetric triacylglycerols with the two primary groups of glycerol chain esterified by fatty acids that are optical isomers.

The number of isomers to separate can be related to the number of esters which could be obtained by transesterification of triacylglycerols n. The number of triacylglycerols corresponding to the first five classes described is $N_1 = (n^3 + 3n^2 + 2n)/6$. Moreover, if the sixth and seventh triacylglycerol classes are taken into account, the number becomes $N_2 = 0.5(n^3 + n^2)$ and $N_3 = n^3$. Due to the large number of possible isomers, their overall separation constitutes a challenge for analysts.

Transesterification followed by a gas chromatographic analysis is one of the most commonly used methods for analyzing triacylglycerols (1,2). It provides the total fatty acid composition but not the partition on the glycerol moiety. From the relative amount of each constitutive fatty acid, the composition in a triacylglycerol can be calculated using a statistical method (3–5). However, it was demonstrated that the real composition can be different (6). Consequently, direct analysis of triacylglycerols is vital for knowledge of the composition of oils and fats. The main difficulty of direct analysis comes from the insolubility of these compounds in hydroorganic solvent, the absence of ultraviolet chromophores in their molecular structure and their low volatility. The answers to the overall separation problems come from the simultaneous use of highly efficient and selective columns, either with a nonaqueous mobile phase and evaporative light scattering detector in high-performance liquid chromatography (HPLC [7,8]) or high-temperature stable stationary phase in capillary gas chromatography (GC [9]).

For every form of chromatography, there are two ways of understanding the analytical process: the first consists of an adaptation of literature information by the analyst. It could be done using optimization techniques or software (10). But in most cases the choice of the stationary phase and mobile phase compositions were not justified. The second way can be developed from the concepts established by analytical chemists. Their goal is to model the molecular phenomena which occurs during the separation process by using standard solutions, stationary and mobile phases in many experiments. The knowledge from the second method helps to choose better experimental conditions to separate a complex mixture.

Triacylglycerol analysis by Reverse Phase Liquid Chromatography (RPLC) has been the subject of many studies, a number of which have been reviewed recently (11,12). One objective of this article is to compile information from experiments with homologous series and triacylglycerols to get an overview on the retention effects of the chain length, structure of the chains bonded to the silica, and nature of the solvent. A second objective is to determine the experimental conditions that permit a comparative study of oils and fats using a single column percolated with an isocratic mobile phase composition. A corresponding challenge is to separate the fourth and possibly the fifth triacylglycerol classes previously described. Such separations have not been performed using these simple conditions. Approximately 50 fats and oils of different biological origins were qualitatively analyzed and a unique spectra for each of them has been established (Héron and Tchapla, O.C.L. (1995), in press [13]).

The last problem considered in this paper concerns the identification of triacylglycerols. Most of standards in existence are either homogeneous (class 1 or 2) or mixed but primarily P, O, L, Ln, and S residues (class 4). Consequently, in the absence of mixed triacylglycerol standards containing acid residues other than P, O, L, Ln, or S, identification is made either by using the diagram from Goiffon (14–16), unambiguous literature data, or by using different chromatographic methods successively such as silver-ion chromatography and RPLC (17,18). Attributes can be determined by examining the results from four hyphenated methods (19): gas chromatography-mass spectrometry (by electron impact and ammonia chemical ionization modes) and liquid chromatography-mass spectrometry (with two modes of ionization).

Results and Discussion

Elaboration of Molecular Interaction Occurring in RPLC.

Many theories have been proposed to account for molecular retention in RPLC (20–22). All of them correlate the retention factor (k') to peculiar molecular properties of the solute that interacts with the stationary and mobile phases. From these theories, the authors tried to develop a structural model of molecular interactions that are developed during the chromatographic process, the final goal being to choose better experimental conditions for the analysis of variable structure solutes.

The first problem arises from the structure of the stationary phase. According to the mono, di, or trifunctional nature of the silanizing agent and the possible presence of traces of water during the reaction, the final grafted silica can be monomeric or polymeric (23–24). In addition to structural differences, these phases also have great differences in separation power (25,26). However, both phases are identically named. Consequently, the model must account for the real "monomeric or polymeric-like" nature of the stationary phase.

Studies were made using homologous series as test solutes. Based on the molecular structure, these kinds of compounds are not representative of the most analyzed solutes. However, they are useful for evaluating the respective roles of the stationary and mobile phases (26). By the regular increases of chain length, it is possible to correlate chromatographic retention to the hydrocarbon skeleton of the solute. Moreover, comparison of the retention of homologues having the same carbon atom number but differing by the nature of the polar head leads to improved knowledge of specific interaction development (27–29). Homologous series are good models to increase the understanding of triacylglycerol retention. Indeed, triacylglycerols from classes 1 and 3 constitute a homologous series. From another point of view, the behavior of alkanes, carboxylic acid methyl esters, and phenyl alkanes or alkenes allows modeling of the chromatographic retention of triacylglycerols included in classes 3, 4, and 5.

Role of the Nature of the Bonded Phase. In a general way, the retention factor of a compound is related to the change of the partial molar Gibbs free energy associated with the transfer of the solute (assumed to be in standard state at infinite dilution) from the mobile phase to the stationary phase, $\Delta G°$ can be calculated according to:

$$\log k' = \log K\Phi = \log [-\Delta G°/2.3 \, RT] + \log \Phi$$

(16.1)

where Φ is the phase ratio and K is the partition coefficient. $\Delta G°$ depends on the stationary and mobile phases, solute, and temperature. In the particular case of a homologous series, a simple approach consists of assuming that $\Delta G°$ is the sum of the individual contributions of each methylene group of the alkyl chain added to the contribution from the polar head. Consequently, for homologous series a linear relation between $\Delta G°$ or $\log k'$ and the carbon atom number of the solutes is observed (20). This linear relationship between $\log k'$ and n has been verified by many authors for homologous series (30) and for triacylglycerols (14–16,31). But generally the chain length of the investigated solute was less than or equal to chain length of the bonded silica. The evaluation of relative chain-length effects needed complementary experiments.

The leveling effect of the logarithmic function shows that by using quadratic methylene selectivity α (Equation 16.2) it is possible to detect subtle phenomena that cannot be observed easily on $\log k'$ vs. n plots (30,32,33).

$$\alpha_n = \sqrt{\frac{k'_{n+1}}{k'_{n-1}}}$$

(16.2)

where: k'_x refers to the retention factor of homologues with x carbon atoms in the chain.

Close examination of both $\log k'$ and α vs. n plots (Figure 16.1) leads to the proposal of two types of retention mechanisms due to the monofunctional-like (usually named monomeric) or polyfunctional-like (usually named polymeric) nature of bonded silica. They are described in a recent review that summarizes all of the aspects of the retention mechanism in RPLC (26).

The first mechanism fits two observations: $\log k'$ vs. n curves exhibit a break point (Figure 16.1*b*), and α vs. n exhibits a discontinuity and a change of slope (Figure 16.1*a*). The observations have been explained by a retention mechanism corresponding to insertion of the hydrocarbonaceous chain of the solute between the grafted chains (Figure 16.2 [26,30]). This phenomenon is independent of the nature of the mobile phase with any homologue including triacylglycerols. It only

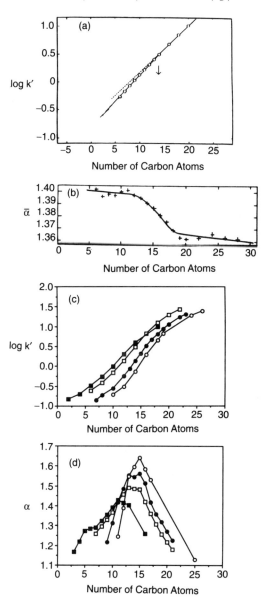

Figure 16.1. (a) log k' vs. n for alkanes on C_{18}-bonded silica. Mobile phase pure MeOH. (b) α vs. n on C_{18}-bonded silica; Mobile phase: MeOH/H_2O 90/10 (v/v). (from Tchapla et al. Ref. 30, with permission). (c) log k' vs. n on Supelcosil LC-PAH column; Mobile phase: pure MeOH; T = 25°C. ■, Phenyl alkanes; □, Alcanes; ●, Methyl esters; ○, Alcohols. (d) α vs. n on Supelcosil LC-PAH column; Mobile phase: pure MeOH; T = 25°C. (Same conditions as for Figure 16.1c).

depends on the bonded chain length (30). The plots of log k' vs. 1/T for each homologue (33) and each triacylglycerol (32) have been examined. The change in partial free molar enthalpy ΔH_0 and entropy ΔS_0 associated with the transfer of the solute from the mobile phase to the stationary phase was also determined. The plots ΔH_0 vs. n and ΔS_0 vs. n exhibited a discontinuity when $n = n_{crit}$ and a change of slope similar to α vs. n plots. However, after the break an increase in entropy and a decrease in the enthalpy of transfer was observed for homologues with $n > n_{crit}$ (32,33). These last results were interpreted as thermodynamic proof of alkyl chain penetration of the solute within the bonded monofunctional silica.

The second mechanism explains two other observations: log k' vs. n plots are sigmoidal, and α vs. n plots exhibit a maximum (Figures 16.1c, 16.1d). These observations suggest that the retention mechanisms are not the same. It has been interpreted by imbedding the homologues completely in the stationary phase with a molecular interaction mechanism (Figure 16.3). However, the phenomenon is not so simple. The molecular insertion mechanism occurs at ambient temperatures with a monofunctional bonded alkyl chain having a chain length equal to or less than 18

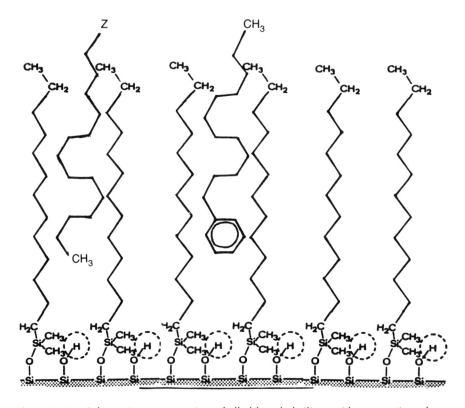

Figure 16.2. Schematic representation of alkyl-bonded silicas with penetration of the solute between the grafted chains.

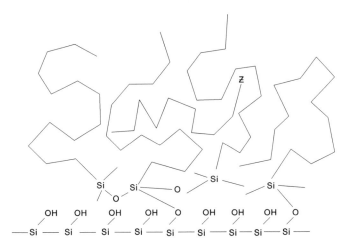

Figure 16.3. Schematic representation of alkyl-bonded silicas with total solubilization of the solute in the stationary phase.

carbon atoms. It is also effective with polyfunctional bonded silicas used at either temperatures above the ambient temperature or subambient temperatures if the mobile phase consists of a solvent with a low dielectric constant, such as methylene chloride, chloroform, tetrahydrofuran, or acetone (19,34). Total solubilization occurs at ambient temperature with polyfunctional bonded silicas or if monofunctional bonded silicas are used below their phase transition temperature (19,34). Since the two mechanisms are not independent, one can pass from one to the other by changing the temperature or the nature of the mobile phase, in spite of the initial monomeric or polymeric nature of grafted silica (Figure 16.4).

Experimental conditions that take advantage of total solubilization resulted in higher selectivity and improved separation of homologues and triacylglycerols. Often the experiments were run below ambient temperature and therefore steps had to be taken to avoid precipitating higher triacylglycerols in the chromatographic system (35). A compromise between low temperature and mobile phase composition must be established to maintain enhanced selectivity compared to analysis using monomeric C_{18}-bonded silica.

The results gained from the study of homologous series behavior have shown the importance of the stationary phase during the separation process. With the same generic denomination (C_{18}, ODS, RP18, etc.), an analyst can use a liquid chromatographic column under the same experimental conditions which gives very different separations due to the silica used.

Role of the Mobile Phase Composition. In RPLC mode, the mobile phase is usually a binary mixture, but sometimes it may be a ternary mixture of solvents. The insolubility of triacylglycerols in hydroorganic solvents makes it impossible to use water as a component of the mobile phase, as is usually done in RPLC. For this rea-

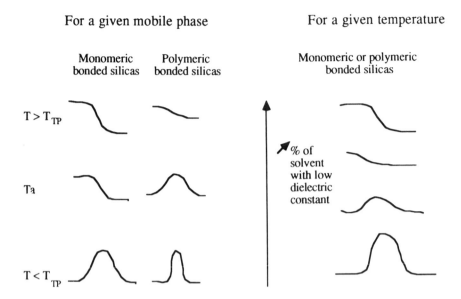

Figure 16.4. Evolution of the α vs. n plots with the experimental conditions. T_{TP}: phase transition temperature; T_a: ambient temperature.

son, Nonaqueous Reversed Phase (NARP) mode is employed to analyze these compounds (12,19,36). Usually the mobile phase consists of a mixture of at least two organic solvents, the weak and the strong solvents. The strong solvent permits simultaneous retention modulation and compound solubilization. The weak solvents most commonly employed are acetonitrile, methanol (12), and propionitrile (37–39), and the strong solvents are methylene chloride, chloroform, tetrahydrofuran, acetone, hexane, (12,19,31,40–42) and benzene.

In the literature there are recent studies on triacylglycerol analysis with monomeric C_{18}-bonded silica and elution gradient comparing the separation effect of methylene chloride and acetone mixed with acetonitrile (43), but few systematic approaches governing the choice of these solvents have been described. The goal of this chapter was to systematically study the two separation mechanism theories to improve solvent choice. Indeed, solvophobic (20) and partition theories (21) correlate solute retention to its hydrocarbonaceous area as well as surface tension of the mobile phase. Some studies have been carried out for homologues, including triacylglycerols, to verify the surface-tension effect, and permit discrimination of the influence of a strong or a weak solvent. Since these influences are dependent on the solute structure, each of these effects will be discussed separately.

Role of Weak Solvents on Saturated Solute Retention. For each saturated solute, a correlation between log k' vs. mobile phase composition curves and the surface tension vs. mobile phase composition curves has been established (28). The variation of log k' for homogeneous saturated triacylglycerols with the composition of ternary mobile phases, such as methanol/acetonitrile/strong solvent (29) and methanol/nitromethane/strong solvent (27), with a constant percentage of strong solvent has been investigated. The effect of the addition of acetonitrile is increased retention due to the surface-tension effect of the solvent.

Role of Strong Solvents on Saturated Solute Retention. Systematic studies of log k' vs. the percentage of a strong organic solvent for many homologues showed different behavior from the weak solvents. When adding solvents with higher surface tensions than those of weak solvents (MeCN, MeOH, or MeNO$_2$) but with low dielectric constants ($\varepsilon < 20$), a systematic decrease in the retention of any homologues (including triacylglycerols) is observed compared to retentions from pure weak solvents. Retention decreases as the percentage of strong solvent is increased. For triacylglycerol analysis, CH$_2$Cl$_2$, CHCl$_3$, THF, pyridine, benzene, and heptane produce similar effects that are stronger than that observed for acetone (27). This phenomenon is due to an effect this type of solvent produces on the conformation of the stationary phase hydrocarbon chains and is demonstrated by FTIR spectroscopy (29). Moreover, the strong solvent conformational effect on the bonded chains is so drastic that when triacylglycerols are eluted by an exclusion mechanism hydroorganic mobile phase containing over 95% THF is used (28). In this case, longer triacylglycerols (or alkanes) have shorter retention times than shorter chain triacylglycerols (or alkanes). This property could be used to perform size discrimination on triacylglycerols by using small size gel permeation chromatography. The major effects of these solvents are to modify the geometrical structure of the stationary phase and to increase the solubility of the solutes.

Role of Unsaturation in Solute, Solvent, and Stationary Phase Structures. When a solute contains double bonds in its structure, the retention time is shorter than expected independent of the organic or hydroorganic nature of the mobile phase. This has been demonstrated by comparing alkane, carboxylic acid methyl ester, and saturated homogeneous triacylglycerol retention with that of phenyl alkanes, unsaturated carboxylic acid methyl esters, and unsaturated homogeneous triacylglycerols (27). All these observations can be rationalized by considering the effect of specific Π–Π interactions between the Π electron systems of the solute and those of acetonitrile in the mobile phase (29).

Thévenon et al. (29) have shown that for triacylglycerols having 18 carbon atoms in each of the three chains, the presence of one, two, or three double bonds has the same effect as the withdrawal of 2.6, 4.8, or 6.5 methylene groups for each solute chain in acetonitrile-chloroform eluents or the withdrawal of 1.9, 3.3, or 4.2 methylene groups from each solute chain in a methanol-chloroform mixture. The effect of two double bonds is not double that of one double bond and depends on the chromatographic conditions. Moreover, the chromatographic selectivity between saturated and unsaturated triacylglycerols is modified according to the amount of acetonitrile (or nitromethane) in the mobile phase and is different for alkyl-bonded silicas or stationary phases possessing Π electrons in their structure, for example PRP1, ACT1, ACT2, phenyl, and cyano.

Combined Role of Unsaturation in Solute, Solvent, and Stationary Phase Structures. Whatever the solute saturation level, the effect from octadecyl-bonded silicas is also observed when the stationary phase contains Π electrons, such as polyvinylbenzenepolystyrene (29), phenyl-1-propyl (19,44), *n*-octyl/cyano-1-propyl or cyano-1-propyl/*n*-octadecyl-bonded silicas (19). The same result is obtained using a porous graphic carbon packing (Hypercarb column) and a mobile phase containing acetonitrile (19). On the aromatic support, the corresponding numbers of methylene units to be withdrawn are less than or equal to 0.5, 1.6, and 2.2 (for phenyl-1-propyl-bonded silicas) and 0.45, 0.9, and 2.2 for polymeric support. The decrease in retention due to unsaturation of the solute molecular structure is larger in octadecyl-bonded phases than in Π-electron-containing stationary phases (29).

The retention effect of the mobile phase and the alkene-bond presence in solutes have been used with the goal to optimize the mobile phase qualitative composition. Log k' of homogeneous unsaturated triacylglycerols varies as a function of the composition of ternary mobile phases, such as methanol/acetonitrile/strong solvent (27,29) and methanol/nitromethane/strong solvent (27) where the strong solvent percentage is constant. The retention variation has a minimum at an acetonitrile percentage that is dependent on the number of solute double bonds. However, maximum selectivity between saturated and unsaturated triacylglycerols is found for the binary mixture acetonitrile/strong solvent (Figure 16.5). This is true independent of the monomeric or polymeric stationary phase; temperature at which experiments have been performed (Figure 16.6); and nature of the strong solvent; methylene chloride, chloroform, THF, or acetone (12). Strong solvents can be separated into two classes; one consisting of acetone and THF, the other consisting of methylene chloride and chloroform (27). The conclusion that acetonitrile produces better resolutions than methanol is due to higher ratios in acetonitrile mixtures and a lower viscosity than methanol, which leads to a better chromatographic efficiency for a given value of k'.

To separate triacylglycerol classes 4 and 5, stationary phases possessing Π electrons leads to modified selectivity compared to alkyl-bonded silicas. Due to the low carbon content, these phases are not retentive enough, and the resulting separation is worse than those currently observed with classical RPLC (19).

In the case of porous graphitized carbon support, the reduction from the first double bond is more than that from the second double bond which in turn is more than that from the third double bond present in the acid residue. However, because of eventual low diffusion between the planes of graphitic carbon atoms or in the pores, chromatographic peaks become large even for small k' values and the result is low resolution (19).

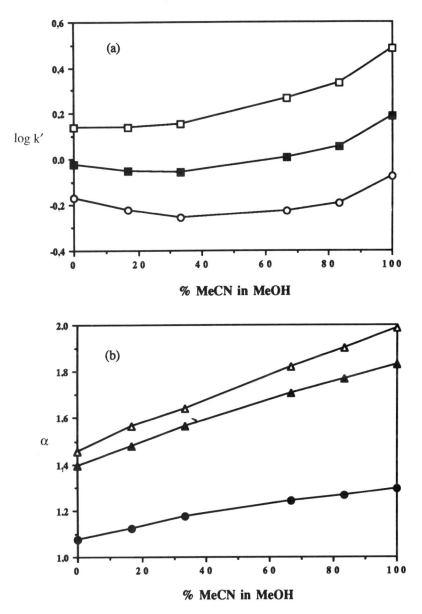

Figure 16.5. (a) log k' vs. (%MeCN in MeOH = x/60); Column: Ultracarb 3 ODS 20; T = 21°C; Mobile phase: MeOH/MeCN/CHCl$_3$. 60-x   x   40
□, OOO; ■, LLL; ○, LnLnLn. (b) α vs. (%) MeCN in MeOH = x/60); Column: Ultracarb 3 ODS 20; T = 21°C; Mobile phase: MeOH/MeCN/CHCl$_3$.
60-x   x   40
●, k'_{PPP}/k'_{OOO}; △ k'_{OOO}/k'_{LLL}; ▲, k'_{LLL}/k'_{LnLnLn}.

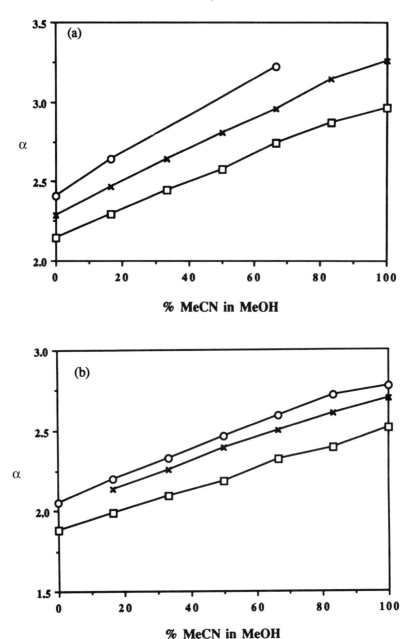

Figure 16.6. α vs. (% MeCN in MeOH); Mobile phase MeOH/MeCN/Acetone 60–x, x, 40; (a) α = k'_{OOO}/k'_{LLL}. (b) α = k'_{LLL}/k'_{LnLnLn}. □, LiChrospher 100 CH 18:2; T = 21°C; ○, Chromegabond C_{22}; T = 12.6°C; ×, Brownlee Spheri 5 ODS; T = 21°C. (From Héron and Tchapla, Ref. 12, with permission).

Optimization of Triacylglycerol Separation

To effectively separate triacylglycerols of different biological origins, oils and fats have been divided into four different classes based on the function of their acid residues (45). There are oils rich in short-chain-length acid residues, such as lauric acid; monounsaturated or diunsaturated acids with more than 20 carbon atoms, such as erucic acid; or 18 carbon atom acids independent of the degree of unsaturation degrees, the most common class of oils that the authors have investigated; and polyunsaturated acids with more than 20 carbon atoms, such as DHA or DPA.

In a recent paper, Hierro et al. (43) showed that using an elution gradient with a single monomeric C_{18}-bonded silica, and a mobile phase of acetonitrile/acetone is better to resolve the short- and medium-chain triacylglycerols but acetonitrile/dichloromethane is better to separate long-chain triacylglycerols. However, in these analyses triacylglycerol classes 4 and 5 were not resolved. Perrin et al. separated these classes using two columns of C_{18}-bonded silica coupled in series and using acetone/methylene chloride/acetonitrile mobile phase to produce elution gradient (40). Stolhywo et al. reported some separation of LLLn and OLnLn using an acetone/acetonitrile isocratic mobile phase with two columns of monomeric C_{18}-bonded silicas (31). Finally, Nikolova-Damyanova et al. using two columns of C_{18}-bonded silicas and a dichloromethane/dichloroethane/acetonitrile mobile phase in isocratic mode have reportedly separated class 5 triacylglycerols (46).

The main objective of this paper was to separate triacylglycerol of classes 3, 4, and 5 using a single column and an isocratic mobile phase. A number of experiments using soybeans as a reference are reported in Table 16.1. The retention time of a class 3 triacylglycerol that eluted later was arbitrarily set with a capacity ratio equal to 20, which corresponds to 1 h. A second criterion was to have a maximum number of solutes in this time (12), considering all peaks with a height greater than 1/250 of the highest peak. In cases where many isocratic mobile phase compositions give the same total number of separated triacylglycerols, the choice is based on chromatograms with a normalized resolution product closest to one (Table 16.2 [47]). This test is weighted toward the less well resolved peak pairs, permitting the analyst to select the best experimental conditions. The overall process is described by Héron, and Tchapla [12], (O.C.L. in press 1995). Chromatograms of two oils analyzed in the best conditions obtained are reported in Figure 16.7. In the case of oils from classes 1, 2, and 4, the mobile phase composition was adapted to obtain the best separation considering their triacylglycerol composition.

Identification of Mixed Triacylglycerols

The identification of triacylglycerols is a difficult problem to solve, due to the extreme complexity of the mixture in natural samples. This is particularly true when

TABLE 16.1
Experimental Conditions Tested

Column	Temperature (°C)	Mobile phase (v/v)		
		MeCN / CH_2Cl_2	MeCN / $CHCl_3$	MeCN / Acetone
Brownlee Spheri 5 ODS	21	40/60	73/27	25/75
		50/50	75/25	30/70
		55/45	80/20	35/65
		60/40		40/60
		63/37		
		65/35		
		68/32		
		69/31		
		73/27		
		75/25		
		77/23		
Chromegabond C22	12.6	55/45	65/35	25/75
		60/40	67/33	27/73
		65/35	69/31	
		68/32		

TABLE 16.2
Selection of the Experimental Conditions (Soybean Oil)

Column	Temperature (°C)	Mobile Phase	tr(mn) POO	k' POO	Number of Peaks (POO included)	Number of Pairs of Peaks Poorly Resolved	r
Brownlee Spheri 5 ODS	21	MeCN / CH_2Cl_2					
		65/35	30	12	15	—	0.98
		68/32	48	20	16	4	
		69/31	53	21	16	4	
		MeCN / $CHCl_3$					
		73/27	44	18	16	6	
		75/25	55	22	16	6	
		MeCN / Acetone					
		30/70	27	11	15	—	
		35/65	37	15	15	—	
		40/60	50	20	14	—	
Chromegabond C22	12.6	MeCN /CH_2Cl_2					
		60/40	41	14	16	5	0.975
		65/35	65	23	16	5	0.84
		MeCN / $CHCl_3$					
		65/35	38	13	16	5	0.95
		67/33	46	16	16	5	0.91
		69/31	60	21	16	5	0.85
		MeCN / Acetone					
		25/75	43	15	14	—	
		27/73	49	17	14	—	

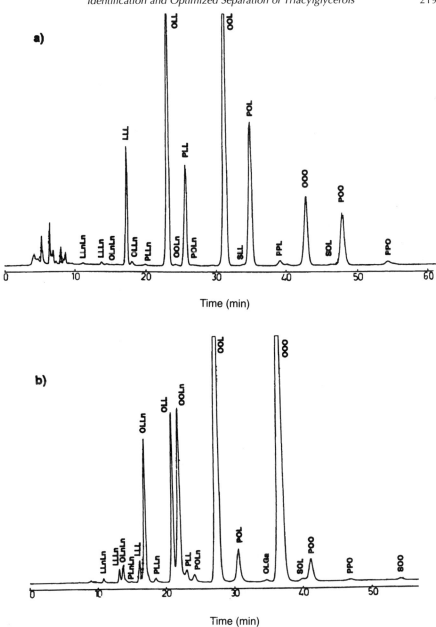

Figure 16.7. HPLC chromatograms. (*a*) Oats-seed oil: Column: Brownlee Spheri 5 ODS (250 × 4.6 mm); Mobile phase: MeCN/CH$_2$Cl$_2$ 68/32; flow rate 1 mL/min; Temperature = 21°C; Light-scattering detection. (*b*) Rapeseed oil: Column: Chromegabond C$_{22}$ (250 × 4.6 mm); Mobile phase: MeCN/CH$_2$Cl$_2$ 60/40; flow rate 1 mL/min; Temperature = 12.6°C; Light-scattering detection.

there is a new sample to analyze and also occurs when known oil or fat is analyzed under new experimental conditions.

The first method is to inject a standard of known structure and compare its retention time (or capacity ratio) with that of the analyzed mixture. This is possible for all homogeneous and some mixed triacylglycerols, but it is not possible to prepare all the standards that would be necessary (15). The second method occurs under particular circumstances when there are conjugated polyunsaturated acid residues in the structure, UV detection can be used to detect corresponding triacylglycerols (48). By using two detection types in series (UV detector and LSD) a partial identification of peaks from the LSD chromatogram can be made by comparing it with the one from UV detection (19). The third method, uses the GC results of corresponding derivatized fatty acids. However, the observation of an acid residue as a main component in GC does not signify that the corresponding homogeneous triacylglycerol is a main component observed in the direct analysis of fats. Stolhywo has reported that the analysis of a high-erucic acid rapeseed oil does not present the trierucin as a triacylglycerol (31). The phenomenon is not always so drastic. Predicting the composition of triacylglycerols from a statistical treatment of acid composition must be used cautiously. For example, Figure 16.8 shows the packed GC analysis of two types of peanut oil that differ in composition (45) and the corresponding RPLC analysis (Figure 16.9). As shown in the chromatograms, oleic acid is the main component of the two samples. According to the relative abundance of acid residues, the first sample can be attributed to a peanut oil from Africa or China (45) and the RPLC chromatogram shows that the main triacylglycerols possess oleic acid in their structures. In this case triolein is the major component, as it was deduced from statistical treatment. The situation is not as clear with examination of the second sample. In this case, the abundance of oleic and linoleic acids is approximately the same, and the origin of the sample could be attributed to South America (45). However, the abundance of four triacylglycerols, LLL, OLL, OOL, and OOO, is not the same. The observed peak of OOL is smaller than the three other triacylglycerols. This proves that the biosynthesis of oils is a rather selective process.

Another aspect of the problem is the presence of minor components detected in capillary GC analysis using methyl ester acids. In the absence of standards, the corresponding triacylglycerols cannot be identified by a single direct chromatographic analysis due to either the small quantity and possible confusion with another minor component, or coelution with a main component. A solution is to first separate the triacylglycerols by class using silver-ion HPLC or RPLC, then characterizing the oils and fats using GC analysis of the different esterified triacylglycerol fractions (46). The only inconvenience is the time required. Another solution is to use the general scheme of Goiffon, based on the additivity of Gibbs free energy for solutions. Stolhywo et al. (31) reported using such a diagram and its performance. In particular, they show that the retention prediction is inaccurate when the linear relationship is invalid for the retention data and the number of double bonds. Thus, they

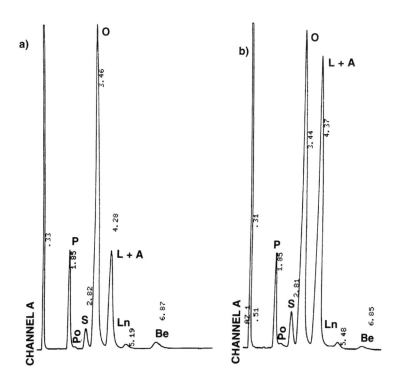

Figure 16.8. GC chromatograms of peanut oils after transesterification. Column: 6 FT 10% Silar 10C WHP 80/100; Gas vector He; Temperature of the analysis: 190°C; Temperature of the injector: 250°C; Temperature of the FID detector: 230°C. (a) Peanut oil from China or Africa. (b) Peanut oil from South America.

report that "the use of this diagram makes it very easy to find out a retention time of a known triacylglycerol. It is more complex however to identify an unknown." Considering that in the experimental conditions described, the linear relationship between log k' and the unverified number of carbon atoms in the saturated triacylglycerols, the authors have not used such a graphical method of identification. Identification can be done by comparing literature data if one can be sure of the published data (49).

For most of the oils and fats that the authors have analyzed (19), characterizing the triacylglycerols has been done by comparing chromatograms to those of many other authors (12,45). However, it is possible that many triacylglycerols are not separated by any known analytical conditions.

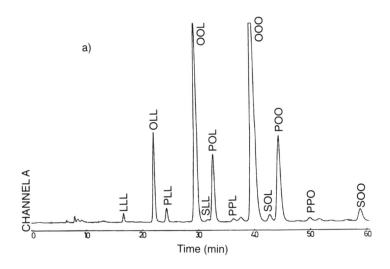

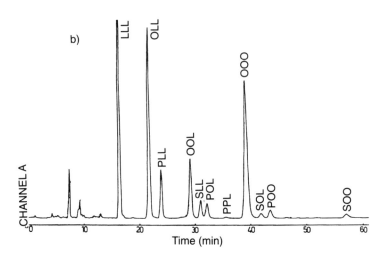

Figure 16.9. HPLC chromatograms of peanut oils. Column: Brownlee Spheri 5 ODS (250 × 4.6 mm); Mobile phase: MeCN/CH$_2$Cl$_2$ 68/32; flow rate 1 mL/min; Temperature = 21°C; Light-scattering detection. (a) Peanut oil from China or Africa. (b) Peanut oil from South America.

The retention order for GC using a polar stationary phase increases as the number of carbon atoms increases. For the same total number of carbon atoms as the unsaturation increases, so does the retention. The retention order in NARP Liquid chromatography is not the same; for the same total number of carbon atoms, retention decreases as unsaturation increases. Therefore, the two techniques are complementary. Some triacylglycerols poorly resolved by liquid chromatography can be well resolved by GC. Moreover, by using hyphenated techniques, such as chromatography coupled with mass spectrometry, some triacylglycerols can be identified more easily.

Optimizing triacylglycerol separation by capillary GC is described by Geeraert (9) and Geeraert and Sandra (50,51). As recommended by these authors, this study used an OV17 column having high temperature stability and hydrogen as the carrier gas. After preliminary results obtained with a FID detector (Table 16.3, Figure 16.10 [19]), the final experimental conditions used a mass spectrometer as the detector. A loss of separation was observed when compared to the chromatograms obtained from GC-FID (Figure 16.11). When two peaks are not well resolved in classical GC analysis, they can appear as an individual peak in GC-MS. Mass spectrometry enables the analyst to characterize individual chromatographic peaks in single or multiple solute systems. The beginning, middle, and end of a homogeneous peak must give the same qualitative information since they are produced by a single solute, only the total intensity will change. For instance, if the peak corresponds to two solutes and if the mass spectrum of each is different, the mass spectra at the beginning, middle, or end of the peak must be different. Failures of this

TABLE 16.3
Experiments in GC with a FID Detector[a]

Initial Temperature (°C)	Temperature Gradient (step) (min)	Rate (°C/min)	Final Temperature (°C)	(step) (min)	Oils or Fats
300		1	355		Coconut
280		1	355		Coconut, saturated TG
275		1	290	then	Coconut
290		2	355		
260 (5)		1	355		Coconut
250		1	355		Coconut, palm kernel, butter, soybean
340		1	365		Meadowfoam
250		1	350		Castor
250		3	350		Castor
320		1	355		Soybean, chicken, goose, duck, carapa, lard, calophyllum, ox's foot, palm, fish, spermaceti, olive, linseed

[a]Column: OV-17; Gas vector: H_2
Injector temperature: 340°C; Detector temperature: 360°C; Split 2%.

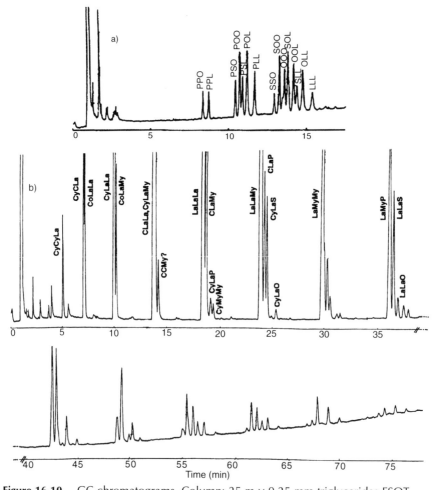

Figure 16.10. GC chromatograms. Column: 25 m × 0.25 mm triglycerides FSOT OV-17, 0.1 μm (Alltech); Gas vector H_2; Temperature of the injector: 340°C; Temperature of the detector: 360°C; Split 2%. (a) Calophyllum oil from 320 to 355°C at 1°C/min. (b) Coconut oil from 250 to 355°C at 1°C/min.

assertion could be due to the presence of triacylglycerols, such as XYY and YXX where X and Y are two different acid residues having the same retention factor. By using efficient chromatographic systems such pairs are well separated today. Another failure is the result of detection limits and is related to the detector sensitivity and the amount of analyzed species. Finally, if the retention time of two solutes is identical, their corresponding peaks will be superimposed, consequently mass spectrometry does not provide analytical data on the structure of either of them. Identical results can be achieved using a diode array detector for determination of peak purity in liquid chromatography. In this chapter, the authors combined

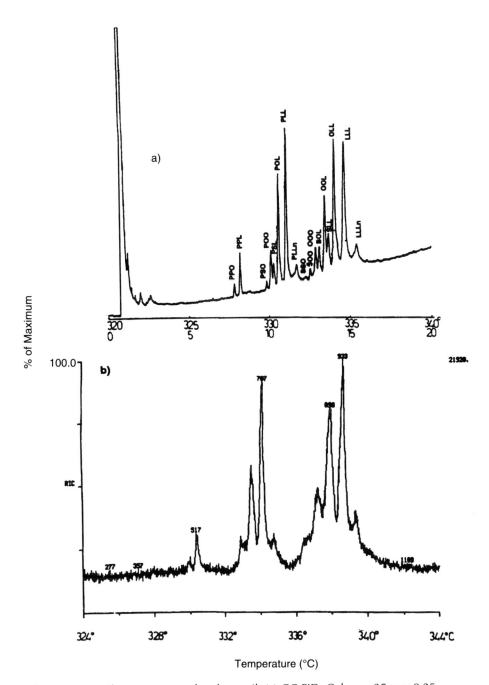

Figure 16.11. Chromatograms of soybean oil. (*a*) GC-FID; Column: 25 m × 0.25 mm triglycerides FSOT OV-17, 0.1 μm; (Alltech); Gas vector H$_2$; From 320 to 355°C in 1°C/min; Temperature of the injector: 340°C; Temperature of the detector: 360°C; Split 2%. (*b*) GC/MS in electron impact mode; Column: OV-17; Gas vector H$_2$; From 320 to 355°C in 1°C/min; Temperature of the injector: 340°C; Temperature of the transfer line: 360°C; Ion source temperature: 160°C; Split 2%.

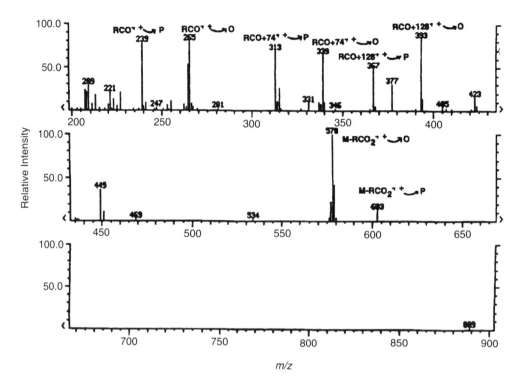

Figure 16.12. Mass spectrum of POO in electron impact mode.

the results of liquid chromatography and GC which has cancelled all the resolution problems except detection sensitivity.

Interpretation of Triacylglycerol Mass Spectra in Electron Impact Mode and Chemical Ionization Mode

Electron Impact Mode. From a triacylglycerol with a molecular mass M, it is possible to observe peaks at m/q, where m is the mass of the fragment and q the single electric charge; corresponding to ions or ion radicals having the structures: $[RCO]^+$ the most abundant; $[M-RCO_2]^+$; $[RCO+74]^+$; $[RCO+128]^+$; $[RCO_2]^+$; $[RCO_2H]^{o+}$; $[M-RCO_2H]^{o+}$; and $[M-RCO_2CH_2]^+$ (52–55) (Jamet, J., Rapport of DEA, University of Dijon, France, unpublished data). The molecular ion radical M^{o+} is never observed under these conditions. For example, Figure 16.12 shows the mass spectrum of POO. The abundance of each ion depends on the chain length of fatty acid residue (Jamet, J., unpublished data). Consequently, mass spectra are not the same for all triacylglycerols analyzed. This is supported by the fact that for a triacylglycerol XXY, the abundance of a fragment corresponding to X is not twice that of Y. Consequently, if the two acid residues X and Y are identified by mass spectrometry, the analyst cannot know if the corresponding triacylglycerol was XXY or XYY. This problem can be more difficult when the acid residues have similar structure. In addition, there is a lack of information due to the absence of peaks corresponding to molecular radicals. To overcome this inconvenience, it is possible to use mass spectrometric analysis by chemical ionization mode.

Chemical Ionization Mode. To characterize the triacylglycerols ammonia has been used as a reactant. In this case the characteristic ions are $[M+NH_4]^+$, corresponding to the molecular adduct; $[MH-RCO_2H]^+$; $[RCO]^+$; $[RCO_2]^+$; and $[RCO+74]^+$ (56–59).

As in electron impact mode, the relative intensity of each ion is related to the structure of the acid residues constituting the triacylglycerol (chain length and unsaturation number).

The observation of a peak at a mass corresponding to $[M+NH_4]^+$ leads to the determination of the total number of carbon atoms in the triacylglycerol (which can also be deduced from the retention factor in Gas or Liquid Chromatography). It permits the determination of the total number of unsaturations present in the molecule considering that one unsaturation decreases the molecular mass of two amu. The presence of fragments having the structure $[MH-RCO_2H]^+$, and $[RCO_2]^+$ gives information on the nature of each acid residue in the triacylglycerol. The experimental information used with a table summarizing all possibilities results in identifying a structure with a chromatographic peak.

For example, the process is used on the mass spectrum of triacylglycerols of hun fat (Figure 16.13). The observation of an adduct to m/q equal to 876 amu indicates a molecular mass equal to 876 − 18 = 858 amu. This corresponds to a total number of carbon atoms equal to 55. Accounting for the glycerol, the total number of carbon atoms corresponding to acid residues is 52 and there are two double bonds. The theoretical combinations to have such a structure are 4-24-24; 12-18-22; 6-22-24; 12-20-20; 8-20-24; 14-14-24; 8-20-22; 14-16-22; 10-24-24;14-18-20; 10-20-22; 16-16-20; 12-16-24; and 16-18-18 where the number corresponds to the total number of carbon atoms of each fatty acid residue.

The observation of peaks at m/q corresponding to the mass of the fragments $[RCO_2]^+$ leads to the determination of the nature of R. For instance, the peaks at m/q = 255 amu and 281 amu are characteristic of the residues P and O, respectively. The supplementary peaks at m/q = 578 and 603 are related to the ion $[MH-RCO_2H]^+$, with R = O and P and confirm these identities. Consequently, the structure of the triacylglycerol is POO.

Results of the Analysis in GC-MS and LC-MS. The GC-MS analysis and literature data permit identification or confirmation of the presence of many triacylglycerols in oils and fats, but GC-MS results do not result in complete identification of the observed peaks in liquid chromatography. Several reasons can be cited.

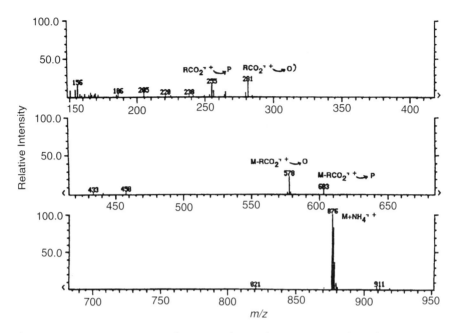

Figure 16.13. Mass spectrum of POO in chemical ionization mode with ammonia.

Triacylglycerols that are highly unsaturated or contain some hydroxyl groups have large GC peaks resulting in a loss of resolution. In mass spectra the characteristics of several triacylglycerols are superimposed and are not the same from one scan to the next. Thus, the gas chromatogram of castor oil which contains triacylglycerols with ricinoleic acid is not easily interpreted in GC-MS. A similar but less drastic situation occurs, with fish oils which contain polyunsaturated acid residues (DHA; $C_{22:6}$ and EPA, $C_{20:5}$) in their structure. In these cases, the information gained by mass spectrometry provides some idea about the potential triacylglycerols present in these oils.

Even if the oil triacylglycerol composition is well determined by GC-MS, characterizing each triacylglycerol to the observed peaks in liquid chromatography is not always simple. The Goiffon diagram only details regions of the chromatogram where a given triacylglycerol is eluted. Thus some characterizations can be uncertain due to superimposing saturated triacylglycerols with higher carbon content polyunsaturated triacylglycerols.

A loss of resolution has been observed in chromatograms obtained with NARP Liquid Chromatography and followed by a mass spectrometer with a particle beam interface. Considering that the mobile phase is purely organic, interface efficiently produces similar chromatograms to those obtained with evaporative light scattering detector (DDL or ELLSD). The ions which are observed in LC-MS are identical to those observed in GC-MS with electron impact ionization mode and chemical ionization mode. Thus, the combination of information from GC-MS and LC-MS enable us to complete identification (cherry nut oil, borage oil, etc.).

Mass spectra show also that there are significant amounts of diacylglycerols present in some oils and fats (cherry nut oil, borage oil, linseed oil, "ox foot" oils, wheat oil, calophyllum oil, spermaceti, etc.). However, the retention time or capacity ratio of these diacylglycerols is short, and often poorly resolved due to inappropriate experimental conditions to separate triacylglycerols.

Conclusions

Several conclusions can be drawn from the results presented in this paper.

When using monomeric alkyl-bonded silicas ($n < 18$), a triacylglycerol alkyl chain is inserted inside the bonded chains. When using polymeric bonded silicas, there is complete triacylglycerol solubilization in the bulk stationary phase, and results in enhanced selectivity compared to monomeric silicas. The first mechanism can be converted to the second by decreasing the analytical temperature or the amount of strong solvent in the mobile phase. Maximum selectivity between homogeneous triacylglycerols is obtained with a polymeric C_{18}-bonded silica at ambient temperature with a binary mixture of acetonitrile/strong solvent. The amount of strong solvent modulated the retention time.

Identification of triacylglycerols has been accomplished by comparison of known standards, literature data, and using chromatographic techniques coupled with mass spectrometry. No single technique such as GC-MS or LC-MS permits unambiguous characterization of triacylglycerol structure to chromatographic peak.

Taking into account of the nature of constituting triacylglycerols of coconut oil, palm oil or butter, the analysis and the identifications seem to be more easy in GC than in LC. To the opposite, for high content polyunsaturated triacylglycerols oils (such as these which contain acid residus such γLn, Ln, DHA, EPA...), or containing hydroxylated triacylglycerols (castor oil), LC analysis gives better results than GC analysis to identify and separate a maximum of solutes. For oils whose acid residues are mainly saturated and monounsaturated with a chain length between 16 and 20, such as P, Po, S, O, L, there are different points of view. If the analyst wants to select the efficiency of separation, capillary GC analysis gives very good results as demonstrated by Geeraert and Sandra, but if the selectivity of separation of critical pairs of peaks is chosen as first criterion, the liquid chromatography gives the best results.

References

1. Entressangles, B., Chazan, J.B., Parmentier, J., and Mordret, F. (1992) in *Manuel des corps gras,* Karleskind, A., Technique et Documentation, Lavoisier, Paris, pp. 531–682 and 1033–1432.
2. Christie, W.W. (1989) *Gas Chromatography and Lipids—A Practical Guide,* The oily

press, Ayr, Scotland.
3. Blank, M.L., and Privett, O.S. (1966) *Lipids 1:* 27–30.
4. Merritt, C., Jr., Vajdi, M., Kayser, S.G., Halliday, J.W., and Bazinet M.L. (1982) *JAOCS 59:* 422–432.
5. Naudet, M. (1992) in *Manuel des corps gras,* Karleskind, A., Technique et Documentation, Lavoisier, Paris, pp. 85–87.
6. Zeitoun, M.A.M., Neff, W.E., Selke, E., and Mounts, T.L. (1991) *J. Liq. Chromatogr. 14:* 2685–2698.
7. Stolywho, A., Colin, H., and Guiochon, G. (1983) *J. Chromatogr. 265:* 1–18.
8. Stolywho, A., Colin, H., Martin, M., and Guiochon, G. (1984) *J. Chromatogr. 288:* 253–275.
9. Geeraert, E. (1987) in *Chromatography of Lipids in Biomedical Research and Clinical Diagnosis,* Kuksis, A., J. Chromatography Library, Elsevier, Amsterdam, vol. 37, pp. 48–75.
10. Tchapla, A. (1992) *Analusis 20,* M71–M81.
11. Barron, L.J.R., and Santa-Maria, G. (1987) *Chromatographia 23:* 209–214.
12. Héron, S., and Tchapla, A. (1994) *Analusis, 20:*114–125.
13. Héron, S., and Tchapla, A. (1994) in *Fingerprints of Triacylglycerols from Oils and Fats by HPLC Isocratic Elution and Evaporative Light Scattering Detector,* Sedere, Paris, France.
14. Goiffon, J.P., Reminiac, C., and Olle, M. (1981) *Rev. Fr. Corps Gras 28:* 167–170.
15. Goiffon, J.P., Reminiac, C., and Furon, D. (1981) *Rev. Fr. Corps Gras 28:* 199–207.
16. Perrin, J.P., and Naudet, M. (1983) *Rev. Fr. Corps Gras 30:* 279–285.
17. Christie, W.W. (1987) *J. High Res. Chromatogr. Chromatogr. Commun. 10:* 148–150.
18. Hudiyono, S., Adenier, H., and Chaveron, H. (1993) *Rev. Fr. Corps Gras 40:* 131–141.
19. Héron, S. (1992) Ph.D. Thesis, Contribution à l'etude des mécanismes d'interactions moleculaires en Chromatographie Liquide à Polarité de Phases Inversée Application á la separation de triglycérides. University of Paris VI, France.
20. Melander, W.R., and Horvath, Cs. (1980) in *High Performance Liquid Chromatography: Advances and Perspectives,* Horvath Cs, Academic Press, New York, Vol. 2, pp. 113–319.
21. Dorsey, J.G., and Dill, K.A. (1989) *Chem. Rev. 89:* 331–346.
22. Héron, S., and Tchapla, A. (1994) *Analusis, 22:*161–177.
23. Unger, K.K. (1990) in *Packings and Stationary Phases in Chromatographic Techniques,* Unger, K.K., Chromatographic Sciences Series, Marcel Dekker Inc., New York, vol. 47, pp. 353–470.
24. Héron, S., and Tchapla, A. (1993) *Analusis 21:* 327–347.
25. Sander, L.C., and Wise, S.A. (1987) *CRT Crit. Rev. Anal. Chem. 18:* 299–415.
26. Tchapla, A., Héron, S., Lesellier, E., and Colin, H. (1993) *J. Chromatogr. 656:* 81–112.
27. Héron, S., and Tchapla, A. (1993) *Analusis 21:* 269–276.
28. Héron, S., and Tchapla, A. (1991) *J. Chromatogr. 556:* 219–234.
29. Thévenon-Emeric, G., Tchapla, A., and Martin, M. (1991) *J. Chromatogr. 550:* 267–283.
30. Tchapla, A., Colin, H., and Guiochon, G. (1984) *Anal. Chem. 56:* 621–625.
31. Stolywho, A., Colin, H., and Guiochon, G. (1985) *Anal. Chem. 57:* 1342–1354.

32. Martin, M., Thévenon, G., and Tchapla, A. (1988) *J. Chromatogr. 452:* 157–173.
33. Tchapla, A., Héron, S., Colin, H., and Guiochon, G. (1988) *Anal. Chem. 60:* 1443–1448.
34. Héron, S., and Tchapla, A. (1993) *Chromatographia 36:* 11–18.
35. Tsimidou, M., and Macrae, R. (1985) *J. Chromatogr. Sci. 23:* 155–160.
36. Parris, N.A. (1978) *J. Chromatogr. 149:* 615–626.
37. Podlhaha, O., and Toregard, B. (1982) *J. High Res. Chromatogr. Chromatogr. Commun 5:* 553–558.
38. Podlhaha, O., and Toregard, B. (1989) *J. Chromatogr. 482:* 215–226.
39. Frede, E. (1986) *Chromatographia 21:* 29–36.
40. Perrin, J.L., and Prévot, A. (1986) *Rev. Fr. Corps Gras. 33:* 437–445.
41. Pauls, R.E. (1983) *J. Am. Chem. Soc. 60:* 819–822.
42. Barron, L.J.R., Santa-Maria, G., and Diez-Masa, J.C. (1987) *J. Liq. Chromatogr. 10:* 3193–3212.
43. Hierro, M.T.G., Najera, A.I., and Santa-Maria, G. (1992) *Rev. Esp. Cienc. Technol. Aliment. 32:* 635–651.
44. Thévenon, G. (1986) Ph.D. Thesis, Mecanisme de retention en chromatographie liquide à polarité de phases inversée en milieu non aqueux. Elûde de la retention de triglycérides. University of Paris VI, France.
45. Merrien, A., Morice, J., Pouzer, A., Morin, O., Sultana, C., Helme, J.P., Bockelee Morvan, A., Rognon, F., Wuidard, W., Pontillon, J., Monteuuis, B., Uzzan, A., Foures, C., Sebedio, J.-L., and Chambon, M. (1992) in *Manuel des corps gras* Karleskind, A., Technique et Documentation, Lavoisier, Paris, pp. 115–280.
46. Nikolova-Damyanova, B., Christie, W.W., and Herslof, B. (1990) *J. Am. Oil Chem. Soc. 67:* 503–507.
47. Schoenmakers, P.J. (1986) *Optimizanon of Chromatographic Selectivity. A Guide to Method Development,* J. Chromatography Library, Elsevier, Amsterdam, vol. 35, pp. 116–169.
48. Comes, F. (1989) Ph.D. Thesis, Etude comparative des lipides de graines de quelgues rosaceés, prunoïdées et rosoïdées. Institut National Polytechnique de Toulouse, France.
49. Kuksis, A. (1987) in *Chromatography of Lipids in Biomedical Research and Clinical Diagnosis,* Kuksis, A., J. Chromatography Library, Elsevier, Amsterdam, vol. 37, pp. 1–42.
50. Geeraert, E., and Sandra, P. (1984) *High Resol. Chromatogr. and Chromatogr. Commun 7:* 431–432.
51. Geeraert, E., and Sandra, P. (1985) *High Resol. Chromatogr. and Chromatogr. Commun 8:* 415–422.
52. Hites, R.A. (1970) *Anal. Chem. 42:* 1736–1740.
53. Murata, T., and Takahashi (1973) *Anal. Chem. 45:* 1816–1823.
54. Myher, J.J., Kuksis, A., Marai, L., and Sandra, P. (1988) *J. Chromatogr. 452:* 93–118.
55. Lauer, W.M., Aasen, A.J., Graff, G., and Holman, R.T. (1970) *Lipids 5:* 861–868.
56. Murata, T., and Takahashi, S. (1977) *Anal. Chem. 49:* 728–731.
57. Murata, T. (1977) *Anal. Chem. 49:* 2209–2213.
58. Schulte, E., Hohn, M., and Rapp, U. (1981) Fresenius *Z. Anal. Chem. 307:* 115–119.
59. Duffin, K.L., Henion, J.D., and Shieh, J.J. (1991) *Anal. Chem. 63:* 1781–1788.

Chapter 17

Gas Chromatography-Fourier Transform Infrared Spectrometry in the Analysis of Fatty Acids

Jean-Luc Le Quéré

INRA. Laboratoire de Recherches sur les Arômes, 17 rue Sully, 21034 Dijon, France.

Introduction

Infrared spectroscopy is a dedicated spectroscopic method to distinguish *cis* (Z) and *trans* (E) isomers of unsaturated compounds. It has also been used extensively in the lipid field, especially to determine the total amount of *trans* (E) fatty acids in fats and oils. This constitutes an official method of the American Oil Chemists' Society.

The main advantage of Fourier transform infrared spectrometry (FT-IR) compared with conventional dispersive infrared spectroscopy is the coupling capability with separation techniques. Gas chromatography-Fourier transform infrared spectrometry (GC-FTIR) is the only way to obtain infrared analysis of individual isomers in complex mixtures. The first example of a such hyphenated technique mentioned uses a light-pipe interface. An FT-IR instrument is basically a single-beam infrared instrument, an interferometer, delivering a modulated infrared beam (1). The interface, the "light-pipe," is basically an infrared gas cell, and the effluent coming from the gas chromatograph passes continuously through the cell.

The cell itself consists of a glass capillary tube with alkali halide windows transparent to infrared light. The inside of the cell is coated with a gold layer, allowing multiple reflections of the modulated infrared beam coming from the interferometer. Infrared spectra are measured continuously by signal averaging as the effluent from the column passes through the heated light-pipe interface. This technique is an on-line technique, with infrared spectra collected on-the-fly, in the gas phase. For detailed reviews, two dedicated treatises may be consulted (1,2).

Analysis of Fatty Acids with a Light-Pipe Interface

Figure 17.1 shows the GC-FTIR spectra of the methyl esters of oleic and elaidic acids. They provide useful structural information about functional group (characteristic ester absorptions) and the main features are significant differences in the fingerprint region (600–1000 cm^{-1}, out-of-plane CH deformations) as well as in the CH-stretching region around 3000 cm^{-1}. The *trans* isomer is characterized by a medium absorption band at 968 cm^{-1} (δ CH) and the *cis* isomer by weak to medium absorptions around 700 cm^{-1} (δ CH) and at 3013 cm^{-1} (ν CH).

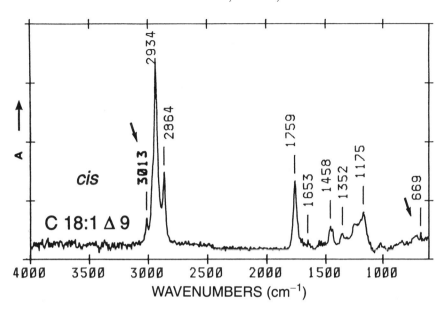

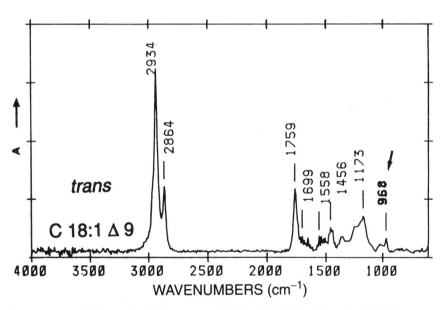

Figure 17.1. GC/light-pipe/FTIR spectra of oleic (cis-C18:1 Δ9) and elaidic (trans-C18:1 Δ9) acid methyl esters. The arrows show characteristic absorption bands discussed in the text.

The GC-FTIR method was employed to characterize geometrical isomers of polyunsaturated fatty acids isolated from liver lipids of rats fed high levels of geometrical isomers of linolenic acid in heated linseed oil (3). These polyunsaturated fatty acid isomers (mainly eicosapentaenoic and docosahexaenoic acids) probably resulted from desaturation and elongation of a geometrical isomer of the linolenic acid present in the dietary oil.

Thus, a mono*trans* geometrical isomer of eicosapentaenoic acid could be characterized. For instance, the spectrum of the methyl ester of all-*cis* eicosapentaenoic acid (Figure 17.2) is characterized by a band at 719 cm^{-1} and an intense CH-stretching band at 3022 cm^{-1}, while the presence of one *trans* double bond in an isomer gives rise to a band at 966 cm^{-1} (Figure 17.2).

The technique was also applied to the analysis of cyclic fatty acid mixtures isolated from heated linseed and sunflower oils (4). Figure 17.3 displays the spectrum of a diunsaturated propylcyclopentenyl-alkenoate with unique features at 712 and 966 cm^{-1}. The first band characterizes the *cis* double bond inside the ring, the second one reveals the *trans* configuration of the double bond in the carboxylate moiety.

The spectrum of a propyl-cyclohexenyl-nonenoate is shown in Figure 17.4. It displays characteristic features at 662 and 968 cm^{-1}. The first band is characteristic of the *cis* double bond in the cyclohexene ring, as in cyclohexene itself, and the sec-

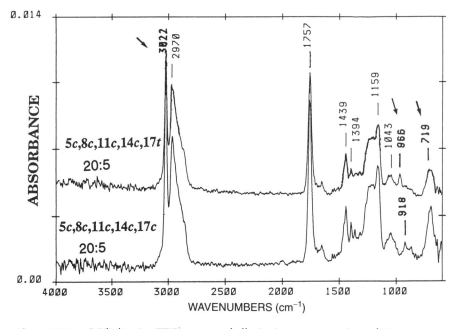

Figure 17.2. GC/light-pipe/FTIR spectra of all *cis*-eicosapentaenoic and 17-*trans*-eicosapentaenoic acid methyl esters. Characteristic features discussed in the text are indicated by arrows.

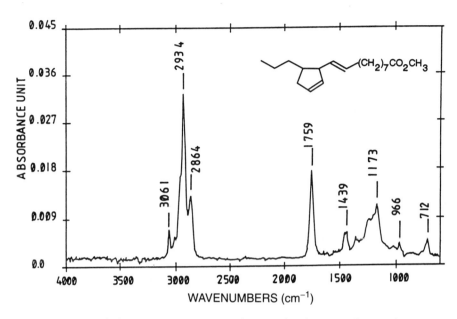

Figure 17.3. GC/light-pipe/FTIR spectrum of a propylcyclopentenyl-*trans*-decenoic acid methyl ester isolated from heated linseed oil (4).

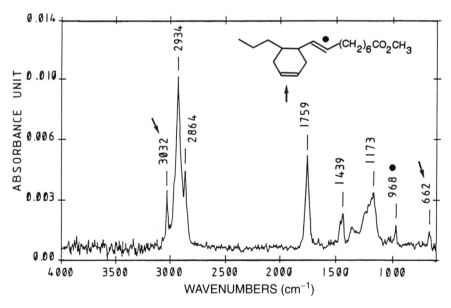

Figure 17.4. GC/light-pipe/FTIR spectrum of 9-(2'-propyl-4-cyclohexenyl)-*trans*-8-nonenoic acid methyl ester isolated from heated linseed oil. (4).

ond one reveals also a *trans* double bond in the carboxylate chain. The sizes of the unsaturated cyclic fatty acid rings were thus established by differentiating between the *cis* double bonds in a five-membered ring (712 cm^{-1}) and the ones in a six-membered ring (662 cm^{-1}) [4].

The main drawback of the GC-FTIR technique of using a light-pipe as the infrared cell is the poor sensitivity. The detection limit is in the ng range, and the identification limit for fatty acid esters is in the range of a few tens of ng per analyte. Another important disadvantage resides in the dead volume of the light-pipe. Mixing of incompletely separated solutes may occur within the interface, leading to degradation of the chromatographic resolution.

Technique Developments

Greater sensitivity, to the pg range, was recently provided by GC/matrix isolation/FTIR (GC/MI/FTIR) in which the GC effluent, containing a small amount of argon, is cryogenically trapped (at a temperature of about 12K under vacuum) onto the surface of a slowly rotating gold-plated disk, forming a microscopic solid argon matrix for subsequent analysis by FT-IR in reflection mode (1).

This technique was successfully applied to fatty acid analysis by Mossoba et al. (5). For example, *trans* fatty acid methyl esters in hydrogenated soybean oil and margarines were identified and quantified by GC/MI/FTIR (6). Figure 17.5 shows the MI/FTIR spectra of the *trans,trans*-octadecadienoate and of the linoleate methyl esters. The *trans,trans*-18:2 ester exhibited an intense characteristic CH deformation band at 972 cm^{-1} and two weak CH-stretching absorptions at 3005 and 3035 cm^{-1}. These absorptions were replaced by medium bands at 730 cm^{-1} and 3018 cm^{-1}, respectively, for the *cis,cis*-isomer.

However, the main drawback of this technique is that it operates at 12K with a solid argon matrix under a high vacuum (10^{-5} torr). Moreover, the argon matrix spectra are different from the conventional spectra obtained in condensed phase and different from the gas-phase spectra.

A new and encouraging development for GC-FTIR is the direct deposition interface (commercially available as the Tracer from the BioRad Corporation, Digilab Division) which operates at 77K (liquid nitrogen temperature) with the same pg sensitivity level (7). The interface is a direct deposition device situated in a vacuum chamber (Figure 17.6). The analytes in the effluent are directly deposited on a rectangular infrared window maintained at the temperature of liquid nitrogen. Sample deposition produces small frozen spots on the window. The window is moved continuously step by step, and each spot passes through the infrared beam after a short time delay. The infrared beam is transmitted through the sample and window, and collected by a microscope attachment. Postrun infrared measurements are possible, as they are with the matrix isolation system (in both cases the sample peaks are frozen), leading to improved signal-to-noise ratios. Moreover, the spectra are obtained in the condensed phase, in the traditional transmittance mode.

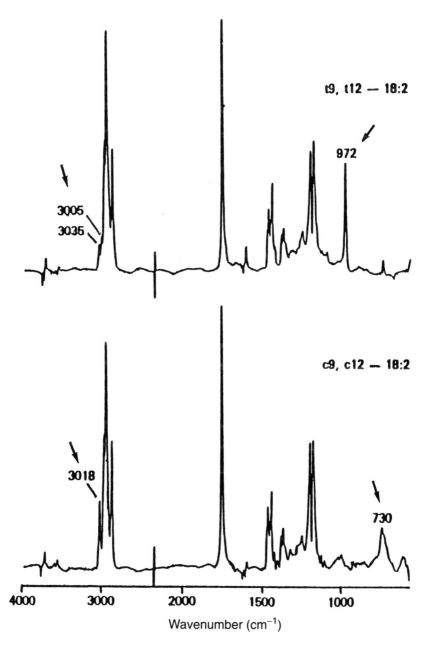

Figure 17.5. GC/MI/FTIR spectra with 4 cm^{-1} resolution of methyl *trans,trans*-octadecadienoate and linoleate showing absorption bands characteristic of the geometry of the double bonds. Reproduced from Mossaba et al. (6) by permission of the authors and the American Chemical Society.

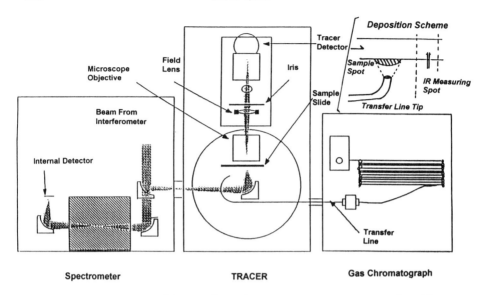

Figure 17.6. Schematics of the GC/direct deposition/FTIR system. Reproduced by permission of BioRad S.A.

Hydrazine reduction of geometric isomers of linolenic acid isolated from heated vegetable oils leads to a mixture of monounsaturated C_{18} fatty acids. They were analyzed in gas chromatography as methyl esters, and two were poorly separated, even on highly polar columns. Analysis of these esters with a light-pipe GC-FTIR system resulted in poor detection. Mixing of the two compounds inside the light-pipe occurred when higher amounts were injected. As a result, the double-bond configurations could not be determined (Le Quéré, J.-L., and Sébédio, J.-L., unpublished data).

Analysis of the same mixture with a direct deposition interface system produced an unambiguous determination, with no degradation of the chromatographic resolution. Both isomers were identified as *trans* fatty acid methyl esters (FAME) by observing in their GC/direct deposition/FTIR spectra intense absorption bands at 966 and 967 cm^{-1}, respectively, characteristic of *trans*=CH deformation (Figure 17.7).

Another development, illustrated by two important papers (8,9), is the coupling of supercritical fluid chromatography with FTIR (SFC/FTIR). One of the main advantages of SFC is that it requires minimal sample preparation. Moreover, free fatty acids, esters, and triglycerides can be analyzed using low temperatures and similar chromatographic conditions. Use of supercritical CO_2 permits component detection in the fingerprint and CH-stretching infrared regions.

The on-line SFC/FTIR spectra of FAME, triglycerides, and free fatty acids are similar, with characteristic absorbances for carbonyl and unsaturation geometries (methyl palmitate, linoleic acid, and trilinolein infrared spectra are shown in Figure 17.8). The carbonyl absorptions of methyl palmitate and of trilinolein are found at 1747 and 1748 cm^{-1}, respectively, whereas the linoleic acid carbonyl gives rise to a

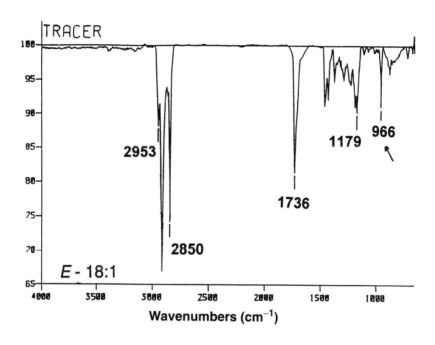

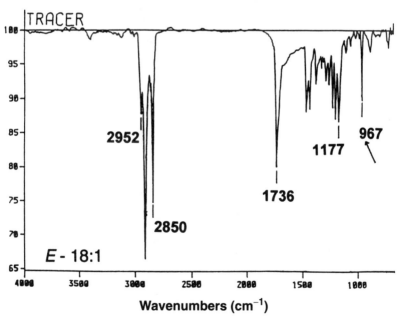

Figure 17.7. GC/direct deposition/FTIR spectra of two *trans*-octadecenoic acid methyl esters resulting from the hydrazine reduction of geometrical isomers of linolenic acid.

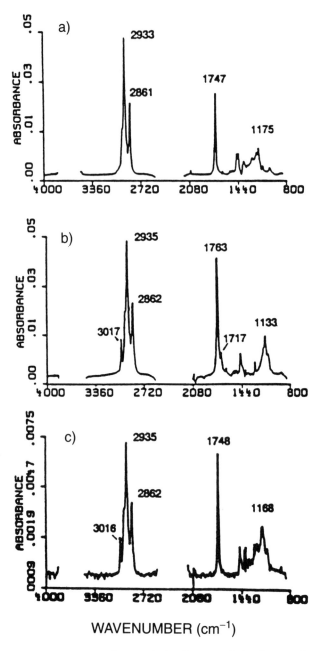

Figure 17.8. SFC/FTIR spectra of (a) methyl palmitate, (b) linoleic acid, (c) trilinolein. Reproduced from Calvey et al. (9) by permission of the authors and the American Chemical Society.

band at 1763 cm^{-1}. The characteristic *cis* CH-stretching bands of linoleic acid and of trilinolein are found at 3017 and 3016 cm^{-1}, respectively. This technique allowed the determination of the relative level of unsaturation and the extent of isomerization in partially hydrogenated soybean oils (9).

Conclusion

The use of FTIR provides sensitive detection of species eluting from a gas chromatograph and produces positive identification through the observation of characteristic spectral information. Particularly, GC-FTIR reveals the geometry of double bonds of individual unsaturated fatty acids occurring in complex mixtures. Sensitivity and chromatographic resolution problems associated with the light-pipe interface are solved in the new techniques, particularly the direct deposition interface. The latter, however, needs to be carefully evaluated for the analysis of complex fat and oil mixtures containing a large number of geometric isomers.

Acknowledgments

The author is indebted to Pierre Delmont, BioRad S.A., for arranging access to a Tracer instrument, and to Jean-Louis Sébédio for providing the *trans*-octadecenoate samples.

References

1. Griffiths, P.R., and de Haseth, J.A. (1986) *Fourier Transform Infrared Spectrometry*, p. 656, Wiley-Interscience, New York.
2. Herres, W. (1987) *HRGC-FTIR: Capillary Gas Chromatography-Fourier Transform Infrared Spectroscopy. Theory and Applications*, p. 212, Hüthig, Heidelberg.
3. Grandgirard, A., Piconneaux, A., Sébédio, J.-L., O'Keefe, S.F., Sémon, E., and Le Quéré, J.-L. (1989) *Lipids 24:* 799–804.
4. Sébédio, J.-L., Le Quéré, J.-L., Sémon, E., Morin, O., Prévost, J., and Grandgirard, A. (1987) *J. Am. Oil Chem. Soc. 64:* 1324–1333.
5. Mossoba, M.M. (1993) *INFORM 4:* 854–859.
6. Mossoba, M.M., McDonald, R.E., Chen, J.Y.T., Armstrong, D.J., and Page, S.W. (1990) *J. Agric. Food Chem. 38:* 86–92.
7. Bourne, S., Haefner, A.M., Norton, K.L., and Griffiths, P.R. (1990) *Anal. Chem. 62:* 2448–2452.
8. Hellgeth, J.W., Jordan, J.W., Taylor, L.T., and Ashraf Khorassani, M. (1986) *J. Chromatogr. Sci. 24:* 183–188.
9. Calvey, E.M., McDonald, R.E., Page, S.W., Mossoba, M.M., and Taylor, L.T. (1991) *J. Agric. Food Chem. 39:* 542–548.

Chapter 18

Contribution of Grignard Reagents in the Analysis of Short-Chain Fatty Acids

Michel Pina[1], Catherine Ozenne[2], Gilles Lamberet[3], Didier Montet[1], Jean Graille[1]

[1]CIRAD-CP, BP 5035, 34032 Montpellier, France, [2]ULN, 50890 Condé Sur Vire, France, and [3]INRA, 78350 Jouy en Josas, France

Introduction

Alkylmagnesium bromides react in ether and at low temperatures on triglyceride ester groups. The result is the transformation of fatty acid chains into tertiary alcohols where primary and secondary ester bonds are broken at the same rate in spite of the nature of the fatty acid chain.

Application of Grignard Reagents to Oils and Fats Analysis

Two applications successfully used Grignard reagents in oils and fats analysis. One was analyzing the fatty acid regiodistribution in triglycerides (1,2). The nonselectivity regarding fatty acid chains and nonregioselectivity of Grignard reagents was used to start operations through partial deacylation. The brief, controlled degradation resulted in a mixture of diglycerides. The second involved analyzing wax constituents, particularly jojoba (3). Ethylmagnesium bromide attacked the ester bonds, and in a single step completely transformed jojoba wax rapidly and completely into a mixture of tertiary alcohols that corresponded to the fatty acids and native primary alcohols. The mixture of alcohols was analyzed directly in gas chromatography (GC) with a single injection.

The Grignard reagents methyl-, ethyl-, and butylmagnesium bromides were chosen for two reasons. First, they are highly reactive when compared to their chlorine equivalents, which do not form complete reactions. They are also available commercially, apart from butylmagnesium bromide which can be prepared by reacting 1-bromobutane with magnesium.

The reaction occurs according to the simplified mechanism in Figure 18.1. The reaction begins with R' MgBr, the Grignard reagent, reacting with one of the ester groups of a triglyceride adding onto the carbonyl and producing intermediate compound A. This compound is unstable and is spontaneously transformed. The ketone B, generated in situ, reacts in the presence of magnesium to give a new intermediate C. This intermediate is transformed in the presence of water into a tertiary alcohol with three radicals, the fatty chain R of the original ester and two alkyl radicals R' of the Grignard reagent.

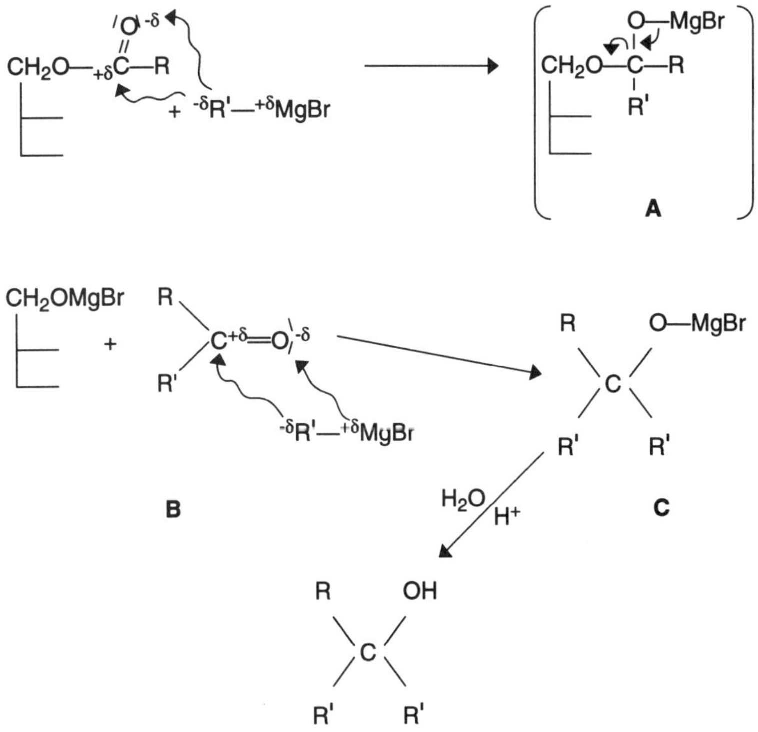

Figure 18.1. A simplified mechanism of the reaction that Grignard's reagent has on an ester group.

The procedure used utilized different Grignard reagents to react with short-chain fatty acid oils and fats, such as lauric oils, milk fat, and on a normal reference oil, palm oil. For the same fat or oil, the tertiary alcohol molar compositions were compared with those obtained through conventional conversion into methyl esters (ME) or isopropylic esters (IPE), in the case of milk fat. The aim was to find a solution to the many problems encountered in short-chain fatty acid analysis. To identify tertiary alcohols chromatographically, standards from a mixture of copra, palm, rapeseed, and groundnut triglycerides were synthesized for carbon chains from 6–24 carbons in length. This demonstrated the effectiveness of Grignard reagents for analyzing short-chain oils and fats.

Analytical Problems Inherent to Short-Chain Fatty Acids

Given their volatility, the conversion of short-chain fatty acids to ME and subsequent analysis by gas–liquid chromatography (GLC) has a number of problems that must be considered (4). During conversion to methyl esters, some sources of error are difficulty in obtaining quantitative methylation, secondary saponification reaction, and difficulty quantitatively extracting short-chain esters from the aqueous phase. Sources of error during chromatographic injection are losses in the syringe, losses through selective evaporation of the short-chain fatty acid esters, and molecular discrimination dependent on the type of injection. While processing the results there may be difficulty in applying adequate response coefficients with a flame ionization detector (FID). All these difficulties are encountered during the analysis of milk fat, which is rich in butyric acid. Lauric oils are easier to analyze because the shortest constituent, caproic acid (C_6), is a minor constituent.

Derivation techniques have been developed to minimize the problems by converting the fatty acids into ME (in the cold state [5]); or into higher molecular weight esters, such as isopropylic (6) or butylic esters (7), but these have long derivation times. Conversion into tertiary alcohols offers the advantage of rapidly increasing molecular weight in the cold state.

Effects of Tertiary Alcohol Conversion on Molecular Weight Increase

Such conversion leads to an increase in the carbon condensation of $2nC$ fatty acid chains, with n being the number of carbons in the alkyl group of the Grignard reagent used (Table 18.1). For a fatty acid of molecular weight x, the molecular weight gain and increase in the number of carbons is dependent upon the alkylmagnesium bromide used. With methylmagnesium bromide a dimethylalkylcarbinol (DMAC) is obtained, having a molecular weight of $x + 14$ and 2 additional carbons, that is the same molecular weight as the ME, but with one more carbon. A diethylalkylcarbinol (DEAC) is obtained with ethylmagnesium bromide, resulting in a molecular weight of $x + 42$ and 4 additional carbons, in other words, the same molecular weight as the IPE, but with one more carbon. Finally butylmagnesium bromide produces a dibutylalkylcarbinol (DBAC) with a molecular weight of $x + 98$ and 8 additional carbons. In the latter case, for the butyric chain at C_4 the number of carbons is tripled, and the molecular weight increases from 88 in the fatty acid to 186 in the dibutylated tertiary alcohol, a more than twofold increase. The molecular weight of the ME at C_{10} is 186.

Effect of Tertiary Alcohol Conversion on GC Correction Coefficients

Conversion into a tertiary alcohol eliminates the carbonyl group. This is a fundamental structural modification which allows no response by flame ionization detection (FID). In order to determine the theoretical response coefficients for FID, the contribution of the atoms of a molecule is calculated in terms of the number of effective carbons, adding 1 for an aliphatic carbon, 0.95 for an olefinic carbon, 0 for the atoms of a carbonyl, and subtracting 0.25 for an oxygen of a tertiary alcohol or an ester (8,9).

TABLE 18.1
Molecular Weights and Carbon Condensations

	FA	R-COOH	x	
CH_3MgBr	DMAC	$R-C(CH_3)_2-OH$	x + 14	+ 2C
	ME	$RCOOCH_3$	x + 14	+ 1C
C_2H_5MgBr	DEAC	$R-C(C_2H_5)_2-OH$	x + 42	+ 4C
	IPE	$R-COOCH(CH_3)_2$	x + 42	+ 3C
C_4H_9MgBr	DBAC	$R-C(C_4H_9)_2-OH$	x + 98	+ 8C

For the Butyric Chain	FA	DBAC		
nC	4	12	(×3)	
mw	88	186	(×2.11)	
		MW of ME (C_{10})		

The response coefficients for the shortest terms were calculated and compared to the saturated term at C_{18} (intentionally the worst choice). In routine analysis correction factors are not normally taken into account for values less than 1.1 (Table 18.2). The theoretical correction factor was only 1.06 for the dibutylated tertiary alcohol when the term from C_4 was compared to that for C_{18}, which is remarkable when compared to the ME at 1.54 and the IPE at 1.33. For the diethylated tertiary alcohol the correction factor was already below 1.1 from the C_6 term (1.08). The

TABLE 18.2
Theoretical Response Coefficients (Compared with 18:0)

Fatty Chains	ME	IPE	DMAC	DEAC	DBAC
4:0	1.54[a]	1.33	1.17	1.12	1.06
6:0	1.33[a]	1.21	1.11	1.08	1.05
8:0	1.19[a]	1.14	1.07	1.05	1.035
10:0	1.12	1.09	1.05	1.04	1.025
16:0	1.020	1.015	1.008	1.006	1.004

[a]Advisable experimental response coefficients.

same applied for the term at C_8 with a dimethylated derivative (1.07). By converting fatty acids into tertiary alcohols the problem of correction coefficients becomes secondary. Nevertheless, this paper utilized correction coefficients in the tables.

Preparation of Tertiary Alcohols

The preparation was carried out at room temperature in a perfectly anhydrous ether solution using 10–15 mg triglyceride and a large stochiometric excess of Grignard reagent. In all cases even with the least reactive reagent used, complete conversion took less than 10 min. There was a thin-layer chromatography (TLC) control to verify the conversion. It was necessary to cool the reaction mixture to 0°C before destroying the excess Grignard reagent with HCl. Water added at the end of the reaction released the tertiary alcohol and led to two phases. The upper ether phase, containing the derived products, was recovered and diluted with an appropriate quantity of solvent for GC.

GC Conditions

The constituents were separated on a 20M Carbowax column (25 m length, 0.28 mm i.d., and 0.25 µm film thickness), using a split injection method. We were unable to use on-column injection, which probably would have been preferable in this case. The carrier gas used was helium with a flow rate of 2 mL/min and a 1/30 split ratio. The injector and detector were maintained at 220 and 250°C, respectively. Temperature programming was adapted to the type of tertiary alcohol mixture to be analyzed for DMAC and ME (the lightest), from 50–190°C with a temperature increase of 7.5°C/min; for DEAC and IPE, from 80–200°C with a temperature increase of 7.5°C/min; for DBAC (the heaviest), from 100–210°C with a temperature increase of 7.5°C/min.

Results

Palm Oil

Palm oil was used as the reference oil (Table 18.3). Any results diverging from this oil, the lightest constituent of which is myristic acid, would have ended the study. It was encouraging to determine that the molar compositions of the three tertiary

TABLE 18.3
Fatty Acid Composition of the Different Derivatizations for Palm Oil (mol%)

Fatty Chains	ME	DMAC	DEAC	DBAC
12:0	0.3	0.4	0.3	0.3
14:0	1.2	1.3	1.2	1.3
16:0	45.6	45.9	45.8	45.3
18:0	4.2	4.3	4.3	4.3
18:1	38.8	38.3	38.8	38.9
18:2	9.3	9.2	9.0	9.2
Others	0.6	0.6	0.6	0.6

TABLE 18.4
Fatty Acid Composition of the Different Derivatization for Coconut Oil (mol %)

Fatty Chains	ME	DMAC	DEAC	DBAC
6:0	0.9	1.1	1.2	1.3
8:0	9.8	10.6	10.7	10.9
10:0	7.6	7.7	7.9	7.8
12:0	50.3	49.6	49.7	49.5
14:0	17.3	17.3	16.9	17.0
16:0	7.6	7.8	7.5	7.3
18:0	2.2	2.3	2.3	2.3
18:1	4.9	4.9	4.6	4.6
18:2	1.3	1.3	1.3	1.3

alcohols derived were identical to each other, and to the derivatization into ME, proving that derivatization by organic magnesium compounds was effective and valid for analysis of a common fat or oil.

Lauric Oils

Palm kernel and babassu oils were analyzed also, but only the results obtained from copra oil will be discussed, since it is the riches in caproic acid (Table 18.4). The similarity of results between the three derivatizations with the magnesium compounds was remarkable, even though the value of the dimethylated alcohol of C_6 was slightly lower. Compared to ME, the low values recorded for the light constituents could be attributed to the nonoptimal preparation or to the application of theoretical correction coefficients.

Milk Fat

A comparison was made with FPE and considering the complexity of this fat, only components up to C_{12} and the major constituents are indicated in Table 18.5. The results recorded show that for butyric derivatives, DMAC derivatization is not appropriate (only 11.7%), while DEAC derivatization is not quite satisfactory

TABLE 18.5
Fatty Acid Composition of the Different Derivatizations for Milk Fat (mol %)

Fatty Chains	IPE	DMAC	DEAC	DBAC
4:0	12.5	11.7	12.3	13.6
6:0	5.2	4.9	5.3	5.8
8:0	2.3	2.1	2.2	2.4
10:0	4.1	3.9	3.8	4.1
12:0	3.9	4.0	4.0	14.1
14:0	11.8	11.9	11.7	11.5
16:0	25.3	25.5	25.1	24.9
18:0	9.7	9.9	9.5	9.2
18:1	20.1	20.4	19.9	19.2
Others	5.1	5.5	6.2	5.2

TABLE 18.6
Volatility of Short-Chain Tertiary Alcohols[a]

	DMAC	DEAC	DBAC
Butyric chain	65	15	—
Caproic chain	20	tr	—
Caprylic chain	2	—	—

[a]Losses are expressed in comparison with the myristic chain.

(12.3%) and DBAC derivatization leads to a value greater than that of IPE (13.6 compared to 12.5%). Therefore, it seems that the butylmagnesium bromide method is more effective in minimizing short-chain fatty acid losses. In fact, it is the first time that a C_4 has been transformed into a C_{12} for analytical purposes.

Volatility of Short-Chain Tertiary Alcohols

In order to determine the volatility of short-chain tertiary alcohols, an aliquot of the solutions of milk fat was evaporated to dryness under a nitrogen stream, then diluted to the same volume and rechromatographed (Table 18.6). Losses of the shortest constituents caused by this treatment were assessed taking the C_{14} as a reference. The dibutylalkylcarboniols were nonvolatile, while the DEAC were not volatile from C_6. The volatility of the DMAC was negligible for C_8, which indicates that it should be possible to analyze milk fat with butylmagnesium bromide, lauric oils with ethylmagnesium bromide, and possibly all other oils with methylmagnesium bromide.

Conclusions

There are two major disadvantages to using this method. A neutral fat must be used since there is no FFA derivatization. Gas chromatography analyses take longer, although this is not the case with DMAC, for which the analytical time is comparable to that of ME. Furthermore, for the same fatty acid chain, tertiary alcohol branching partially compensates for the increase in carbon condensation regarding the retention times. For example, isothermal GC analysis of a groudnut oil at 195°C takes 22 min to elute C_{24} as ME and 23 min as DMAC.

If the alkyl in the Grignard reagent chosen is appropriate for the fat or oil to be analyzed, the advantages far outweigh the disadvantages. The derivatization is a reaction which is simple, rapid, and complete (even with the least reactive Grignard reagents); performed in the cold state (which prevents loss due to volatility); and problem-free for extraction. During GC analysis fatty acids are converted into sufficiently heavy tertiary alcohols that are nonvolatile; the response correction coefficients are highly uniform for all the constituents; and solutions that are too dilute can be concentrated.

References

1. Brockerhoff, H. (1965) *J. Lipid Research 6:* 10.
2. Muderhwa, J.M., Dhuique-Mayer, C., Pina, M., Galzy, P., Grignac, P., and Graille, J. (1987) *Oléagineux 42:* 207.
3. Pina, M., Pioch, D., and Graille, J. (1987) *Lipids 22:* 358.
4. Ackman, R.G., and Sipos, J.C. (1964) *J. Chromatogr. 16:* 298.
5. Christopherson, S.M., and Glass, R.L. (1969) *J. Dairy Science 52:* 1289.
6. Wolff, R.L., and Castera-Rossignol, F.M. (1987) *Rev. Fse Corps Gras 34:* 123.
7. Iverson, J.L., and Sheppard, A.J. (1986) *Food Clam. 21:* 233.
8. Tranchant, J. (1982) *Manuel Pratique de Chromatographie en Phase Gazeuse,* 3rd edn., pp. 68–69, Masson, Paris.
9. Bading, H.T., and De Jong, C. (1988) *J. Am. Oil Chem. Soc. 65:* 659.

Chapter 19
Information About Fatty Acids and Lipids Derived by ^{13}C Nuclear Magnetic Resonance Spectroscopy

Frank D. Gunstone

School of Chemistry, The University, St. Andrews, Fife, KY16 9ST, United Kingdom.

Introduction

The analysis of lipid mixtures is dominated by chromatographic procedures and it is appropriate that several chapters in this book should be devoted to procedures such as thin-layer chromatography (TLC), gas chromatography (GC), and high-performance liquid chromatography (HPLC). These are used mainly for the quantitative measurement of known compounds. On the other hand, spectroscopic procedures are used chiefly for the identification of lipid compounds—both familiar and unfamiliar—often in conjunction with a chromatographic procedure. In these hyphenated techniques the separating power of chromatography is combined with the facility of identification found in spectroscopy. It is, therefore, not surprising to find other chapters in this book devoted to gas chromatography–Fourier Transform Infrared (GC-FTIR) and gas chromatography–mass spectrometry (GC-MS).

This chapter is devoted to the application of high-resolution ^{13}C NMR spectroscopy to the study of lipids and will illustrate what can be achieved by this procedure. Useful information derived from the study of single compounds and of concentrates of one component has been applied to natural and synthetic mixtures. The author has already prepared a full review of this topic (1) and also several shorter reviews (2–4).

A ^{13}C NMR spectrum provides two kinds of information for each signal: the chemical shifts and the relative intensities. The former is of qualitative value and permits identification of important structural features; while the latter provides quantitative information of analytical value. Chemical shifts may vary slightly with the concentration of the solution under study and with the nature of the solvent. Most measurements are made with solutions of about 1M concentration and $CDCl_3$ is the solvent most commonly employed. Other solvents, such as $CD_3OD/CDCl_3$ mixtures, $(CD_3)_2SO$, D_2O, and C_6D_6 have also been used.

To obtain quantitative data, attention has to be given to the protocol for obtaining the spectrum. In particular, the relaxation problem has to be overcome by adding a relaxation agent, such as chromium acetonylacetonate ($Cr[acac]_3$), and/or by including a delay time between successive scans of the spectrum. This will add to the time required to collect the spectrum. Spectrometers are now available at a frequency for ^{13}C of 68 MHz or more and spectra are generally obtained using an nuclear overhauser enhancement suppressed, inverse-gated, proton-decoupled technique. Excitation pulses have a 45–90° pulse angle, and acquisition times (including delay times) are 1–20 sec per scan. The number of scans is usually 1000 or more. The sample size for a routine ^{13}C NMR spectrum is 50–100 mg, but high-quality spectra can be obtained from 1 mg with a suitable investment in a large number of scans.

Oils Containing Only the Common Fatty Acids.

This survey begins with oils containing only the common fatty acids such as palmitic, stearic, oleic, linoleic, and sometimes α-linolenic. A typical spectrum is shown for safflower oil in Figure 19.1. The printout of chemical shifts and intensities which accompanies the spectrum provides more detailed information. In such a spectrum it is easy to recognize the carbon atoms listed in Table 19.1. The complex group of signals (the methylene envelope) with a chemical shift of 29.0–29.8 ppm belong to those carbon atoms which are not significantly influenced by any functional group and have similar or identical chemical shifts. With individual compounds these can sometimes be assigned (5–9), but this becomes difficult with mixtures and useful assignments can seldom be made. Therefore, these signals are generally ignored. However, this still leaves many other signals which can be recognized and assigned to carbon atoms in a particular environment.

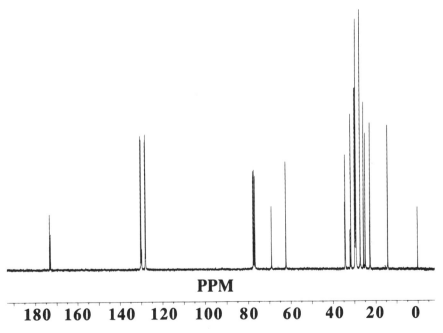

Figure 19.1. ^{13}C NMR spectrum of safflower oil.

TABLE 19.1
Easily Recognized Carbon Atoms Present in Most Triacylglycerols and Their Approximate Chemical Shifts (ppm)

Carbon atoms close to the acyl group	C-1 (173), C-2 (34), C-3 (25)
Carbon atoms close to the methyl group	ω-1 (14), ω-2 (23), ω-3 (32)
Glycerol carbon atoms	G-1 and G-3 (62), G-2 (69)
Olefinic carbon atoms	130
Allylic carbon atoms	27, 25.6
Methylene envelope	Remaining carbon atoms (29.0–29.8)

The chemical shifts of carbon atoms are strongly influenced by a nearby acyl group (acid or ester), and in single species this effect can be recognized along the acyl chain to C-6. In mixtures, only carbon atoms 1–3 are easily recognized (Table 19.2). They generally appear as clusters of several signals with similar but not identical shifts, since the shift is influenced by whether the acyl chain is attached to a primary (α) or secondary (β) hydroxyl group of glycerol, and by the presence of a nearby double bond. The effect of a double bond in the $\Delta 9$ position (as in oleate, linoleate, etc.) is quite small and can only be observed in the C-1 to C-3 signals when the spectrum is collected with great care over a long period of time (10). These changes increase as the double bond gets closer to the carboxyl group. Examples with acids having $\Delta 4$–$\Delta 6$ unsaturation are discussed in later sections of this chapter. The observed chemical shifts change slightly with experimental factors, such as solvent, concentration, and temperature, but the differences between members of a cluster remain more constant. The difference for C-3 α and β signals is quite small and may be observed only after resolution enhancement. For the C-1 and C-2 atoms the α and β signals are easily recognized. Since there are two α chains and only one β chain in a glycerol triester, the two peaks appear in a 2:1 ratio and the members of each pair can be recognized easily.

The ω-1 to 3 carbon atoms also have easily identified signals with the shift for saturated acyl chains being changed by double bonds in the n-9, n-6, and n-3 positions. Although the presence of such double bonds is easily demonstrated by the ω-1 to 3 signals, other evidence is required before positive identification of the fatty acid occurs. The n-3 signal in a vegetable oil probably comes from α-linolenic acid, but in a fish oil the signal would be indicative of all n-3 acids (docosahexaenoic acid [DHA], eicosapentaenoic acid [EPA], etc.). It is sometimes possible to recognize an ω-3 signal corresponding to n-7 acids. This is most likely produced by 9c-16:1 and/or 11c-18:1 (Table 19.3).

TABLE 19.2
Chemical Shifts (ppm) for C-1–C-3 Signals in the α (sn-1 and sn-3) and β (sn-2) Chains of Tripalmitin

	C-1	C-2	C-3
α	173.27	34.07	24.89
β	172.86	34.24	24.94
Difference	0.41	0.17	0.05

TABLE 19.3
Chemical Shifts (ppm) for the ω-1-3 Signals in Saturated and Several Unsaturated Glycerol Esters

	ω-1	ω-2	ω-3
Saturated	⎫	⎫	⎫ 31.95
n-9	⎬ 14.13	⎬ 22.60	⎬
n-7	⎭	⎭	⎭ 31.83
n-6	14.09	22.71	31.55
n-3	14.29	20.57	[a]

[a]Olefinic signal.
Source: Gunstone (1).

In general each olefinic carbon atom has its own signal, although signals with very similar chemical shifts may overlap. Thus oleic, linoleic, and linolenic acids and esters have two, four, and six signals in the region 127–132 ppm. Mixtures of oleate and linoleate often show only five signals instead of six because of overlap of one pair while mixtures with α-linolenate usually have only 10 signals with two pairs overlapping (Table 19.4). Careful studies show that in some cases there are small differences in the olefinic chemical shifts for the α and β chains (10). The previous statements and figures relate to natural acids with *cis* configuration; the *trans* isomers have their own distinct chemical shifts (Table 19.5).

TABLE 19.4
Chemical Shifts (ppm) for Olefinic Signals in Oleic, Linoleic, and α-Linolenic Glycerol Esters

Carbon Atom	Oleate	Linoleate	α-Linolenate
9	129.70	[129.98][a]	[130.19][a]
10	[129.98][a]	128.09	127.77
12	—	127.91	128.24
13	—	[130.19][a]	128.29
15	—	—	127.13
16	—	—	131.93

[a]Overlapping signals are shown in brackets.
Source: Gunstone (3).

TABLE 19.5
Chemical Shifts (ppm) for Olefinic Carbon Atoms in 18:1 Isomeric Acids

Δ	cis		Difference[a]	trans		Difference[a]
8	130.17	129.65	(0.52)[a]	130.67	130.13	(0.54)[a]
9	130.09	129.78	(0.31)	130.54	130.23	(0.31)
10	130.00	129.83	(0.17)	130.47	130.31	(0.16)
11	129.96	129.89	(0.07)	130.43	130.34	(0.09)
12	129.94		—	130.14		—

[a]Difference in chemical shift between the two olefinic carbon atoms.
Source: Gunstone et al. (8) and Gunstone (11).

TABLE 19.6
Chemical Shifts (ppm) for Some Allylic Carbon Atoms

9-18:1 esters	c	27.19 (8)[a], 27.24 (11)[a]
	t	32.58 (8)[a], 32.63 (11)[a]
9,12-18:2 esters	c,c	27.32 (8)[a], 25.77 (11)[a], 27.32 (14)[a]
	c,t	27.18 (8)[a], 30.55 (11)[a], 32.64 (14)[a]
	t,c	32.67 (8)[a], 30.56 (11)[a], 27.22 (14)[a]
	t,t	32.59 (8)[a], 35.68 (11)[a], 32.59 (14)[a]
9,12,15-18:3 esters	c,c,c	25.65 (11)[a], 25.55 (14)[a]

[a]Numbers in parentheses refer to carbon atoms.
Source: Gunstone et al. (8) and Gunstone (11).

TABLE 19.7
Influence of *cis* or *trans* Double Bonds on the Chemical Shift of ω-1–3 Signals When These Are γ to the Double Bond

	Normal Signal	Signal γ to Double Bond		18:1 Isomer
		cis	*trans*	
ω-1	14.12	13.80	13.66	$\Delta 14$
ω-2	22.72	22.37	22.21	$\Delta 13$
ω-3	31.95	31.56	31.43	$\Delta 12$

Source: Gunstone (1,11) and Gunstone et al. (8).

The double bond has a marked influence on the chemical shift of its α-carbon atoms (allylic). There are distinct signals for CH_2 groups allylic to one or two double bonds and these vary with the configuration of the double bond(s) (Table 19.6). The unsaturated center also has a smaller, but still significant, effect on carbon atoms in the γ-position. The effect of this γ-shift for *cis* and *trans* double bonds on the easily recognized ω-1 to -3 signals is demonstrated in Table 19.7. These values have been exploited in the study of partially hydrogenated oils (11). Several of the features discussed are apparent in the spectrum of safflower oil (Figure 19.1, Table 19.8).

Oils Containing Acids With a Double Bond Close to the Carboxyl Group

Some acids with a double bond close to the carboxyl group are listed in Table 19.9 along with some common oil sources and references that have information concerning their ^{13}C spectra. The content of these acids and their distribution between the α and β chains of the glycerol triesters can be determined using intensity and chemical shift data. The latter has been measured previously by regiospecific enzymatic hydrolysis, but this method is not appropriate for acids with double bonds at $\Delta 7$ or closer to the carboxyl group (16).

For example, it is possible to recognize DHA and EPA in fish oils and to calculate the percentage of each of these in the α and β chains. Docosahexaenoic acid is characterized by C-1 signals at 172.50 and 172.10 ppm (γ to the double bond) and EPA by C-2 signals at 33.54 and 33.35 ppm (γ to the double bond). Typical results are collated in Table 19.10.

TABLE 19.8
Chemical Shifts (ppm) of ^{13}C NMR of Safflower Oil

C-1α	173.21		Gβ	68.89
C-1β	172.80	Glycerol	Gα	62.10
C-2β	34.19	Olefinic	L13	130.20
C-2α	34.03		O10,L9	129.98
			O9	129.70
C-3	24.85		L10	128.07
			L12	127.89
ω-1	14.09			
ω-2	22.71[a]	Allylic	cis	27.21[c]
	22.60[b]		L11	25.64
ω-3	31.93[a]			
	31.54[b]	Methylene-envelope		29.11–29.78 (11 signals)

Abbreviations: L, linoleic; O, oleic.
[a]n-9, sat.
[b]n-6.
[c]Carbon atoms allylic to one double bond (O8, O11, L8, L14)
Source: Gunstone, unpublished data.

TABLE 19.9
Structure of Fatty Acids with a Double Bond in the Δ4–Δ6 Positions

	Acid	Structure	Source	Ref.
Δ4	DHA	(22:6) 4, 7, 10, 13, 16, 19	Fish oils	12, 13
Δ5	EPA	(20:5) 5, 8, 11, 14, 17	Fish oils	12, 13
	Columbinic	18:3 5t,9c,12c	Seed oils	14
Δ6	Petroselinic	18:1 6c	Seed oils	14
	γ-linolenic	18:3 6c,9c,12c,	Seed oils	15

TABLE 19.10
Distribution of DHA and EPA Between α and β Chains by NMR Spectroscopy

	EPA (C-2)			DHA (C-1)		
	wt%	α	β	wt%	α	β
S. Af. anchovy	23.3	86	14	8.6	36	64
Chilean	11.6	62	38	13.0	31	69
Menhaden	13.7	67	33	12.9	39	61
HBF[a] fish oil	10.9	53	47	10.9	23	77

Source: Gunstone, F.D., and S. Seth, Chem. Phys. Lipids 72:119–126 (1994).
[a]A commercial fish oil from HBP Ltd. (Fraserburgh, Scotland) [See JAOCS 70:133–138 (1993)].

Seed oils of the *Umbelliferae*, characterized by the presence of high proportions of petroselinic acid (6c-18:1), are also amenable to study by ^{13}C NMR spectroscopy, and some typical results are given in Tables 19.11 and 19.12. Table 19.11 contains the important C-1–C-3 chemical shifts for one petroselenic acid containing oil and Table 19.12 gives the content of this acid and its distribution between the α and β chains for a number of *Umbelliferae* oils. These results are based on the intensities of the signals for carbon atoms C-1–C-3 (14).

TABLE 19.11
Chemical Shift (ppm) for Glycerol Esters of Petroselinic and Other Acids (Saturated and Δ9) in Carrot Seed Oil

	C-1		C-2		C-3	
	α	β	α	β	α	β
Petroselinic	173.05	172.66	33.95	34.11	24.51	24.54
Other	173.16	172.76	34.05	34.21	24.87	24.90

Source: Gunstone (14).

TABLE 19.12
Content (%) of Petroselinic Acid and Its Distribution Between α and β Chains in Seed Oils of the *Umbelliferae*

	Carrot	Celery	Parsnip	Parsley	Celery	Coriander
Petroselinic acid (mol%)						
C-1	70.9	69.3	60.1	79.8	38.9	75.8
C-2	70.3	67.9	60.0	77.7	41.6	nd[b]
C-3	70.5	68.1	58.2	79.0	nd[b]	74.8
Olef[a]	70.5	66.7	58.7	78.5	36.8	75.0
Petroselinic acid (mol% in α and β chains from C-1 and C-2 signals)						
α {C-1	83	78	82	82	34	90
α {C-2	81	73	78	75	34	nd[b]
β {C-1	48	54	16	77	47	49
β {C-2	51	59	25	81	54	53

[a]P-6/O-9 ratio.
[b]Not determined.
Source: Heimermann et al. (16).

Oils Containing Short-Chain Acids

In the common C_{16} and C_{18} acids, the methyl and carboxyl ends of the chains are too distant to influence each other so that, for example, there is no significant difference in the chemical shifts for the C-1 to C-3 and ω-1 to 3 carbon atoms in myristic (C_{14}), palmitic (C_{16}), and stearic (C_{18}) esters. However, with esters of the C_8 acid and lower homologues the two ends of the molecule influence each other and produce signals which differ from those in the longer chain compounds. This is illustrated with figures taken from a paper by Lie Ken Jie et al. (Table 19.13 [9]) and from a spectrum of butter fat containing C_4, C_6, and C_8 glycerol esters (Table 19.14 [20]). Through the use of the chemical shifts observed in butter fat for the C_4–C_8 acids, it is possible to detect the presence or absence of this material in fat blends (17).

Oils Containing Oxygenated Acid

The presence of an oxygenated group (hydroxy, acetoxy, oxo, or epoxy) in a fatty acid chain produces signals characteristic of the carbon atom(s) bearing the oxygenated function and affects the chemical shifts of nearby carbon atoms. Some values from the literature are summarized in Table 19.15. It is usually possible to recognize the oxygenated function and to fix its position in the acyl chain.

TABLE 19.13
Chemical Shifts (ppm) of Some Synthetic Short-Chain Glycerol Esters and Glycerol Tristearate

	$(4:0)_3$	$(6:0)_3$	$(8:0)_3$	$(18:0)_3$
C-1	173.14	173.31	173.29	173.27
	172.74	172.90	172.89	172.88
C-2	35.94	34.03	34.08	34.07
	36.09	34.19	34.24	34.24
C-3	—	24.56	24.88	24.91
		24.59	24.92	24.94
C-4	—	—	29.09	29.16
			29.05	29.12
C-5	—	—	28.94	29.32
			28.96	29.34
ω-3	—	31.26	31.68	31.98
		31.22	31.70	
ω-2	18.37	22.31	22.63	22.73
	18.40			
ω-1	13.63	13.90	14.07	14.13
	13.57			

Source: Lie Ken Jie et al. (9).

TABLE 19.14
Chemical Shifts (ppm) and Intensities of Signals from Butterfat

C-1α	173.26	3.04	ω-1 n-9, sat.	14.14	8.51
C-1α 4:0	173.06	0.57	ω-1	14.03	0.35
C-1β	172.85	1.84	ω-1 6:0	13.91	0.71
			ω-1 4:0	13.63	1.26
C-2α 4:0	35.91	1.51			
C-2β	34.22	3.90	ω-2 n-9, n-6, sat.	22.72	9.17
C-2α	34.05	5.73	ω-2 6:0	22.32	0.73
			ω-2 4:0	18.36	1.50
C-3α,β	24.88	7.25			
C-3 6:0	24.55	0.81	ω-3 n-9, sat.	31.96	8.34
			ω-3 n-7	31.79	0.65
9c	130.00	1.69	ω-3 8:0	31.69	0.41
9c	129.70	1.43	ω-3 6:0	31.26	0.67
11t	130.30	0.38			
11t	130.29	0.40	allylic t	32.64	1.04
			c	27.24	2.44
			c	27.19	2.32
			cc	25.63	0.33

Source: Gunstone (1,17).

When the oxygenated function occurs with a double bond then the two functional groups act together to produce several unique references (18). This is illustrated in Table 19.16 for ricinoleic ester (C8–C14) in castor oil (~90%) and for vernolic ester (C8–C18) in *Vernonia galamensis* seed oil (70–75%) (Table 19.16) (19).

TABLE 19.15
Changes in Chemical Shift of an Alkyl Chain Resulting from the Presence of an Oxygenated Function

CHX[a]	CHX	α	β	γ
CH—CH (epoxide)				
cis	+27.4	−1.71	−2.93	−0.38
trans	+28.7	+2.57	−3.50	−0.23
CHOH	+42.2	+7.8	−4.0	+0.06
CHOAc	+44.7	+4.4	−4.4	−0.2
C = O	+182.1	+13.1	−5.8	−0.4

Source: Gunstone (1).
[a]Carbon atom(s) llinked to the oxygenated functions.

TABLE 19.16
Chemical Shifts (ppm) in Ricinoleate and Vernolate Different from Nonoxygenated Esters

Carbon Atom	Ricinoleic Ester from Castor Oil	Vernolate Ester from *Vernonin galamensis* seed oil
8	27.38	27.79[a]
9	125.39	124.05
10	133.00	132.40
11	35.37	26.33[b]
12	71.44	57.05
13	36.85	56.40
14	25.73	27.40[a]
15	—	26.28[b]
16	—	31.77
17	—	—
18	—	14.01

[a,b]These pairs may be interchanged, chemical shifts are not significantly different from other esters also present.
Source: Gunstone (2,19). (Later work suggests that some of these assignments may be in error—Lie Ken Jie, M.-S.F., and A.K.L. Cheng, *Nat. Prod. Letters* 3:65–68 (1993) and unpublished work from the same laboratory.)

Epoxidized vegetable oils (palm super-olein, soybean oil, and linseed oil) are produced on a large scale for use as plasticizers and stabilizers in polyvinyl chloride (PVC). When epoxidized, the oleic, linoleic and linolenic acids in these oils are converted to mono-, di-, and triepoxides. From the ω-1 to 3 signals in the spectra of these epoxidized oils, it is possible to distinguish acyl chains with epoxide groups at 9,10-; 12,13-; and 15,16- positions (Table 19.17). A comparison of the chemical shifts in Table 19.17 with those for the effect of a double bond on the ω-1 to 3 signals (Table 19.3) shows that the effect of the epoxide function is different from (and usually greater than) that of the *cis* double bond.

TABLE 19.17
Chemical Shifts (ppm) for ω–1–3 Carbon Atoms in Nonepoxidized and Epoxidized Oils

	ω-1	ω-2	ω-3
Saturated	} 14.12	27.70	31.93
9,10-ep		22.68	31.86
12,13-ep	13.99	22.58	31.68
15,16-ep	} 10.61[a]	21.25[a]	[b]
	10.50	21.16	

[a]Diastereoisomers.
[b]Epoxide carbon atoms.
Source: Gunstone (19).

TABLE 19.18
Chemical Shifts for Selected Carbon Atoms in Unbranched, Iso-, and Anteiso Acids Present in Wool Wax Fatty Acids[a]

	Unbranched	Iso	Anteiso
ω-1	14.14	22.68	11.42
ω-2	22.68	27.98	—
ω-3	31.95	39.08	34.41
ω-4	—	27.45	36.66
ω-5	—	—	27.14
ω-6	—	—	30.06
brhr Me	—	22.68	19.23

[a]The intensities of these signals indicate the presence of unbranched acids (~30%), isoacids (~32%), anteisoacids (~32%), and others (6%).
Source: Adapted from Gunstone (20).

Cyclic and Branched Chain Acids

The ^{13}C NMR spectrum of a branched chain acid frequently provides information about the nature and position of the branched groups. The author has reported data for a wide range of substituted acids (20). When the results of this study were applied to wool-grease fatty acids, it was possible to recognize and quantitate the n, iso-, and anteisoacids (Table 19.18 [20]).

Kapok seed oil contains two cyclopropene acids, sterculic and malvalic in Figure 19.2. The characteristic chemical shifts of these acids resulting from the cyclopropene

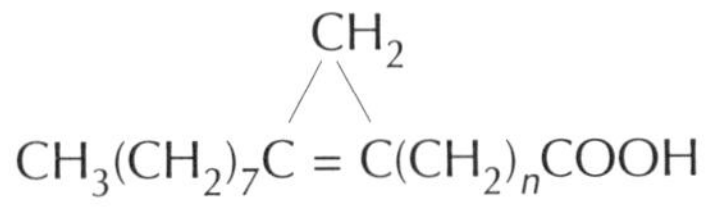

$$CH_3(CH_2)_7 \overset{\overset{\displaystyle CH_2}{\diagup \diagdown}}{C} = C(CH_2)_n COOH$$

Figure 19.2. Structure of sterculic and malvalic acids. For sterculic acid n = 7 and malvalic acid n = 6.

TABLE 19.19
Chemical Shifts (ppm) in Sterculate and Malvalate Observed in Kapok Seed Oil Different from Common Saturated and Unsaturated Acids

Sterculic	C-7	C-8	C-9	(CH$_2$)	C-10	C-11	C-12
ppm	27.37	26.01	109.20	7.42	109.47	26.07	27.37
Malvalic	C-6	C-7	C-8	(CH$_2$)	C-9	C-10	C-11
ppm	27.43	25.96	109.11	7.42	109.56	26.07	27.43

Source: Gunstone (4).

unit are listed in Table 19.19. The difference in the chemical shift for the olefinic carbons is 0.27 for sterculic acid and 0.45 for malvalic acid. These values are indicative of the position of the cyclopropene system (compare with Table 19.5).

Partially Hydrogenated Vegetable Fats

Unsaturated vegetable oils are commonly used in a partially hydrogenated form. Hydrogenation with a heterogeneous catalyst is a complex reaction in which stereomutation and double-bond migration accompany reduction. The product therefore contains *cis* and *trans* forms of 18:1 isomers with double bonds in several positions and is difficult to analyze completely. Total *trans* unsaturation is an important parameter in the specification of partially hydrogenated fats and can be measured by infrared spectroscopy or less satisfactorily by gas chromatography. There is no easy way to assess double-bond migration. This latter requires oxidative fission and recognition of all the mono- and difunctional scission products. This aspect of the analytical problem is frequently ignored. The author has recently demonstrated that ^{13}C NMR spectroscopy can contribute to the study of partially hydrogenated fats in three ways (11), the ω-1 to 3 signals provide information on the *cis* and *trans* isomers of the Δ15–Δ12 18:1 esters (see Table 19.7), allylic signals distinguish between *cis* and *trans* isomers and can be used to determine the proportion of these two classes of unsaturated esters (see Table 19.6), and the olefinic signals give information about and can furnish a semiquantitative estimate of most of the 18:1 isomers (see Table 19.5). For example, a study of the olefinic signals of a rapeseed oil hydrogenated to melting point 43°C indicates the presence of the following 18:1 isomers 5*t* (2.3%), 6*t* (4.8), 7*t* (7.6), 8*t* (13.8), 9*t* (13.7), 10*t* (15.5), 11–13*t* (18.3), 14–15*t* (3.4), 8*c* (3.4), 9*c* (7.2), 10*c* (7.2), 11–13*c* (2.8); total *trans* 79.4%, total *cis* 20.6%. Further details are found in Gunstone (11).

Glycerol and Propylene Glycol Esters

Certain esters of glycerol and of propylene glycol are frequently used as food emulsifiers. These are generally analyzed by gas chromatography, probably after silylation of free hydroxyl groups, but it has been shown that ^{13}C NMR spectra also provide useful insights into the nature of the mixtures present in commercial products (Gunstone, [21] and unpublished data).

Glycerol can have five different ester forms, two monoacylglycerols (1- and 2-), two diacylglycerols (1,2- and 1,3-), and one triacylglycerol. Each has signals with characteristic chemical shifts corresponding to the C-1 and C-2 carbon atoms and to the three glycerol carbon atoms. The last three of these are generally the most useful for analytical purposes. The three "symmetrical" esters (2-mono-, 1,3-di-, and the triacylglycerol) show only two glycerol signals in a ratio of 2:1 corresponding to two α and one β carbon atoms, while the "nonsymmetrical" esters (1-mono- and 1,3-diacylglycerols) have three different glycerol signals in a 1:1:1 ratio. The chemical shifts of these glycerol carbon atoms in $CDCl_3$ and in a $CDCl_3$, CD_3OD (2:1) solution are listed in Table 19.20. The intensity data can be used in a semiquantitative manner to analyze mixtures of glycerol esters, and two examples are given in Table 19.21. The spectroscopic data give results on a mol% basis, but these can be adjusted to the more common wt% basis.

Acetylated monoglycerides are also produced on a commercial scale for use as food emulsifiers. The study of a number of samples by ^{13}C NMR spectroscopy indicated the presence of six components. The structures found are listed in Table 19.22 along with eight significant chemical shifts. These are assigned to the three glycerol carbon atoms, C-1 in the long acyl chain, and to the two acetate carbon atoms in each of the α and β chains.

Propylene glycol (propane-1,2-diol) forms three types of esters, the 1- and 2-monoacyl compounds and the 1,2-diacyl compound. Mixtures of these three show important signals corresponding to the three-carbon propane unit (P-1–3) and C-1–3 of the acyl chains. The chemical shifts listed for these in Table 19.23 and Table 19.24 indicate the composition of two commercial examples. These were mainly monoester and diester, respectively. The data in Table 19.24 show the composition

TABLE 19.20
Chemical Shifts (ppm) for Glycerol Carbon Atoms in Glycerol Esters in $CDCl_3$ and in $CDCl_3$-CD_3OD (2:1) Solution

Glycerol Ester		$CDCl_3$	$CDCl_3$-CD_3OD[a]
β-carbon atom			
	2-mono-	74.97	75.36
	1,2-di-	72.25	72.40
	1-mono-	70.27	70.24
	tri-	68.93	69.41
	1,3-di-	68.23	67.71
α-carbon atom			
	1-mono-	65.04	65.58
	1,3-di-	65.04	65.36
	1-mono-	63.47	63.46
	1,2-di-	62.20	62.92
	tri-	62.12	62.50
	2-mono-	62.05	61.06
	1,2-di-	61.58	60.82

[a]Chemical shifts for glycerol in $CDCl_3$-CD_3OD are 72.78(β) and 63.66(α).
Source: Gunstone (3,21).

TABLE 19.21
Composition of Two Mixtures of Glycerol Esters Determined by NMR and Compared with Specifications

Glycerol Ester	Sample A (wt%)		Sample B (wt %)	
	NMR	Specification	NMR	Specification
1-mono	46.2 } 47.4	48.9	42.9 } 46.2	45.3
2-mono	1.2		3.3	
1,2-di	0.5 } 5.7	7.5	14.0 } 43.5	40.5
1,3-di	5.2		29.5	
Tri	46.9	42.4	10.3	13.2
Other	—	1.2	—	—

Source: Gunstone (21).

TABLE 19.22
Chemical Shifts (ppm) of Acetylated Monoacylglycerols in $CDCl_3$

Carbon Atom	1	2	3	4	5	6
G-1	65.15	62.07	65.00	62.00	62.46	62.35
G-2	70.26	72.39	68.14	69.16	72.03	68.76
G-3	63.39	61.40	65.26	62.33	61.32	62.35
C-1	174.31	173.84	173.96	173.38	173.52	173.01
$COCH_3(\beta)$ {	—	21.00	—	20.88	—	—
	—	170.98	—	170.16	—	—
$COCH_3(\alpha)$ {	—	—	20.79	20.69	20.74	20.69
	—	—	171.10	170.57	170.65	170.57
	OCOR	OCOR	OCOR	OCOR	OAc	OAc
	OH	OAc	OH	OAc	OCOR	OCOR
	OH	OH	OAc	OAc	OH	OAc

Source: Gunstone (4), unpublished data.

TABLE 19.23
Chemical Shifts (ppm) for Propylene Glycol Esters

	P-1[a]	P-2[a]	P-3[a]	C-1[a]	C-2[a]	C-3[a]
1-monoacyl	69.46	66.13	19.19	173.97	34.23	24.97
2-monoacyl	65.92	71.77	16.25	173.97	34.58	25.03
1,2-diacyl	65.92	67.98	16.54	{ 173.51	34.51	24.97
				173.24	34.19	25.03

[a]P-1–3 refer to carbon atoms in the propane unit, C-1–3 refer to carbon atoms in acyl chain(s).
Source: Gunstone (4), unpublished data.

TABLE 19.24
Composition of Propylene Glycol Monoester and Diester Samples by ^{13}C NMR Spectroscopy

	Monoester		Diester (~80%)		
Results Based On	1-Acyl	2-Acyl	1-Acyl	2-Acyl	1,2-Diacyl
C-1	—	—	—— 22 ——		78
C-2	69	31	—	—	—
C-3	66	34	—	—	—
P-1,P-2	71	29	17	7	70[a]
P-3	72	28	18	7	75
Average	70	30	17	7	74

[a]Also 6% unidentified.
Source: Gunstone (4), unpublished data.

of the mixtures based on different NMR signals. The results do not vary greatly, and quantitation could be improved by an adjustment of the protocol for collecting the spectra and by a study to determine which signals give the best results.

Phospholipids

The important signals in the phospholipid spectra are those associated with C-1 and C-2 of the acyl chains and with the carbon atoms of the head groups. These include the carbon atoms in glycerol and in the choline or ethanolamine unit. In phospholipids, the signals for the *sn*-3 glycerol carbon and for OCH$_2$ are split by phosphorus and show two-bond coupling, while the *sn*-2 glycerol carbon atom and CH$_2^+$,N show three-bond coupling. The former pair have lower coupling constants than the latter pair. As the chemical shifts in Table 19.25 show, it is possible to distinguish between triacylglycerols, phosphatidylcholines, and phosphatidylethanolamines.

TABLE 19.25
Chemical Shifts (ppm) for Selected Signals Observed in a Mixture of Triacylglycerols, Phosphatidylethanolamines, and Phosphatidylcholines Derived from Soybean Oil

	Triacylglycerols	Phosphatidylethanolamines	Phosphatidylcholines
C-1α	173.91	174.11	174.14
C-1β	173.49	173.74	173.79
G-1	62.46	62.86	62.86
G-2	69.38	70.72 / 70.62	70.67 / 70.62
G-3	62.46	64.06 / 63.99	63.88
CH$_2$O	—	61.98 / 61.91	59.44 / 59.38
CH$_2$N$^+$	—	40.87	66.65
Me$_3$N$^+$	—	—	54.30

Source: Gunstone (2), unpublished data.

However, these compounds are probably better studied by ^{31}P NMR spectroscopy, since each phospholipid class contains only one phosphorus atom with a characteristic chemical shift (1,22).

Acknowledgment

The spectroscopic studies were supported financially by Karlshamns.

References

1. Gunstone, F.D. (1993) in *Advances in Lipid Methodology—Two,* Christie, W.W., The Oily Press, Dundee, pp. 1–68.
2. Gunstone, F.D., (1994) *Progress in Lipid Research, 33:*19–28.
3. Gunstone, F.D., *Applications of Magnetic Resonance to Food Science,* Webb, G., in press.
4. Gunstone, F.D., *Developments in the Analysis of Fats and Other Lipids,* Tyman, J.M.P., and M.H. Gordon, Royal Society of Chemistry, Cambridge (UK), pp. 109–122.
5. Bus, J., Sies, I., and Lie Ken Jie, M.S.F. (1976) *Chem. Phys. Lipids, 17:* 501–518.
6. Bus, J., Sies, I., and Lie Ken Jie, M.S.F. (1977) *Chem. Phys. Lipids, 18:* 130–144.
7. Gunstone, F.D., Pollard, M.R., Scrimgeour, C.M., Gilman, N.W., and Holland, B.C. (1976) *Chem. Phys. Lipids, 17:* 1–13.
8. Gunstone, F.D., Pollard, M.R., Scrimgeour, C.M., and Vedanayagam, H.S. (1977) *Chem. Phys. Lipids, 18:* 115–129.
9. Lie Ken Jie. M.S.F., Lam, C.C., and Yan, B.F.Y. (1992) *J. Chem. Research, S:* pp. 12–13, *M:* pp. 250–272.
10. Wollenberg, K.F. (1990) *J. Am. Oil Chem. Soc., 67:* 487–494.
11. Gunstone. F.D. (1993) *J. Am. Oil Chem. Soc., 70:* 965–970.
12. Aursand, M., Rainuzzo, J.R., and Grasdalen. H. (1993) *J. Am. Oil Chem. Soc., 70:* 971–981.
13. Gunstone, F.D. (1991) *Chem. Phys. Lipids, 59:* 83–89.
14. Gunstone, F.D. (1991) *Chem. Phys. Lipids, 58:* 159–167.
15. Gunstone, F.D. (1990) *Chem. Phys. Lipids, 56:* 201–207.
16. Heimermann, W., Holman, R.T., Gordon, D.T., Kowalshyn, D.E., and Jensen, R.G. (1973) *Lipids, 8:* 45–47.
17. Gunstone, F.D. (1993) *J. Am. Oil Chem. Soc., 70:* 361–366.
18. Pfeffer, P.E., Sonnet, P.E., Schwartz, D.P., Osman, S.F., and Weisleder, D. (1992) *Lipids, 27:* 285–288.
19. Gunstone, F.D. (1993) *J. Am. Oil Chem. Soc., 70:* 1139–1144.
20. Gunstone, F.D. (1993) *Chem. Phys. Lipids, 65:* 155–163.
21. Gunstone, F.D. (1991) *Chem. Phys. Lipids, 58:* 219–224.
22. De Koch, J. (1993) *Fat Sci. Technol., 95:* 352–355.

Chapter 20
Sensory Assessment of Fats and Oils

Renée Raoux and Odile Morin

Institut des Corps Gras, Iterg, Rue Monge, Parc Industriel, F33600 Pessac, France.

Introduction

Sensory analysis has significantly improved in recent years due to the development of new methodologies such as descriptive and quantitative sensory profiles, and data analysis through sophisticated statistical and mathematical methods leading to more accurate appraisals. These new methodologies include Factorial Correspondence Analysis (FCA), Principal Component Analysis (PCA), and Canonical Variate Analysis (CVA) (1); standardization of methods; standardization of testing conditions, such as sensory evaluation, laboratory, and sample presentation; and use of computers to help sensory operation management record and process scores and descriptions. Sensory analysis can be a valuable tool and, when correctly used, the ultimate one to assess food, particularly fat and oil quality. Depending on the fat products considered: refined oils; virgin olive oils; or elaborate fats, such as margarines, spreads, and low-fat butters, the purpose will be different. Since sensory assessment is based on product tasting and analysis of the sensory ratings from the panel, working sessions should obey standardized conditions to obtain objective results.

Practical Organization of Sensory Measurement

Panel

Sensory assessment of fats and oils requires assembling a motivated team to differentiate typical flavors and measure their intensity. Assessors should not have negative feeling toward the products they are testing. This implies training and selection of the panelists according to the directives of the international standard methods ISO 8586-1 and 2 (23), ASTM STP 758 (4), or AOCS Cg 2-83 (5). This training will allow the assessors to become familiar with the products and will develop more accurate perceptions (qualitative and quantitative) of the specific odors and flavors they are asked to detect.

Test Rooms: Panel Room

The dedicated room or laboratory must avoid any possibility of error caused by an unfavorable environment: noise, light, or other panelist influence. The installation of individual booths must follow the guidelines described in standard recommendations. The specific facilities must include regulated temperature equipment for sample presentation, as in the case of refined oils, virgin olive oils, and solid fats such as tallow, lard, and palm oil; and an area regulated at 20°C for elaborate fats, such as margarines and spreads, in order to perform spreadability evaluations.

Working Session

To produce the best conditions for the assessors, working sessions must be organized in midmorning and must not last more than 30 min to avoid saturation symptoms. During the session, the panelists are asked to drink some warm water before tasting and between each sample; this helps to maintain equal sensitivity for each product tested. A maximum number of 6–8 test samples can be proposed in the case of difference tests but only three in descriptive tests.

Samples Presentation

The test portions must be codified and all presented at the same temperature, in the same quantity, and the same type of vessel. They are randomized and offered in a different order to each panelist. In case of a replicate test, each judge will receive the samples in a different order than in the first test. For refined oils and virgin olive oils, the best conditions for sensory evaluation are the use of "brandy" or balloon glasses; a sample volume of 20 mL; and the use of an aluminum foil lid (or clock-glass) to close the glass, allowing the concentration of volatiles in the headspace (Figure 20.1). The tasting temperature is $45 \pm 2°C$ for refined oils and $28 \pm 2°C$ for virgin olive oils. Elaborate fats are presented in small pieces, all of the same shape and size, and kept at 7°C in a refrigerator until tasting.

Data Collection and Processing

Data collection and processing can be done either on score sheets with data analysis (statistical software); or by means of a computerized system dedicated to sensory assessment and research, where individual screens and/or keyboards are used by the judges for recording scores and descriptions, with the results analyzed through a central computer. Such systems significantly improve sensory analysis management and contribute to panel motivation. They simplify panel management and training, sample management, organization of the working sessions, and obtaining and processing the data.

For each kind of test, obtaining specific statistical processing is required.

1. Difference tests: Difference tests (ISO 6586) are based on application of the binomial expansion. Statistical tables determine whether tests are significant or not, depending on the confidence level (95%), the number of assessors, and the number of answers expected to make the test valid.

Sensory Assessment of Fats and Oils

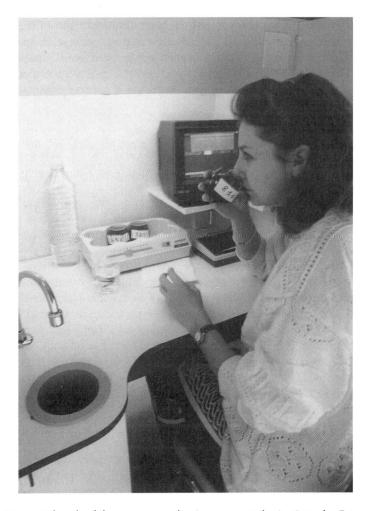

Figure 20.1. A booth of the sensory evaluation room at the Institute for Fats and Oils (ITERG), with blue balloon glasses for olive-oil assessment, a device for sample heating, and an individual computer (screen and keyboard).

2. Ranking tests: French standard method (NF ISO 8587) recommends the Friedman test.
3. Intensity tests: In this presentation, the profiling of quantitative data will be considered only if they follow a gaussian distribution. The interpretation of results is done by parametric tests. The Student's "t" test is used to compare two products and the analysis of variance (ANOVA) is used to compare several products. After an analysis of variance, the products can be gathered in subgroups by using the Newman-Keuls test.

A product is usually characterized by several attributes, each of them being rated (intensity scale) independently. Multivariate analysis, such as FCA and PCA, is used to make a complete analysis of an attribute/product matrix. These methods can also be used during the attribute-selection stage when establishing a sensory profile.

Difference Tests

Difference tests are particularly well adapted to the assessment of fats during their processing, because in this case taste differences are slight. They are used for controlling the refining efficiency and the stability of manufactured fats, evaluating the shelf-life of the products, testing a new technology or a change in raw materials, and controlling raw materials that enter the formulation of elaborate products. These methods have the advantage of being standardized, easy to run, and requiring no calculation thanks to the statistical tables available in all sensory manuals.

Considering the low number of test portions a person is able to taste during a session, triangular tests, duo-trio and pair-comparison tests are used most frequently for fat and oil assessment. Assessors must give an answer: indeed, since the differences are slight, not answering would be too easy a choice. A panel of 20 assessors is generally required, because the probability of uncertainty is 50% in the case of pair-comparison or duo-trio tests and 30% in the case of a triangular test. Panelists do not need to be specially trained (only a few sessions are necessary to become accustomed to the product), but they must show ability to detect the differences between samples.

Descriptive Quantitative Analysis: Sensory Profile

This method is used to get an exhaustive description of the products. Assessment of a complex sensory magnitude, such as appearance, flavor, odor, and texture, is hard to describe through instrumental analysis, and requires a specific methodology. The methodology is based on the research of appropriate descriptors. The aim is to detail all characteristics of the samples (or groups of samples) with as few and pertinent words as possible. The result is a precise identification of the product, reproducible and readable by anybody. The profiling method is used from qualitative and quantitative points of view.

Identification of Descriptors for Refined Oil Flavors

Identification of descriptors for establishing a sensory profile for refined oil flavors whatever the oxidation stage is based on the methodology described by Barthelemy (6), which is about to be standardized (ISO/Dis 11035). The experiment includes different stages.

The first stage is the selection of the refined oils to enable the assessors to distinguish all possible qualitative differences detectable in products for which the profile is to be constructed. Three types of oils were selected: peanut, sunflower, and rapeseed; each oil type had three oxidation levels.

In the second stage, the assessors are asked to generate the maximum number of terms to describe the sensations produced by these oils whether visual, tactile, olfactory, or gustatory. The ITERG panel established large lists of terms to describe the oils. They included 16 terms for appearance, 34 for odor, 31 for taste, and 9 for texture. The descriptors were too numerous. Some terms, such as hedonic, quantitative, and irrelevant attributes, might be eliminated. These lists also included synonyms.

In the third stage, two successive reductions of the number of descriptors are made for each group of terms (aspect, odor, taste, and texture). In the first reduction, the terms were classified by intensity and frequency. In the second, multivariate analysis was used to group synonymous descriptors or to eliminate descriptors which contributed little to show differences between the oils. After these two reductions, two terms for aspect, six for odor, ten for taste and one for texture, for a total of nineteen terms were obtained. This was still too much. Then PCA (Figure 20.2) and FCA were applied to the overall list of nineteen terms. These statistical analyses enabled the elimination of appearance and texture terms and showed the redundancy of taste and odor terms.

The final list of attributes to describe the flavors of refined oils includes six terms: butter, painty, fishy, rancid, fruity (sui generis, seed), and grassy/beany (7,8). The intensity of each attribute is assessed on a continuous unipolar scale ranging from absent or extremely weak to strong. The sensory profiles of two refined oils obtained by using the generated list of descriptors are given in Figure 20.3.

A corn oil and a rapeseed oil are compared through the Student's "t" test. If the "t" value for the panel is above the theoretical t value for three descriptors, then the fishy, rancid, and grassy intensities are significantly different for the two oils (9). To correctly use this list, the panel must be trained to recognize each flavor by spiking a tasteless oil with either chemical substances or other oils presenting the studied attribute (Table 20.1), and to quantify the flavor on the intensity scale with the help of different dilutions of samples, each of them presenting only one specific descriptor (8).

Organoleptic Assessment of Virgin Olive Oil

The profile sheet is divided in two parts: the list of descriptors which includes qualities, such as olive, fruity, bitter, and pungent; and defects, such as musty, sour, and fusty; and the rating table based on overall marks from nine to one (Table 20.2). The overall mark takes into account both the fruity intensity and the maximum defect intensity. For instance, if the maximum defect intensity is slight and the fruitiness is rather imperfect, the judges must give it five points.

On November 1, 1992, the European Community introduced in its regulation the sensory assessment based on the IOOC method for quality control of olive oils. This law went into effect on May 1, 1993. In the international trade standard apply-

PCA : CORRELATION CIRCLE

Plane 1, 2 :   **68.2 % (axis 1, horizontal ; axis 2, vertical)**

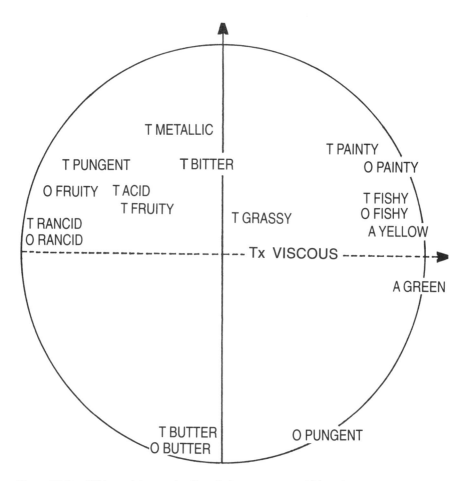

Figure 20.2. PCA applying to the list of nineteen terms. Abbreviations: Appearance (A), two attributes; odor (O), six attributes; taste (T), ten attributes; and texture (Tx), one attribute. *Source:* Raoux et al. (10).

ing to olive oils (2), "the virgin olive oils" are classified in four groups in accordance to their organoleptic characteristics and free acidity (Table 20.3). For example, an olive oil will be qualified as "extra virgin," if the overall mark ≥ 6.5 and free acidity ≤ 1.

The IOOC controlled the efficiency of this method through a collaborative study run in 1992: 10 panels from five countries tested 12 samples of virgin olive

Sensory Assessment of Fats and Oils 271

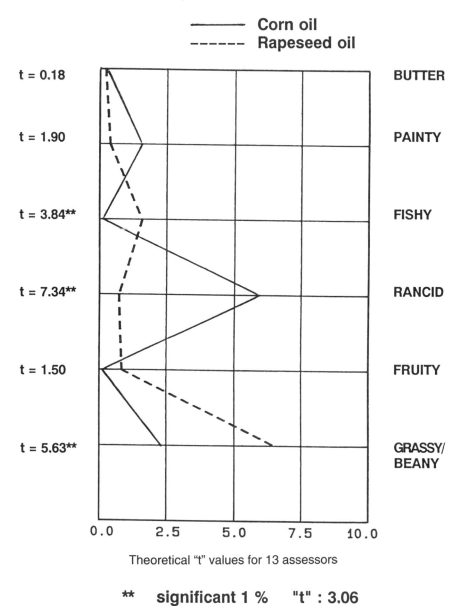

Figure 20.3. Sensory profiles of a corn oil and a rapeseed oil; graphical comparison of mean score/10 through the Student's "t" test.

TABLE 20.1
Training of the Oil Panelists: Reference Standards for Individual Flavors

Descriptors	Chemical Substance or Oil
Butter	200 ppb of diacetyl
Painty	3% of linseed oil
Fishy	5% of cod liver oil
Rancid	500 ppb of hexanal
Fruity	Refined oil + 2% crude sunflower oil
Grassy/beany	Refined soybean oil (after reversion)

TABLE 20.2
Virgin Olive Oil List of Attributes: IOOC Method

Qualities	Defects	Scale
Olive fruity	Sour	0 None whatsoever
Apple	Rough	1 Barely
Other ripe fruit	Metallic	2 Slight
Green (leaves, grass)	Musty	3 Average
Bitter	Muddy sediment	4 Great
Pungent	Fusty	5 Extreme
Sweet	Rancid	

Rating Table

Defects (Intensity)	Characteristics	Overall Mark/9
	Olive fruity	9
None	Fruitiness of other	8
	fresh fruit	7
Barely	Weak fruitiness	6
	of any type	
Slight	Rather imperfect	5
	fruitiness	
Average	Clearly imperfect	4
	odors and tastes	
Great/	Inadmissible odors	3
Extreme	and tastes for	2
	consumption	1

oils. For each sample, statistical analysis of the results obtained by the various panels gave the following margins of error for overall marks:

Overall mark repeatability ± 0.17
Overall mark reproducibility ± 0.51
Overall mark confidence limits ± 1.0

The value of the confidence limits is very important; if a panel gives a mean score of six to an olive oil, the score should fluctuate between five and seven (6 ± 1) or the quality label will be different.

TABLE 20.3
International Trade Standard Applying to Olive Oils

Quality Criteria	Organoleptic Characteristics	Free Acidity (%)
Extra virgin olive oil	≥ 6.5	≤ 1.0
Fine virgin olive oil	≥ 5.5	≤ 1.5
Semifine virgin olive oil	≥ 3.5	≤ 3.3
Lampante virgin olive oil	< 3.5	> 3.3

In order to face the practical difficulties met in the application of this recommended practice (10), the IOOC decided to review the score sheet and clarify it, with the help of panel supervisors of each country. The judges will have to indicate whether the oil shows some defects or not, and quantify their intensity. The panel supervisor then will calculate the overall mark of the oil by the means of each panelist's assessment.

Room Odor Tests

Some ITERG's recent works (11) compared the odor quality of two Canadian varieties of rapeseed oils (Westar, 11% linolenic acid; low-linolenic, 3% linolenic acid) with a French one (Bienvenu, C18:3, 7%) during frying. Room-odor tests enable the evaluation of odors developed while frying potatoes (french fries), under domestic conditions. In this experiment, room-odor tests included eight fryings over a 2-day period with panel assessments on the first, fourth, and eighth frying.

Each panelist was first asked to determine the overall strength of odor on a five point intensity scale ranging from "unnoticeable frying odor" to "very poor odor" (12). Second, the judges were asked to determine what characteristic odors they had perceived among a list of possible descriptors: attribute 1—nutty, sweet, fruity; attribute 2—grassy, beany; attribute 3—buttery, hydrogenated, tallow; attribute 4—burnt, acrid/pungent, rancid; attribute 5—painty, plastic, fishy; and to rate the intensity as none, slight, moderate, or strong (12).

The overall strength of odor (scores) of the three rapeseed oils compared with a sunflower oil used as a reference are shown in Figure 20.4. For French rapeseed, the scores were always below a mean value of five from the first frying. Westar had scores that were similar to those of the French rapeseed. The scores obtained for low-linolenic were significantly higher than the two other rapeseed oils, and these scores were very close to those obtained for the sunflower oil fryings.

The intensities of the odors for the different fryings can be seen in Figure 20.5. For French rapeseed, the painty, plastic, fishy odors and burnt, acrid, rancid ones were predominant from the first frying. These odors increased during the fryings, and the fruity odor was always weak. The odor profile of Westar was quite similar to that of French rapeseed. A fruity odor was dominant in the low-linolenic, and the fishy-painty odors were significantly reduced. The Canadian low-linolenic rapeseed

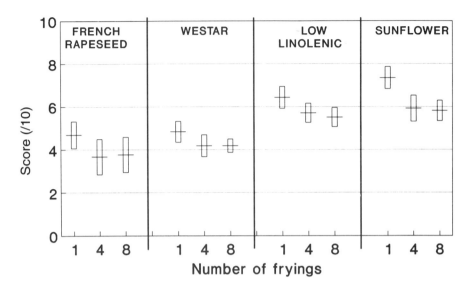

Figure 20.4. Room odor scores of three refined oils compared with sunflower oil. Tests after the first, fourth, and eighth frying. Each bar represents the mean value ± SEM (standard error mean value; $P < 0.05$; n ranges from 15–25). *Source:* Prevot, et al., (13).

oil with 3.1% linolenic acid has a significantly different frying behavior when compared with the other rapeseed oils (with higher C18:3 content). This difference is strongly related to the intensities of the "painty, plastic, fishy" odors and "burnt, acrid, rancid" odors.

Conclusion

Sensory evaluation, as described in this paper, has an important role to play in the fat and oil industry. It implies an environment specific for this kind of product with selected and trained assessors, sufficient facilities, and a common list of attributes. All these requirements are now well described in national and international standards or recommended practices, (2–7, 15, 16), manuals on sensory testing methods, and all publications on oil and fat assessment (17).

Descriptive and quantitative sensory evaluation can be considered as a time- and money-consuming method. Depending on its growth policy, a firm may either develop an internal sensory laboratory, or use the services of specialized organizations. Another solution is to perform the simplest tests (difference tests) and to subcontract the more complex ones (identification of attributes, sensory profile).

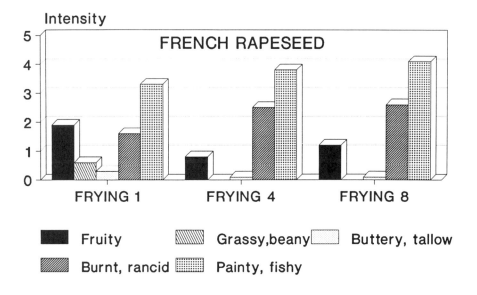

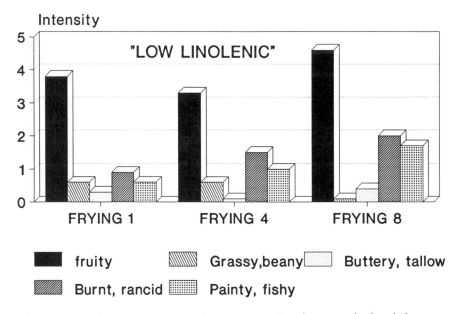

Figure 20.5. Characteristic room odor intensities: French rapeseed oil and "low-linolenic" oil. *Source:* Prevot et al., (13).

Acknowledgments

The authors would like to thank all the members of the panel, Laurence Puchaud, and ITERG's Documentation Department for their collaboration.

References

1. S.S.H.A. (1990) *Evaluation sensorielle manuel methodologique, Technique and Documentation Lavoisier,* Paris.
2. International Olive Oil Council (1993) *International Trade Standard Applying to Olive Oils and Olive-Pomace Oils,* COI/T.15/NC, 1 Rev. 6, June 10, 1993.
3. International Olive Oil Council (1992) *Organoleptic Assessment of Virgin Olive Oil,* COI/T.20/Doc. 3, Rev. 2, May 28, 1992.
4. ASTM (1981) *Guidelines for the Selection and Training of Sensory Panel Members,* STP 758, American Society of Testing and Materials, Philadelphia, PA.
5. ASTM (1968) *Manual on Sensory Testing Methods,* STP 434, American Society of Testing and Materials, Philadelphia, PA.
6. ASTM (1990) *Standard Practice for the Bulk Sampling, Handling, and Preparing of Edible Oils for Sensory Evaluation,* E1346, American Society of Testing and Materials, Philadelphia, PA.
7. AOCS (1989) *Official and Tentative Methods of the Arnerican Oil Chemists' Society,* 4th edn., Methods Cd 8-53, Cg 1-83, and Cg 2-83, The American Oil Chemists' Society, Champaign, IL.
8. Barthelemy, J. (1990) *Evaluation d'une Grandeur Sensorielle Complexe: Description Quantifiée; Evaluation Sensorielle, Manuel Méthodologique,* Tec. and Doc. Lavoisier, Paris, pp. 144–158.
9. Raoux, R. (1992) *Analyse Sensorielle des Huiles Raffinées: Manuel des Corps Gras,* Karleskind, A., Tec. & Doc. Lavoisier, 1992, Paris, 1419–1427.
10. Raoux, R., Diris, J., and Mordret, F. (1989) *First European Lipid Conference,* June 6–9, 1989, Angers.
11. O'Mahony, M. (1986) *Sensory Evaluation of Food. Statistical Methods and Procedures,* Marcel Decker, Inc., New York.
12. Fedeli, E. (1993) *La valutazione organolettica degli oli vergini di oliva,* Riv. Ital. Sostanze grasse, Vol. LXX, 81–85.
13. Prevot, A., Perrin, J.L., Laclaverie, G., Auge, Ph, and Coustille, J.L. (1990) *J. Am. Oil Chem. Soc.* 67: 161–164.
14. Prevot A., Desbordes, S., Morin, O., and Mordret, F. (1988) *Frying of Food, Principles, Changes, New Applications,* Varela, G., Arela, Bender, A.E., and Morton, I.D.
15. A.F.N.O.R. (1991) *Contrôle de la Qualité des Produits Alimentaires—Analyse Sensorielle,* 4th edn.,
16. Canada Department of Agriculture (1978) *Methods for Sensory Evaluation of Food,* Ottawa, Canada.
17. Warner, K. (1991) *Assessment of Fat and Oil Quality: Sensory Methodology, Analyses of Fats, Oils and Lipoproteins,* American Oil Chemists' Society, Champaign, Illinois, p. 344–386.

Chapter 21
Classical Chemical Techniques for Fatty Acid Analysis

J.-L. Sebedio

Institut National de la Recherche Agronomique, Station de Recherches sur la Qualité des Aliments de l'Homme, Unité de Nutrition Lipidique, 17 rue Sully, 21034 Dijon Cedex, France.

Introduction

For a number of years, gas–liquid chromatography (GLC) has been the method of choice to analyze fatty acids from oils or biological samples (1–3). The introduction of fused silica columns (4), as well as new stationary phases, allows the separation and quantification of numerous naturally occurring fatty acids, including geometrical and positional isomers of mono-, di-, and triethylenic fatty acids, mostly of C_{18} and C_{20} chain length (5–10).

However, the retention time of compounds on a given stationary phase, under the same conditions (carrier gas, temperature, etc.) depends on the chain length, the degree of unsaturation, and the position and geometry of the ethylenic bond(s). The introduction of new phases and new column types can lead to excellent separations, especially in the field of fish oil fatty acids where as many as 60 different fatty acids can be separated and quantified (11). At the same time this can be a major drawback of GLC when complex mixtures of *cis* and *trans*, methylene- and nonmethylene-interrupted, dienes and trienes have to be analyzed. Partially hydrogenated oils, especially those of marine origin, come into this category (12–15). The presence of hundreds of peaks results in large overlaps of isomers and gives unreliable qualitative and quantitative analyses even on capillary GLC columns (14,15). It is then necessary to use sophisticated physical and chemical methods along with some isolation techniques to determine the structure of the different molecules.

Fatty acid structural analysis involves determining the number of carbons, the degree of unsaturation, the geometry of the unsaturation(s), and the position of the ethylenic bond(s). While the number of carbons and the degree of unsaturation can be determined before and after total hydrogenation by GLC coupled with mass spectrometry (GC-MS [16]), determination of the position of the ethylenic bonds may involve either sophisticated physical techniques based on mass spectrometry (16–22) and/or complex chemical techniques. Determination of the structures of mono- and diethylenic fatty acids are quite simple, while the determination of polyunsaturated fatty acids, such as those with 4, 5, or 6 ethylenic bonds, is more tedious. This chapter will concentrate on how chemical techniques can be applied to the identification of fatty acids.

Monoethylenic Fatty Acids

The structure of monoethylenic fatty acids can be readily obtained using ozonolysis (23). Two types of ozonolysis can be carried out. Reductive ozonolysis (Figure 21.1) gives two fragments, an aldehyde and an aldehyde ester, while oxidative ozonolysis gives a mixture of acidic fission products (23). Among all the techniques used in reductive ozonolysis, the technique of Stein and Nicolaides (24) using triphenylphosphine for the cleavage of the ozonides into aldehyde fragments has been commonly used. The major advantages of oxidative ozonolysis *vs.* reductive ozonolysis are that a wide range of mono- and diesters are commercially available, which is not the case for the aldehyde esters and mono- and diesters can be separated by GLC readily, which is not the case for mixtures of some aldehydes and aldehyde esters (23).

In practice, oxidative ozonolysis can be executed by two different approaches. The reaction can be carried out in a nonparticipating or participating solvent, such as methanol (23). The different methods were tedious and time consuming (23).

In 1977, a very simple and efficient oxidative ozonolysis procedure was developed by Ackman (25). An important feature of this method was that one handling step integrated three reactions: addition of ozone to the unsaturated fatty acids in 7% BF_3-MeOH medium at room temperature, producing the methoxyhydroperoxide and aldehyde (Figure 21.2); decomposition of the methoxyhydroperoxide and oxidation of the intermediate aldehydes to acid at 100°C in the same solvent; and esterification of the acidic products by MeOH with BF_3 catalysis (26). This method was of particular interest to determine the structures of long-chain monoethylenic

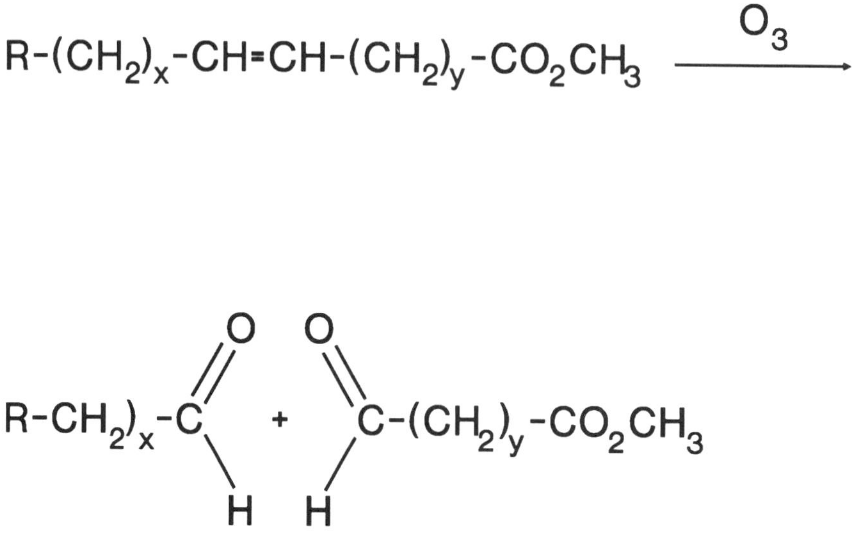

$$R-(CH_2)_x-CH=CH-(CH_2)_y-CO_2CH_3 \xrightarrow{O_3}$$

$$R-(CH_2)_x-\overset{O}{\overset{\|}{C}}\diagdown_H \quad + \quad \overset{O}{\overset{\|}{\underset{H}{\diagup}C}}-(CH_2)_y-CO_2CH_3$$

Figure 21.1. Reductive ozonolysis of unsaturated fatty acids.

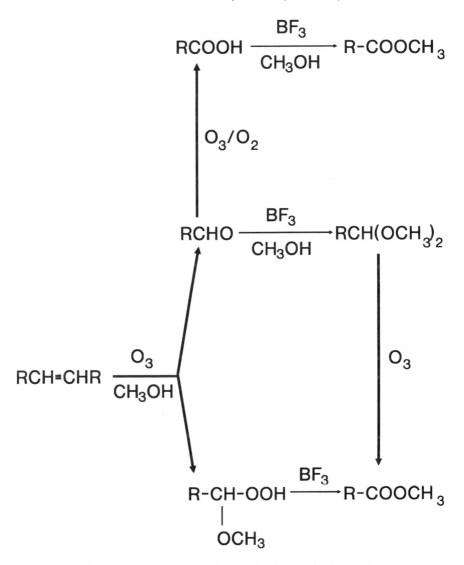

Figure 21.2. Oxidative ozonolysis in BF_3-MeOH. From Sebedio, et al.

acids from partially hydrogenated oils. As has been demonstrated, partial hydrogenation not only results in the formation of geometrical isomers (*trans* fatty acids) but also in the formation of positional isomers. The number and the relative proportions of positional and geometrical isomers depends on hydrogenation parameters, such as temperature, rate of agitation, hydrogen pressure, and type of catalyst (27,28).

An alternative to this method would be to use the mass spectra of the 2-alkenyl-4,4-dimethyloxazolines (21,22). The derivatives of the usual unsaturated fatty acids have good chromatographic properties similar to those of the esters. Simple mixtures of monoethylenic fatty acids could be analyzed and the position of the ethylenic bond determined after analysis of the mass spectra. However, it would be difficult to apply this method to the complex case of partially hydrogenated oils considering the large number of peaks which would have to be separated by GLC.

Diethylenic Fatty Acids

A slightly modified ozonolysis method has been proposed by Ratnayake and Ackman (29) in order to determine the structure of dienoic fatty acids (Figure 21.3). Direct ozonolysis will only locate the ethylenic bond close to the methyl end of the molecule because two dimethyl esters will be formed after ozonolysis. This provides two possibilities to localize the second ethylenic bond. The fatty acid methyl ester (FAME) is first reduced to an alcohol, which is further ozonized. The resulting mixture of monoester, diester, and alcohol ester is then fractionated by thin-layer chromatography (TLC). The alcohol is converted into acetate, and the three fragments are then analyzed by GC or GC-MS. This method has been useful for studying the structures of the dienoic fatty acids present in partially hydrogenated oils (14,15). Mass spectra of picolinyl esters (18) or the dimethyloxazoline derivatives are possible alternatives, as long as the mixture to be analyzed is not too complex.

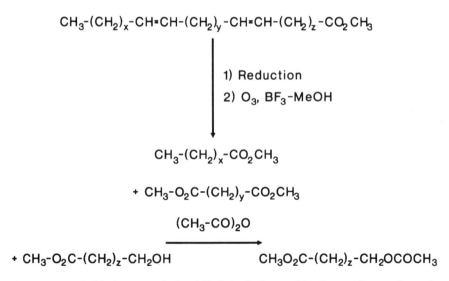

Figure 21.3. Oxidative ozonolysis of diethylenic fatty acids. *Source:* Ratnayake and Ackman (29).

Polyunsaturated Fatty Acids

For polyunsaturated fatty acids having more than two ethylenic bonds, the procedure utilized is much more complex and tedious since it would be impossible to localize more than two ethylenic bonds with the previously described procedures. It is then necessary to transform the polyunsaturated fatty acid into a mixture of monoenes where the ethylenic bond in each monoene would represent one ethylenic bond of the parent molecule with no positional or geometrical change from the original unsaturation. This can be achieved using hydrazine reduction (30–33). The method consists of partial reduction of the polyethylenic acid with hydrazine to give a maximum yield of monoethylenic fatty acids. This is followed by isolation of the monoethylenic isomers using mercuric adduct fractionation (34,35) and determination of the double-bond position using ozonolysis as previously described.

It is now well established that hydrazine reduction occurs with no change in the geometry or the position of the ethylenic bond. However, great care must be taken in choosing reaction parameters, such as time and temperature, to obtain optimal quantities of monoenes without increasing the amount of saturate too much. For example (Figure 21.4), the hydrazine reduction of 18:3Δ9c, 12c,15c will give a mixture of 18:0, three monoenes, 18:1Δ9c, 18:1Δ12c, and 18:1Δ15c, three dienes, 18:2Δ9c, 12c, 18:2Δ9c, 15c, and 18:2Δ12c, 15c, and some unreacted triene. The optimum time (Figure 21.4) to obtain appreciable quantities of monoenes, a low quantity of 18:3, and not too much saturate would be 60–70 min. However, hydrazine reduction should be used in qualitative analysis and not in quantitative analysis due to its selectivity as demonstrated in Figure 21.5, where the quantity of the monoene obtained after the hydrazine reduction of 18:3Δ9c, 12c, 15c depends on the position of the ethylenic bond on the carbon chain.

A more complex case has been reported in Figure 21.6. This is the case of the hydrazine reduction of 20:5 n-3 which produced a mixture of 32 components including 20:0, monoenes, dienes, trienes, tetraenes, and some unreacted pentaene (36). From such a mixture, the major problem is how to isolate all the monoenes in one fraction. This can easily be carried out by $AgNO_3$-TLC (37) when the starting molecule only contains *cis* methylene-interrupted ethylenic bonds. This technique should not be carried out when the presence of *trans* ethylenic bonds is suspected due to the overlap which may occur during $AgNO_3$-TLC. Overlap occurs because the migration is not only influenced by the degree of unsaturation, but by the geometry and position of the unsaturation on the carbon chain (38).

Isolation of the monoenes is carried out using mercuric adduct fractionation (Figure 21.7). As a first step, the unsaturated molecule is converted to the acetoxymercuri-methoxy adduct by reaction of mercuric acetate in methanol. The adduct is then converted to the bromercuri-methoxy derivative by reacting NaBr in MeOH. The migration of these derivatives is carried out in a mixture of hexane-dioxane, and only depends on the degree of unsaturation with no influence by the fatty acid chain length or the geometry of the ethylenic bond(s) unlike $AgNO_3$-

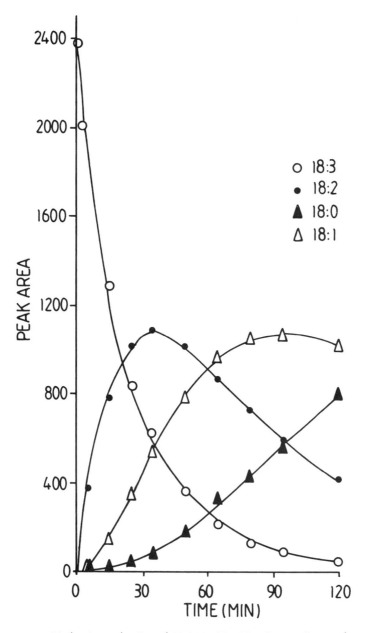

Figure 21.4. Hydrazine reduction of 18:3Δ9c,12c,15c. *Source:* Ratnayake, et al. (33).

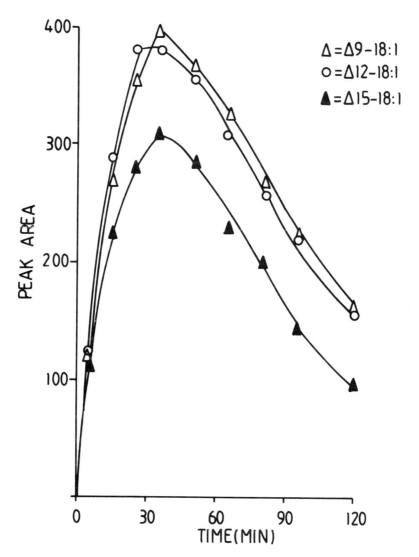

Figure 21.5. Hydrazine reduction of 18:3Δ9c,12c,15c. Influence of the reaction time on the formation of monoenes. *Source:* Ratnayake, et al. (33).

TLC. The major applications of this technique have been reviewed recently (35). The fatty acid methyl esters are then regenerated from the monoene band using HCl in MeOH. These are then separated into *cis* and *trans* monoenes prior to ozonolysis in BF_3-MeOH. The GC-MS analysis of the different fragments then enables structural determination of the parent molecule.

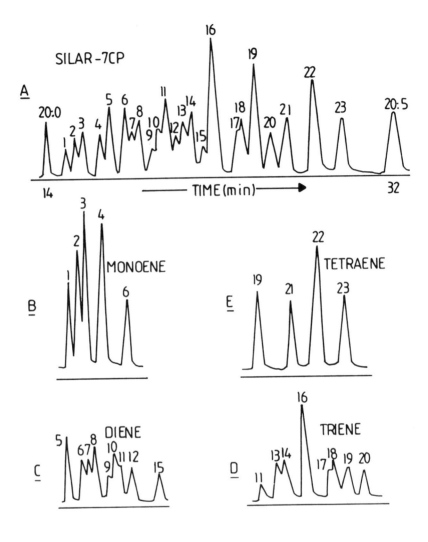

(5) Δ5,8 + Δ5,11
(6) Δ17 + Δ5,14
(11) Δ8,17 + Δ5,8,11
(19) Δ8,14,17 + Δ5,8,11,14
(16) Δ5,8,17 + Δ5,11,17 + Δ8,11,14
(22) Δ5,8,14,17 + Δ5,11,14,17

Figure 21.6. GLC analyses of (A), the hydrazine-reduction product 20:5n-3 and (B,C,D,E) the fractions isolated after mercuric adduct fractionation. *Source:* Sebedio and Ackman (36).

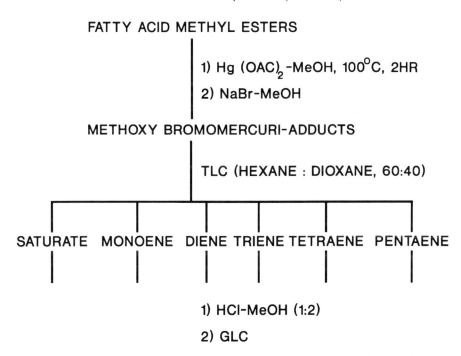

Figure 21.7. Mercuric adduct fractionation of a complex mixture of fatty acids.

These methods were recently applied to structural determination of fatty acids from biological samples. For example, it has been demonstrated that some minor unsaturated fatty acid isomers, such as 18:3Δ9c,12c,15t which is present in refined oils and heat-abused oils (7), could be desaturated and elongated to give geometrical isomers of eicosapentaenoic (EPA) and docosahexaenoic acids (DHA). Using these methods, these were identified as being 20:5Δ5c,8c,11c,14c,17t and 22:6Δ4c,7c,10c,13c,16c,19t (39). Some of the techniques also permit the isolation of sufficient quantities of these fatty acids to look at their potential physiological effects (40,41).

Unusual Fatty Acids

Unfortunately, some of the techniques described above cannot be applied successfully to all types of molecules. This is particularly true with cyclic fatty acid monomers (CFAM). These molecules are usually found in heat-abused fats and oils and are formed from linoleic and linolenic acids. Those formed from linoleic acid are C_{18} disubstituted fatty acids having one ethylenic bond, while those formed from linolenic acid are diunsaturated isomers (42). Mercury adduct fractionation has been used to study the structures of these complex molecules (43).

Thin-layer chromatography of the mercuric adducts of total FAME from a heated linseed oil gave five major bands (Table 21.1), a triene band $R_f = 0.58$, a diene band $R_f = 0.82$, a monoene band $R_f = 0.87$, a saturate band $R_f = 0.90$, and a band which contained a complex mixture of unknown components $R_f = 0.78$. The analysis of each band after total hydrogenation showed that the cyclic fatty acids were mainly distributed in three bands, the triene band, the diene band, and one between the diene and triene bands (Table 21.1). Since the cyclic fatty acids are known to be a mixture of *cis*, and *trans* diethylenic isomers (42), this study indicates that the migration of the adducts of cyclic fatty acids is influenced by more than the degree of unsaturation of the molecule. However, knowledge of the structures of cyclic fatty acids is insufficient to determine which factors influenced the formation and migration of the adducts. Some further experiments may permit the utilization of this fractionation technique in combination with other methods such as silver-ion high-performance liquid chromatography (HPLC), to obtain simple fractions for structural determination of complex cyclic fatty acid mixtures present in heated fats and oils.

Ozonolysis of the cyclic fatty acids is not as simple as it is for the usual polyunsaturated fatty acid. In some cases, such as those represented in Figure 21.8, secondary products are formed. The CFAM *1*, should lead to the formation of the tetraester *2*. Unfortunately, this product goes through a decarboxylation process to give component *3*. In this case, examination of the structure of component *3* would lead to a wrong CFAM structure. In any case, it would be impossible to localize the ethylenic bond on the carbon chain, as this will give a short-chain monoester which could not be detected with the analytical method.

In the other example, component *4* (Figure 21.8) shows the formation of one triester which corresponds to the ozonolysis of the 5-carbon membered ring and of two diesters respectively with 8 and 9 CH_2 units (DMC_8 and DMC_9). As *4* is a synthesized molecule, it is surprising to find such a high quantity of a decarboxylation product (DMC_8). Decarboxylation is known to occur during ozonolysis, but the secondary reaction product does not usually exceed 2% of the major diester (23). Such an analysis indicates the presence of two components. For these types of molecules,

TABLE 21.1
Percentage of CFAM (as total CFAM) in the Methoxy bromoMercuric Adduct Fractions From Heated Linseed Oil

MBM Bands R_f	% CFAM
0.87	11.2
0.82	38.9
0.78	40.6
0.63	4.8
0.58	4.5

Abbreviations: methoxybromomercuric adduct, MBM; cyclic fatty acid monomers, CFAM.
Adapted from Sebedio (43).

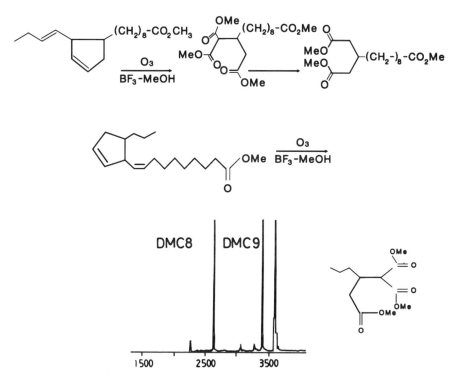

Figure 21.8. Ozonolysis of some synthesized C_{18} CFAM in BF_3-MeOH.

the only alternative to oxidative fission is the GC-MS analysis either on the methyl esters or after derivatization (44,45).

Conclusions

Despite the development of new physical methods for the structural determination of polyunsaturated fatty acids, the utilization of ozonolysis combined with hydrazine reduction, $AgNO_3$-TLC, and mercuric adduct fractionation offers a powerful tool especially for molecules containing *trans* ethylenic bond(s). In fact, GC-MS of polyunsaturated fatty acid derivatives now localizes each ethylenic bond on the carbon chain. Gas-liquid chromatography coupled with infrared spectroscopy detects the presence of a *trans* ethylenic bond on the carbon chain. However, it is impossible to determine the position of the *trans* ethylenic bond on the carbon chain by physical methods if more than one double bond is present. This is the case of fatty acids which are formed during processes, such as hydrogenation, or in vivo by desaturation-elongation of unusual fatty acids, such as the 18:2 and 18:3 isomers, which have been detected in oils used for human consumption.

References

1. Ackman, R.G. (1972) *Prog. Chem. Fats Other Lipids 12:* 167–284.
2. Ackman, R.G., and Eaton, C.A. (1978) *Fette Seifen Anstrichm. 80:* 21–37.
3. Christie, W.W. (1989) *Gas Chromatography and Lipids,* pp. 85–128, The Oily Press, Ayr, Scotland.
4. Ackman, R.G. (1981) *Chemistry and Industry,* 715–722.
5. Ackman, R.G., Hooper, S.N., and Hooper, D.L. (1974) *J. Am. Oil Chem. Soc. 51:* 42–49.
6. Rakoff, H., and Emken, E.A. (1982) *Chem. Phys. Lipids 31:* 215–225.
7. Grandgirard, A., Sebedio, J.-L., and Fleury, J. (1984) *J. Am. Oil Chem. Soc. 61:* 1563–1568.
8. Perkins, E.G., and Smick, C. (1987) *J. Am. Oil Chem. Soc. 64:* 1150–1155.
9. Ratnayake, W.M.N., Hollywood, R., O'Grady, E., and Beare-Rogers, J.L. (1990) *J. Am. Oil Chem. Soc. 67:* 804–810.
10. Wolff, R.L. (1992) *J. Chromatogr. Sci. 30:* 17–22.
11. Ackman, R.G. (1989) in *Marine Biogenic Lipids Fats and Oils,* Ackman, R.G., CRC Press, Boca Raton, Florida, vol. 1, pp. 103–137.
12. Ratnayake, W.M.N., and Beare-Rogers, J.L. (1990) *J. Chromatogr. Sci. 28:* 633–639.
13. Ratnayake, W.M.N., and Pelletier, G. (1992) *J. Am. Oil Chem. Soc. 69:* 95–105.
14. Sebedio, J.-L., and Ackman, R.G. (1983) *J. Am. Oil Chem. Soc. 60:* 1986–1991.
15. Sebedio, J.-L., and Ackman, R.G. (1983) *J. Am. Oil Chem. Soc. 60:* 1992–1996.
16. Ratnayake, W.M.N., and Ackman, R.G. (1989) in *The Role of Fats in Human Nutrition,* Academic Press, New York, pp. 515–565.
17. Anderson, B.A., Christie, W.W., and Holman, R.T. (1975) *Lipids 10:* 215–219.
18. Christie, W.W., Brechany, E.Y., and Holman, R.T. (1987) *Lipids 22:* 224–228.
19. Adams, J., Deterding, L.J., and Gross, M.L. (1987) *Spectros. Int. J. 5:* 199–228.
20. Christie, W.W. (1989) *Gas Chromatography and Lipids,* pp. 161–184, The Oily Press, Ayr, Scotland.
21. Zhang, J.Y., Yu, Q.T., Liu, B.N., and Huang, Z.H. (1988) *Biomed. Environ. Mass Spectrom. 15:* 33–44.
22. Luthria, D.L., and Sprecher, H. (1993) *Lipids 28:* 561–564.
23. Ackman, R.G., Sébédio, J.-L., and Ratnayake, W.M.N. (1981) in *Methods in Enzymology,* Lowenstein, J.M., Academic Press, New York, vol. 72, pp. 253–276.
24. Stein, R.A., and Nicolaides, N. (1962) *J. Lipid Research 3:* 476.
25. Ackman, R.G. (1977) *Lipids 12:* 293–296.
26. Sebedio, J.-L., Ratnayake, W.M.N., and Ackman, R.G. (1984) *Chem. Phys. Lipids 35:*21–28.
27. Sebedio, J.-L., and Ackman, R.G. (1983) *Fette Seifen Anstrichm. 85:* 339–346.
28. Coenen, J.W.E. (1976) *J. Am. Oil Chem. Soc. 53:* 382–389.
29. Ratnayake, W.M.N., and Ackman, R.G. (1979) *Lipids 14:* 580–584.
30. Conway, J., Ratnayake, W.M.N., and Ackman, R.G. (1985) *J. Am. Oil Chem. Soc. 62:* 1340–1343.
31. Privett, O.S., and Nickell, E.C. (1966) *Lipids 1:* 98.
32. Scholfield, C.R., Butterfield, R.O., Mounts, T.L., and Dutton, H.J. (1969) *J. Am. Oil Chem. Soc. 44:* 323.
33. Ratnayake, W.M.N., Grossert, J.S., and Ackman, R.G. (1990) *J. Am. Oil Chem. Soc. 67:* 940–946.

34. White, H.B. (1966) *J. Chromatogr. 21:* 213–222.
35. Sebedio, J.-L. (1993) in *Advances in Lipid Methodology,* Christie, W.W., The Oily Press, Ayr, Scotland, vol. 2, pp. 139–155.
36. Sebedio, J.-L., and Ackman, R.G. (1981) *Lipids 16:* 461–467.
37. Morris, L.J. (1966) *J. Lipid Res. 7:* 717–732.
38. Gunstone, F.D., Ismail, I.A., and Lie Ken Jie, M. (1967) *Chem. Phys. Lipids 1:* 376–385.
39. Grandgirard, A., Piconneaux, A., Sébédio, J.-L., O'Keefe, S.F., Semon, E., and Le Quere, J.-L. (1989) *Lipids 24:* 799–804.
40. Chardigny, J.M., Sebedio, J.-L., and Grandgirard, A. (1993) in *Essential Fatty Acids and Eicosanoids,* Sinclair, A., and Gibson, R., American Oil Chemists' Society, Champaign, pp. 148–152.
41. O'Keefe, S.F., Lagarde, M., Grandgirard, A., and Sebedio, J.-L. (1990) *J. Lipid Res. 31:* 1241–1246.
42. Sebedio, J.-L., and Grandgirard, A. (1989) *Prog. Lipid Res. 28:* 303–336.
43. Sebedio, J.-L. (1985) *Fette Seifen Anstrichm. 87:* 267–273.
44. Sebedio, J.-L., Le Quere, J.-L., Semon, E., Morin, O., Prevost, J., and Grandgirard, A. (1987) *J. Am. Oil Chem. Soc. 64:* 1324–1333.
45. LeQuere, J.-L., Sebedio, J.-L., Henry, R., Couderc, F., Demont, N., and Prome, J.C. (1991) *J. Chromatogr. 562:* 659–672.

Chapter 22
Analysis of Sterol Oxides in Food and Blood

Lars-Åke Appelqvist

Department of Food Science, Swedish University of Agricultural Sciences, P.O. Box 7051, S-750 07 Uppsala, Sweden.

Introduction

Although cholesterol, the predominant sterol in blood and foods of animal origin, has only one double bond, cholesterol oxides are formed at measurable levels in both blood and certain foods. Foods that are dried and then stored dry, as well as those which are heated during industrial manufacture or in the kitchen, are relatively rich in cholesterol oxides (1). Dry plant foods, such as wheat flour, and certain deep-fried foods, such as potato chips, contain oxidized plant sterols, predominantly sitosterol oxides.

The mechanisms of cholesterol oxidation and the resultant reaction products were studied in a few laboratories during the early- to mid-twentieth century following the first report in 1904 (2). An extensive monograph, including 2773 references, was published in 1980 by Smith (3). At that point 66 different primary and secondary oxidation products of cholesterol had been structurally identified. A paper presenting additional literature on the chemistry of cholesterol oxidation up to 1985 has also been published (4). Very little has been published on the isolation of various oxyphytosterols. A few references are included in the reviews on cholesterol oxides. Oxides from cholesterol or plant sterols will be called oxysterols in this chapter. It is beyond the scope of this chapter to present the oxidative mechanisms; this chapter is only meant to give the interested reader key references in the literature (3–5).

Medical Significance of Dietary and Body Oxysterols

In the early 1970s, a number of physiologists, aware of reports on the cytotoxicity, angiotoxicity, and mutagenicity of certain oxysterols, presented data from in vivo animal studies pointing to oxysterols rather than cholesterol as the initiating agent of the atherosclerotic process; see Taylor and Peng (6) for original references. At that time it was also reported that experimental animals which are strongly hypercholesterolemic by means other than dietary cholesterol show no atherosclerotic lesions.

Presently, many research groups are involved in the determination of oxysterols in animal and human blood relating to the initiation of the atherosclerotic process, cancer, and other severe diseases. An increase in human plasma oxysterols after consuming an oxysterol-rich breakfast clearly points to dietary oxysterols as one important source of plasma oxysterols (7). The relative importance of dietary oxysterols compared to oxysterols formed in vivo for the total body burden of oxysterols seems to be completely unknown. To date, no studies on physiological properties, such as resorption and metabolism, of plant oxysterols have been reported in the literature. However, in view of the considerable amounts of oxyphytosterols in certain diets, such studies appear highly warranted.

Major Oxysterols Found in Food or Blood Tissue

The qualitative and quantitative composition of sterol oxides in foods has been reported in the literature (1). Only about 10 of the more than 60 known primary and secondary oxysterols from food have been reported (1). The traditional short names and systematic names for some oxysterols found in foods and the formulas of these are presented in Table 22.1 and Figure 22.1.

All known oxysterols derived from cholesterol probably have counterparts among the oxides derived from plant sterols; the proportion of side chain oxides may be slightly different due to the methyl-, ethyl-, or ethylidene-groups at C_{24}. The 5,6-α- and 5,6-β epoxides and the 7α- and 7-β-hydroxy derivatives of sitosterol were identified by gas chromatography-mass spectrometry (GC-MS) in samples of wheat flour (8). The structures of 4-des-methyl-, 4-monomethyl-, and 4,4-dimethyl sterols and their occurrence in vegetable oils were reviewed by Kochhar (9). The most common des-methyl sterols are presented in Figure 22.2.

Analytical Procedures Used in the Determination of Oxysterols

Since oxysterols are present in very low concentrations compared to cholesterol, the isolation and qualitative and quantitative analysis is delicate and tedious. Some of the steps in the analysis are common for most sources of oxysterols except for pure fats and oils; however, step 1 generally is not applied to plasma or serum samples.

1. Extraction with an organic solvent or solvent mixture for all food matrices except for simple emulsions.

TABLE 22.1
Common Autoxidation Derivatives of Cholesterol

Chemical Name	Trivial Name
Cholest-5-ene-3β,7α-diol	7α-Hydroxycholesterol
Cholest-5-ene-3β,7β-diol	7β-Hydroxycholesterol
3β-Hydroxycholest-5-ene-7-one	7-Ketocholesterol
5α-Cholestane-3β,5,6β-triol	Cholestanetriol
Cholest-5-ene-3β,25-diol	25-Hydroxycholesterol
20S-Cholest-5-ene-3β,20-diol	20α-Hydroxycholesterol
5,6α-Epoxy-5α-cholestan-3β-ol	Cholesterol 5α,6α-epoxide
5,6β-Epoxy-5β-cholestan-3β-ol	Cholesterol 5β,6β-epoxide

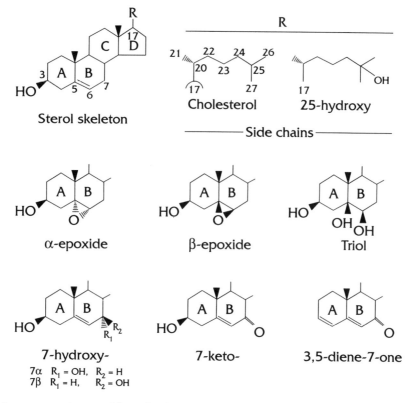

Figure 22.1. Structural formulas for cholesterol and oxysterols.

2. Isolation of sterol oxides from sterols and other lipids.

3. Qualitative and quantitative determination of the individual sterol oxides.

4. Verification of sterol oxide identity, generally by GC-MS.

A recent and complete review of methods used to analyze oxysterols was produced by Park and Addis (14), and includes 89 references. Readers interested in the details of each methodology presented may find that reference a useful place to start.

Extraction

It has generally been assumed that the classical chloroform/methanol extraction procedures according to Folch (10) or according to Bligh and Dyer (11) produce complete extraction of cholesterol from animal tissues. However, recent studies demonstrated that extraction of red blood cells with hexane in combination with ethanol or 2-propanol yielded extracts with a higher cholesterol content than the traditional chloroform/methanol method (12). The author's studies on food oxysterols

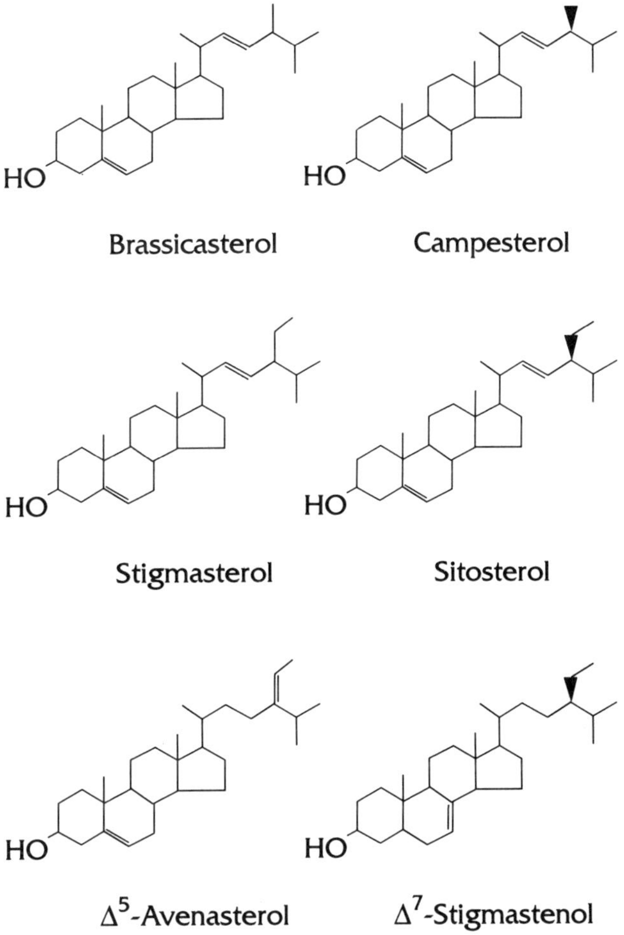

Figure 22.2. Common des-methyl sterols in vegetable oils.

have used the hexane/2-propanol procedure of Hara and Radin (13), but most laboratories seem to use chloroform/methanol; see the review by Park and Addis (14).

While most of the cholesterol of animal tissues is free, with about 10% being esterified in eggs, milk, and meat (15), the sterols in plant tissues can occur as free and esterified sterols, as well as steryl glucosides and esterified steryl glucosides (9,16). Some of these are strongly bound to the matrix. After repeated extraction, considerable amounts of sterols are still released from the defatted plant tissues after heating in strong acids. Such harsh treatment probably destroys any phytosterol oxides present. Obviously the extraction percentage needs to be studied for each food matrix, which is rarely done in papers reporting sterol oxide levels in food.

When indigenous sterols are present in the extracts there is a risk of additional oxidation, before the separation of the sterol oxides from the sterols. According to Park and Addis, the extraction has to be protected by excluding daylight and fluorescent light; adding antioxidants, such as BHT; flushing the solvents with an inert gas, such as argon or nitrogen; or by a combination of any of these precautions (14). In carrying out this step, a compromise between the desire to perform a complete extraction, involving several consecutive homogenizations and filtrations with the disadvantage of large solvent volumes, and the need to minimize oxidation artifacts must be sought in each laboratory working with different matrices.

Enrichment of Sterol Oxides from the Bulk Extract

There are two different routes followed in different laboratories: direct multistep chromatographic enrichment of the total lipid extract, and saponification followed by chromatography. For extracts of animal origin foods, the first method has advantages regarding accuracy, since artifacts are easily formed during saponification, but it is more time consuming. When preparative thin-layer chromatography is used as one of the enrichment steps, artifact formation may occur by oxidation on the plate. The second group of methods has the advantage of accounting for free and esterified sterol oxides, but this may be a minor advantage, because of the relatively minor amount of total cholesterol esterified in eggs, meat, and dairy products (15). Some authors claim that cold saponification (room temperature, overnight) does not generate artifacts, but others have reported that even cold saponification generates substantial artifacts, namely the formation of 3,5-cholesta-diene from 7-ketocholesterol, generation of epoxides, and breakdown of epoxides to cholestanetriol (14).

While most foods of animal origin have a small proportion of cholesterol in esterified form, about two-thirds of the blood cholesterol is esterified. Hence for studies on blood oxysterols, a saponification step has been included. In recent studies comparing cold *vs.* hot saponification, it was found that only 80–90% of added cholesterol oleate labeled with ^{14}C in the cholesterol moiety was hydrolyzed during the time-temperature conditions presented in the literature (17,18). An alternative method to release oxidized sterols from their esterified form without causing artifacts is to transesterify the lipids with sodium hydroxide in methanol-benzene at room temperature for 2 h (19). However, this introduces a lipid-extraction step for plasma or serum samples which usually are saponified directly, without extraction.

It is possible to add cholesterol and oxysterols labeled with deuterium before the saponification step and then calculate the breakdown of certain oxides and the generation of others by analyzing the purified oxides by GC-MS (18,20). However, this is an elaborate way and the deuterium-labeled compounds must be synthesized in the laboratory. Therefore, this elegant analytical technology is hardly applicable to large series of samples.

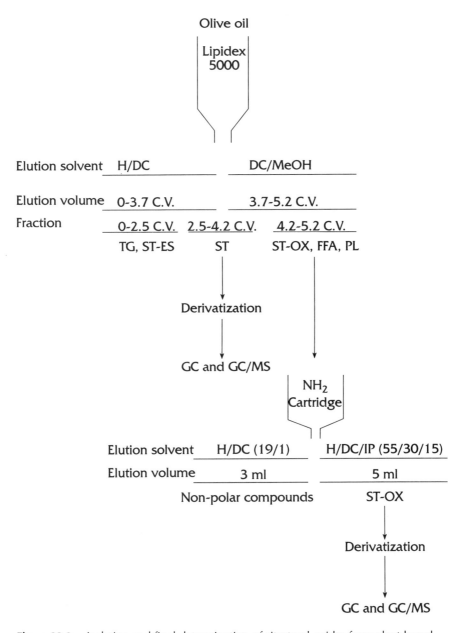

Figure 22.3. Isolation and final determination of sitosterol oxides from plant-based food. Abbreviations: H, heptane; DC, dichloroethane; MeOH, methanol; IP, isopropanol; C.V., column volume; TAG, triacylglycerols; ST-ES, sterol esters; ST, sterols; ST-OX, sterol oxides; FFA, free fatty acids; and PL, phospholipids. *Source:* Nourooz-Zadeh and Appelqvist (8).

When handling plant-lipid extracts, a saponification step generally is unavoidable, since the sterols of vegetable oils often are esterified to a considerable extent and into various derivatives. However, a multistep chromatographic procedure (Figure 22.3) has been developed in the author's laboratory to enrich sterol oxides from vegetable oils with low percentages of esterified sterols (8). The basic concept of utilization of Lipidex® chromatography from egg powders was presented earlier (21).

Qualitative and Quantitative Determination of the Individual Sterol Oxides

This is done either by GC, or HPLC. The necessary verification of the identity of the chromatographic peaks preferably is done by mass spectrometry. Since HPLC-MS generally is not available in most lipid laboratories, and GC-MS is becoming more and more common, the use of capillary GC as tool for quantitation appears to be preferable over HPLC. Data reported in the literature may suffer from serious misidentification when just HPLC is used as the final step; this is probably the case

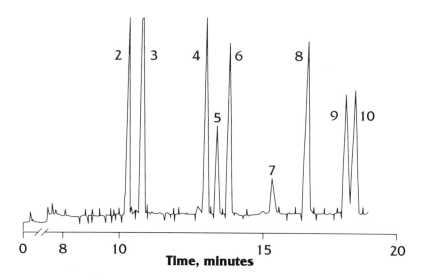

2- Cholest-5-ene-3β, 7α-diol (7α-hydroxycholesterol)
3- Cholest-5-ene-3β-ol (cholesterol)
4- Cholest-5-ene-3β, 7β-diol (7β-hydroxycholesterol)
5- 5,6β-Epoxy-5β-cholestan-3β-ol (5β-,6β-epoxycholestanol)
6- 5,6α-Epoxy-5α-cholestan-3β-ol (5α-,6α-epoxycholestanol)
7- Cholest-5-ene-3β, 20α-diol (20α-hydroxycholesterol)
8- 5α-Cholestane-3β, 5, 6α-triol (cholestane-triol)
9- 3β-Hydroxycholest-5-en-7-one (7-ketocholesterol)
10- Cholest-5-ene-3β, 25-diol (25-hydroxycholesterol)

Figure 22.4. Gas chromatogram of TMS derivatives of synthetic oxysterols. *Source:* Nourooz-Zadeh et al. (21). Reprinted with permission from *J. Food Sci.*

for some data presented for meat products (22). Since there are great similarities in the chromatographic properties of some of the cholesterol oxides, for example, the isomeric 5,6-epoxides and the 7-hydroxy-derivatives, great care has to be taken when choosing separation parameters (3). Generally, nonpolar columns, such as OV-1 or DB-1, or slightly polar columns such as DB-5 are used for capillary GC. Some laboratories use isothermal conditions, while others use temperature programming. Figure 22.4 shows the resolution of eight important cholesterol oxides as TMS-ethers. When analyzing sterol oxides from plant tissues, which can have 3–4 "major" sterols, the large number of sterol oxides present makes the results from capillary GC analysis alone tentative.

Verification of Sterol Oxide Identity, Generally by GC-MS

A valuable introduction into the mass spectrometry of underivatized oxysterols as well as their TMS-ethers (trimethylsilyl-ethers) is presented in Smith (3). The aim of including this point is not to discuss mass spectra, but to emphasize the necessity of verifying the peak assignment, based on relative retention time in GC or HPLC, especially when studying a matrix previously unknown in the laboratory.

References

1. Addis, P.B., and Park, P.S.V. (1992) in *Biological Effects of Cholesterol Oxides*, Peng, S.-K., and Morin, R.J., CRC Press, Boca Raton, Ann Arbor, and London, pp. 71–88.
2. Schultz, E., and Winterstein, E. (1904) *Physiol. Chem. 43:* 316–319.
3. Smith, L.L. (1981) *Cholesterol Autoxidation,* Plenum Press, New York.
4. Smith, L.L. (1987) *Chem. Phys. Lipids 44:* 87–125.
5. Peng, S.-K., and Morin, R.J. (1992) *Biological Effects of Cholesterol Oxides,* CRC Press, Boca Raton, Ann Arbor, and London.
6. Taylor, C.B., and Peng, S.-K. (1992) in *Biological Effects of Cholesterol Oxides,* Peng, S.-K., and Morin, R.J., CRC Press, Boca Raton, Ann Arbor, and London, pp. 2–6.
7. Emanuel, H.A., Hassel, C.A., Addis, P.B., Bergmann, S.D., and Zavoral, J.H. (1991) *J. Food Sci. 56:* 843–847.
8. Nourooz-Zadeh, J., and Appelqvist, L.-Å. (1992) *J. Am. Oil Chem. Soc. 69:* 288–293.
9. Kochhar, S.P. (1983) *Prog. Lipid Rev. 22:* 161–188.
10. Folch, J., Lees, M., and Stanley, G.H.S. (1957) *J. Biol. Chem. 726:* 497–509.
11. Bligh, E.G., and Dyer, W.J. (1959) *Can. J. Biochem. Physiol. 37:* 911–917.
12. Peuchant, (1989) *Analyt. Biochem. 181:* 341–344.
13. Hara, A., and Radin, N.S. (1978) *Analyt. Biochem. 90:* 420–426.
14. Park, P.S.V., and Addis, P.B. (1992) in *Biological Effects of Cholesterol Oxides,* Peng, S.-K., and Morin, R.J., CRC Press, Boca Raton, Ann Arbor, and London, pp. 33–70.
15. Nourooz-Zadeh, J. (1988) *Cholesterol Oxides in Food,* Ph.D. Thesis, Swedish University of Agricultural Sciences, Uppsala, Sweden.
16. Morrison, W.R. (1978) *Cereal Chem. 55:* 548–558.

17. Pie, J.E., Spahis, K., and Seillan, C. (1990) *J. Agric. Food Chem. 38:* 973–979.
18. Breuer, O., and Björkhem, I. (1990) *Steroids 55:* 185–192.
19. Zubillaga, M.P., and Maerker, G. (1988) *J. Am. Oil Chem. Soc. 65:* 780–782.
20. Wasilchuk, B.A., Le Quesne, P.W., and Vouros, P. (1992) *Anal. Chem. 64:* 1077–1087.
21. Nourooz-Zadeh, J., Johanson, B., Sjövall, J., Ryhage, R., and Appelqvist, L.-Å. (1988) *J. Food Sci. 52:* 57–62,67.
22. Higley, N.A., Taylor, S.L., Herian, A.M., and Lee, K. (1986) *Meat Sci. 16:* 175–188.

Chapter 23
Utilization of Stable Isotopes to Study Lipid Metabolism in Humans

B. Descomps

Laboratoire de Biochimie A et, INSERM Unité 58, Faculté de Médecine, Université de Montpellier I, Bd Henri IV, 34000 Montpellier, France.

Introduction

From an ethical viewpoint, stable-isotope-labeled compounds offer an invaluable opportunity to study turnover, oxidation rates, and metabolic pathways in healthy and diseased patients. The nontoxic nature of stable isotopes in the concentrations used makes their utilization quite acceptable in humans. Their analysis by mass spectrometry allows quantitative evaluation of metabolizing processes as well as direct and unambiguous identification of the metabolic pathways (1,2). Moreover, stable isotopes offer peculiar facilities for investigating the metabolism of geometrical or optical isomers; their use is suitable for the simultaneous study of the metabolic behavior of isomers even when not separable by thin-layer chromatography (TLC), high-performance liquid chromatography (HPLC), gas–liquid chromatography (GLC) on capillary columns or by their combination.

The tolerance and toxicity of stable isotopes used in the domain of lipids has been studied extensively. Though isotope effects were frequently detected in enzymatic studies *in vitro*, no biological effect was detectable at the concentrations expected during investigations in humans. The toxicity of ^{13}C was investigated in detail for yeast, algae, plants, and mice; however, growth and reproduction remained normal in these organisms even when 50% enrichment of the body pool was maintained for several months. The tolerance for ^{15}N was also excellent, and the toxicity of deuterium only appeared with considerable doses, such as immersion of fish in more than 30% D_2O-enriched water (3).

In the area of lipid metabolism, most investigators used ^{13}C-labeled fatty acids, mainly at the C-1 position, to determine their oxidation rates. During the last few years, there was growing interest in the synthesis of deuterium (^2H) labeled molecules, mainly for the study of specific metabolic pathways. In the first case (^{13}C-labeled molecules), Isotope Ratio Mass Spectrometry (IR-MS) was used, whereas analytical Gas Chromatography-Mass Spectrometry (GC-MS) with Multiple Ion Detection (MID) was generally preferred for ^2H-labeled compounds. At the present time, the ^{18}O isotope is less commonly used to analyze lipids and ^{15}N is limited to the study of apo lipoprotein metabolism.

Detection Procedures

The main analytical procedures used mass spectrometry in IR-MS, GCMS, and GC-IRMS. Detection by nuclear magnetic resonance (NMR) in man is uncommon.

IR-MS

A simplified schema of the modules involved in an IR-MS device is shown in Figure 23.1. The availability of the sample and reference via switching valves together with the use of simultaneous collection in the detector (reference and sample) gives extreme sensitivity to the system to detect isotope enrichment.

The high sensitivity and precision of IR-MS make this technique well adapted to measure the ^{13}C enrichment in CO_2 exhaled for quantitative studies of oxidative metabolism of ^{13}C-labeled substrates (2). This sensitivity compensates at least in part for the cost of the ^{13}C label. However, larger samples (mg), are needed with IR-MS than with GC-MS (ng) and a pure gas should be introduced into the spectrometer. This requires trapping the volatile contaminants and preparing for isolation of the labeled metabolite if a pyrolytic step is necessary if the metabolite is not CO_2. The pure hydrogen requirement in IR-MS severely limits the use of deuterium because of the difficulties for chemical isolation of the hydrogen element (D_2O decomposition in presence of Zn at high temperature).

GC-MS

Gas chromatography-mass spectrometry requires smaller samples than IR-MS. Most of the separation of analytes is done by the GC module with a minimal loss of material, but the range of atom percent excess (APE) measurement is only 0.5–100% with only 0.5% precision (2).

Figure 23.2 shows the connection of the main modules involved in a GC-MS apparatus. The system admits a wide variety of samples, even complex ones, the only limitation being their acceptability by the gas chromatography column. Methyl esters, oxides, and silyl esters are the principal derivatives used for fatty acids, car-

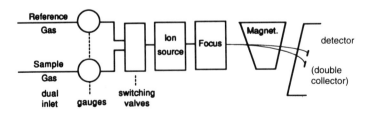

Figure 23.1. Schematic diagram of an IR-MS. Detection with a collector for both reference and sample gases gives extreme precision to APE determination.

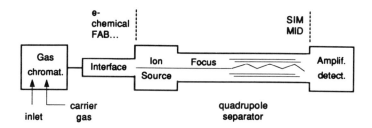

Figure 23.2. Schematic diagram of GC-MS. Three modes of ionization are indicated: electronic ionization (e⁻), chemical ionization, or fast atom bombardment (FAB). SIM and MID refer to specific ion monitoring and multiple ion detection respectively. Abbreviations: specific ion monitoring, SIM; multiple ion detection, MID.

bonyl, and hydroxy compounds, respectively. Chemical ionization (CI) or fast atom bombardment (FAB) avoid fragmenting the molecules which allows for the detection and the quantification of molecular ions, thus giving higher sensitivity than electronic ionization and subsequent fragmentation. Moreover, MID allows simultaneous detection and quantification of molecular ions, such as the natural molecule and the labeled one, or the detection of several isomers of the same molecule labeled differently with the same isotope (for instance m, $m + 1$, $m + 2$).

GC-IRMS

Both IR-MS and GC-MS techniques appear to be complementary. Even though it has been over 10 years since efforts to combine GC-MS and IR-MS in GC-IRMS were started, GC-IRMS is not widespread, and the difficulties encountered for ^2H isolation from ^2H_2O after pyrolysis of deuterium-labeled compounds limits its utilization to the ^{13}C label.

Figure 23.3 shows that the GC combination with IR-MS requires a complex interface system because of the constraints of IR-MS (admission of a pure gas sample). This interface is, in fact, a pyrolytic oven (converting organic samples to CO_2 and H_2O) followed by several traps separating pure CO_2 samples for admission into the IR-MS module.

IR-MS: Practical Aspects and Examples

^{13}C Determination in Expired CO2 from Oxidized Labeled Substrates

An example of the classical IR-MS method as applied to the ^{13}C-labeled fatty acids is the measurement of the oxidation rate of different dietary fatty acids (4). After ingestion of 10–20 mg/kg [1-^{13}C] in stearic, oleic, or linoleic acids (90% enrichment), breath $^{13}CO_2$ enrichment, and label excretion in fecal fat were determined and cumulative absorption and oxidation were calculated.

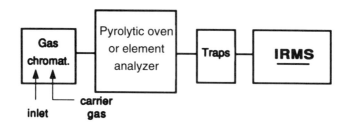

Figure 23.3. GC-IRMS via a pyrolytic interface.

From a practical point of view, $^{13}CO_2$ was trapped from exhaled gases (0.5 mL/min) in 1 N NaOH (10 mL in a 1-m coil) with infrared monitoring of the process. Storage in a freezer is possible at this step. After release of total CO_2 by acidification, water trapping ^{13}C APE was determined in the IR-MS in comparison with the reference sample. The background, determined from the profile of $^{13}CO_2$ exhaled by patients receiving the same diet, during the same period, but without the ^{13}C fatty acids, is subtracted. Figure 23.4 compares the cumulative oxidation rates of stearic, oleic, and linoleic acids calculated from the absorbed ^{13}C exhaled in breath CO_2 after ingestion of the three acids.

The oxidation of stearic acid was less than the other acids. At 9 h exogenous oleic acid was oxidized 14 times faster than stearic acid. It should be observed,

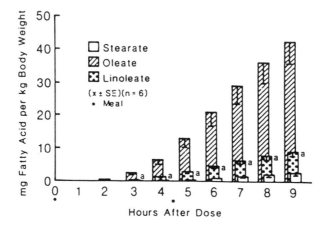

Figure 23.4. Cumulative oxidation rates of ^{13}C-labeled stearic (open bars), oleic (hatched bars), and linoleic (dotted bars) acids calculated from ^{13}C-labeled CO_2 exhaled after ingestion of the three acids. *Source:* Jones et al. (4).

however, that in this type of experiment, the data obtained compare the labeled fatty acids ingested but do not provide direct information on the participation of fatty acids from endogenous pools.

Deuterium Enrichment Measured by IR-MS: Human Lipogenesis Measured Incorporation of D_2O by Plasma Triglycerides (TG)

Human triglyceride (TG) synthesis can be measured from the incorporation of ^2H after the ingestion of ^2H_2O (0.7 g/kg of body water) and 1.4 g ^2H_2O/kg in drinking water for 48 h. Plasma TG were isolated by TLC, pyrolized at 250°C, and the ^2H_2O from ^2H-TG was analyzed by IR-MS after reducing the ^2H_2O with Zn. The background determined from a water standard is subtracted from the experimental values (5). Recently, the same method was used to compare cholesterol and TG, biosynthesis in normolipidemic individuals and hypercholesterolemic patients (apo E_2 associated). In these patients, hypercholesterolemia did not appear to be related to increased synthesis, but the reverse was found for TG (6).

Interesting contributions to our understanding of human lactation were made by two studies. The first one reported the effects of a high-fat or low-fat diet on endogenous fatty acid synthesis by the mammary gland. The second considered incorporation of individual fatty acids in the lipoprotein fractions of human milk as a function of their concentration in the plasma of the lactating mother. In the first study, it was demonstrated that the low-fat diet resulted in a sixfold increase in 16:0 and 18:0 in the plasma, and an increase in medium-chain fatty acids in the milk fat; this was determined by ^2H from ingested ^2H_2O incorporated into the fatty acids (7). The second study is an example of the usefulness of deuterium labeling and GC-MS detection to determine or compare the metabolic pathways of fatty acids.

GC-MS: Practical Aspects and Examples

^{13}C and Lipoprotein Biosynthesis

Surprisingly, ^{13}C-labeled amino acids can be useful to study human lipid metabolism using GC-MS. An example is given in Figures 23.5 and 23.6. In Halliday et al. (8), [1-^{13}C]-leucine was used to measure very low density lipoprotein (VLDL) apo lipoprotein B100 (Apo B100) biosynthesis. After an intravenous priming injection (8 µmol/kg), labeled leucine was infused (7 µmol/kg/h for 9.5 h). Each hour, ^{13}C-labeled VLDL from 12 mL blood samples were separated by ultracentrifugation, isolated, and submitted to HCl hydrolysis. The constitutive amino acids were derived and analyzed by GC-MS for ^{13}C-leucine enrichment. At the same time α-ketoisocaproic acid (KIC), the marker of intracellular ^{13}C-leucine transamination, was determined from plasma extracts after purification and derivatization. Figure 23.5 compares the incorporation of ^{13}C-leucine into apo B100, albumin, and muscle over time. From the data, the authors could demonstrate an increase in the pool

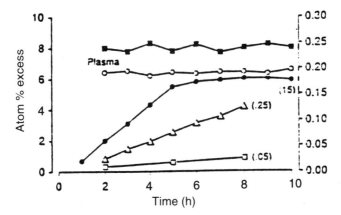

Figure 23.5. Change with time of incorporation of [1-^{13}C]-leucine in apo B100 (•—•), albumin (Δ—Δ), and muscle (□----□) during a constant infusion of the tracer. Calculated fractional protein synthesis rates (%/h) using [^{14}C] KIC as a precursor are given in parentheses. The upper three plots relate to the left-hand axis, lower two plots to the right-hand axis. Source: Halliday et al. (8).

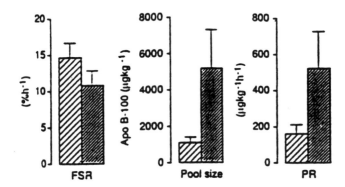

Figure 23.6. Fractional synthetic rate (FSR), pool size, and overall production rate (PR) of VLDL apo B100 in control subjects and patients with familial combined hyperlipidemia. Source: Halliday et al. (8).

size and overall production of VLDL apo B100 in familial combined hyperlipidemia (Figure 23.6).

Deuterium, GC–MS, and Metabolic Pathways of Lipid Components

The human glycerol appearance rate can be determined under steady state or nonsteady state conditions determining deuterium-labeled glycerol with GC-MS (after derivation and with fragmentography). A value of 2.22 ± 0.2 µmol/kg/min was found, and doses of 5–8 µmol/kg and 0.036–0.085 µmol/kg/min were used for bolus and intravenous injections, respectively (9).

Simultaneous administration of differently labeled fatty acids (di-, tetra-, and hexadeuterated molecules) can be used to compare the metabolic fate of the different fatty acids in the same subject. An interesting application compares the incorporation of three labeled fatty acids in plasma lipid fractions of lactating women with their values in human milk. An example of the results reported in Emken, et al. (10) is shown in Figure 23.7. After ingesting equal amounts of palmitic, oleic, and linoleic acids (di-, hexa- and tetra-deuterium-labeled, respectively), it can be observed that palmitic and oleic acids are preferentially incorporated into milk over linoleic acid. The method is especially well adapted to study the selectivity of metabolic pathways of homologous compounds. However, it should be observed that in this type of experiment, the amount of deuterium-labeled material needed is in the range of several grams for each subject. The difficulty of synthesizing multigram quantities of these specific molecules, most of which are not commercially available, remains as a restriction, in spite of the relatively low cost of deuterium

Selective studies of specific metabolic pathways and comparisons of geometric isomer or diastereoisomer metabolism generally are performed by GC-MS and

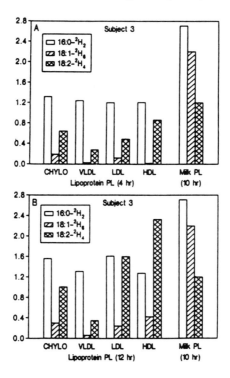

Figure 23.7. Comparison of the distribution of 16:0 $^2H_2$, 18:1 $^2H_6$, and 18:2 $^2H_4$ in chylomicron (CHYLO), very low density lipoprotein (VLDL), low-density lipoprotein (LDL), and high-density lipoprotein (HDL) PL samples to 10-h milk PL data from subject 3. (A) 4-h lipoprotein. (B) 16-h lipoprotein samples. *Source:* Emken et al. (10).

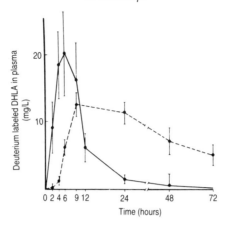

Figure 23.8. Deuterium-labeled 20:3 *n*-6 concentration in plasma triglycerides (●—●) and plasma phospholipids (●---●) of diabetic patients before treatment. The concentration in the plasma is shown after ingesting 2 g of [18,19-^{2}H$_4$] 20:3 *n*-6 at 0 hours (mean values ± SEM, *n* = 4).

especially with deuterium labeling. By measuring the conversion of deuterium-labeled [18,19-^{2}H$_4$] dihomo-linolenate (20:3) into [18,19-^{2}H$_4$] arachidonate (20:4), the reality of this conversion, the impairment of Δ5 desaturase in diabetic patients, and the insulin dependence of this enzyme in humans could be demonstrated (11).

In this study each subject received 2 g of deuterium-labeled dihomo linolenic acid (free acid form) for each test. The distribution of this precursor in the TG and phospholipid (PL) fractions of plasma lipids was determined by GC-MS with MID over time, as shown in Figure 23.8. The change of plasma PL deuterium-labeled arachidonate, a conversion product of Δ5 desaturase, was time-dependent in diabetics both before and after insulin treatment, demonstrating the effect of insulin treatment (Figure 23.9).

Simultaneous analysis of the behavior of geometrical or optical isomers labeled with different deuterium loads is an elegant way to compare the metabolic fate of the isomers under the same biological conditions. This method clearly demonstrated the striking difference in behavior between *cis* and *trans* fatty acids as well as between positional isomers in plasma cholesterol esters (12).

A recent clinical application identified an altered discrimination between stereoisomers of α-tocopherol involved in a familial isolated vitamin E deficiency (13). After simultaneous ingestion of 20 mg of each of the α-tocopherol acetate isomers, (natural isomer [^{2}H$_3$] RRR and diastereo isomer [^{2}H$_6$] SRR), the change of concentration in plasma and lipoprotein with time (Figure 23.10) clearly demonstrated impaired retention of the natural α-tocopherol in the affected patient.

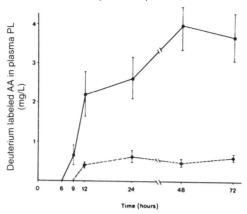

Figure 23.9. Deuterium-labeled 20:4 n-6 (arachidonate) in phospholipids of diabetic patients before (●---●) and after (●—●) insulin treatment and diabetes equilibration. The concentration in the plasma is shown after ingesting 2 g [18,19-^{2}H$_4$] 20:3 n-6.

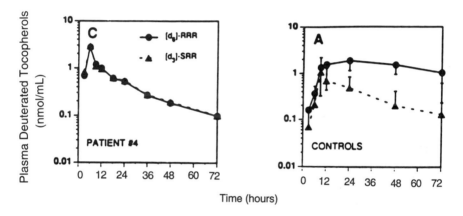

Figure 23.10. Plasma concentrations of [^{2}H$_6$]-RRR and [^{2}H$_3$]-SRR tocopherols in controls. (A) Mean of seven subjects ± SD (B) Patient affected by familial isolated vitamin E deficiency. Each subject received a capsule containing 20 mg of each isomer in the acetate form after an overnight fast. *Source:* Traber et al. (13).

Other Perspectives

Carbon-13 NMR spectroscopy has been used in several model systems, such as model membranes (14) subcellular fractions (15), and microorganisms (16), but the low sensitivity of this method impaired its utilization in vivo and only preliminary but encouraging data appear for "ex vivo" tracer analysis of ^{13}C-labeled fatty acids in mammals (17).

Conclusion

From the data available at the present time, ^{13}C-labeling and IR-MS appear to be most suitable for studies of energetic and oxidative metabolism. Deuterium-labeling and GC-MS detection using MID is convenient for comparative studies of metabolic behavior or pathways of homologous molecules when the analytical separation is cumbersome, difficult, or impossible.

The relatively low cost and the availability of deuterium, the wide variety of samples acceptable in GC, and the high precision of IR-MS makes the combination of GC-IRMS with deuterium-labeled tracers a promising technique for the future. However, this interesting combination could remain theoretical until technical difficulties for automation can be overcome.

References

1. Thompson, G.N., Pacy, P.J., Ford G.C., and Halliday, D. (1989) *Biomed. Mass Spectrom. 18:* 321–327.
2. Wolfe, R.R. (1984) *Laboratory and Research Methods in Biology and Medicine,* Vol. 9, Alan R. Liss Inc., New York.
3. Matwiyoff, N.A., and Ott, D.G. (1973) *Science 181:* 1125–1133
4. Jones, P.J.H., Pencharz, P.B., and Clandinin, M.T., (1985) *Am. J. Clin. Nutr. 42:* 769–777.
5. Leitch, C.A., and Jones, P.J.H. (1991) *Biol. Mass Spectrom. 20:* 392–396.
6. Jones, P.J., Dendy, S.M., Frohlich, J.J., Leitch, C.A., and Schoeller, D.A. (1992) *Arteriosclerosis and Thromb. 12:* 106–113.
7. Hachey, D.L., Silber, G.H., Wong, W.W., and Gurza, C. (1989) *Pediatr. Res. 25:* 63–68.
8. Halliday, D., Vankatesan, S., and Pacy, P. (1993) *Am. J. Clin. Nutr. (suppl.) 57:* 726S–31S.
9. Beylot, M., Martin, C., Beaufrère, B., Riou, J.P., and Mornex, R. (1987) *J. Lipid Res. 28:* 414–422.
10. Emken, E.A., Adlof, R.O., Hachey, D.L., Garza, C., Thomas, M.R., and Brown-Booth, L. (1989) *J. Lipid. Res. 30:* 395–402.
11. El Boustani, S., Causse, J.E., Descomps, B., Monnier, L., Mendy, F., and Crastes de Paulet, A. (1989) *Metabolism 38:* 315–321.
12. Emken, E.A., Rohwedder, W.K., Adlof, R.O., Dejarlais, W.J., and Gulley, R.M. (1986) *Lipids 21:* 589–595.
13. Traber, M.G., Sokol, R.J., Kohlshütter, A., Yokota, T., Muller, D.P.R., Dufour, R., and Kayden, H.J. (1993) *J. Lipid Res. 34:* 201–210.
14. Bhamidipati, S.P., and Hamilton, J.A. (1993) *J. Biol. Chem. 268:* 2431–2434.
15. Jin, S.J., Hoppel, C.L., and Tserng, K.Y. (1992) *J. Biol. Chem. 267:* 119–125.
16. Jung, S., Lowe, S.E., Hollingsworth, R.J., and Zeikus, J.G. (1993) *J. Biol. Chem. 268:* 2828–2835.
17. Cunnane, S.C., McDonagh, R.J., Narayan, S., and Kyle, D.J. (1993) *Lipids 28:* 273–277.

Chapter 24
Utilization of Stable Isotopes to Study the Compartmental Metabolism of Polyunsaturated Fatty Acids: An In Vivo Study Using ^{13}C-Docosahexaenoic Acid

M. Croset[1], N. Brossard[1], J. Lecerf[1], C. Pachiaudi[2], S. Normand[2], V. Chirouze[3], J.P. Riou[2], J.L. Tayot[3], and M. Lagarde[1].

[1]INSERM U352, Chimie Biologique, INSA-Lyon, 69621 Villeurbanne; [2]INSERM U197, Faculté de Médecine A. Carrel, 69008 Lyon; and [3]IMEDEX, 69630 Chaponost, France.

Introduction

N-6 and n-3 polyunsaturated fatty acid (PUFA) families are the two series of essential fatty acids in mammals. Linoleic acid (18:2n-6) is the precursor of the n-6 series, and its deficiency leads to well-known symptoms, such as growth retardation and skin lesions. Linolenic acid (18:3n-3) is the precursor of the n-3 series and is desaturated and chain elongated to eicosapentaenoic acid (20:5n-3), docosapentaenoic acid (22:5n-3), and docosahexaenoic acid (22:6n-3).

N-3 fatty acid deficiency does not produce classical symptoms of essential fatty acid deficiency, since growth is normal (1), but the depletion of 22:6n-3 in some tissues, such as the brain and retina, during periods of growth leads to functional impairment (2–4). 22:6n-3 is weakly present in blood lipids but is concentrated in the specialized neuromembranes, such as the brain synaptosomes and photoreceptor outer segments of the retina. Limited visual acuity and altered physiologic responses of the retina have been observed in animals (5) and preterm infants (6,7), as a result of 22:6n-3 deficiency.

20:5n-3 and 22:6n-3 are found at high levels in fish and marine food (8). Their concentrations are increased in human lipids at the expense of n-6 PUFA, after dietary supplementation with fish oils. These changes are believed to have beneficial effects on thrombosis, probably by reducing the concentration of plasma triglycerides and by fatty acid incorporation into cellular lipids where they interfere with the arachidonic acid cascade (9). Despite the biological functions of docosahexaenoic acid, little has been learned about its dynamic forms of plasma transport for efficient distribution to tissues. In humans, 22:6n-3 enrichment has been analyzed in blood lipid components after daily ingestion of relatively high amounts of oils, but this approach cannot report on the fatty acid turnover.

More information can be obtained on the dynamic fluxes of PUFA with tracer experiments based on isotope substitution. The transport of 22:6n-3 to target tissues via the circulatory system has been studied in rats with ^{14}C-labeling (10). In humans, utilization of stable isotopes has been developed, since the use of radiotracers is limited due to concerns over the risk of radiation (11). The investigation of 22:6n-3 metabolism in humans would be facilitated by the availability of a tracer labeled with stable isotopes that could be detected in the blood with high sensitivity. Looking for these requirements, triglycerides containing 22:6n-3 enriched with ^{13}C (^{13}C-22:6-TG) have been synthesized and fed to rats to monitor the appearance of ^{13}C-22:6n-3 in blood lipids with gas chromatography combustion-isotope ratio mass spectrometry (GCC-IRMS). This technique can detect isotope enrichment (IE) as low as 0.0010 atom ^{13}C percent (At%) and could be easily adapted to humans.

Methods

Blood from healthy volunteers or rats was withdrawn into 1/9 vol of citric acid, citrate, and dextrose (ACD) and plasma obtained by centrifugation at 100 g for 15 min. Platelets were pelleted by plasma centrifugation, and the supernatant was used to prepare lipoprotein classes. Very low density lipoprotein + chylomicrons, HDL, LDL and nonlipoprotein proteins were obtained by ultracentrifugation in a fixed angle rotor at 600,000 g for 2 h on a KBr gradient, similar to that used by Chung et al. in a vertical rotor (12). Total lipids were extracted according to Bligh and Dyer (13), and lipid classes were separated by thin-layer chromatography (TLC). After separation, lipids were scraped off the plate and directly treated with 5% H_2SO_4 in methanol for fatty acid methyl ester preparation. N-3 PUFA methyl esters were purified by reverse phase high-performance liquid chromatography (HPLC) using a 5 µm Supersphere column and an isocratic elution with acetonitrile/H_2O (8:2, v/v [14]).

For feeding experiments, ^{13}C-22:6-TG was synthesized with microalgae grown on [1-^{13}C] glucose for 7 d under heterotrophic conditions. Lipids were extracted from the biomass; then the triglycerides were purified by TLC. Their fatty acid composition was analyzed by GLC using a SP 2380 column (Supelco [15]) and 22:6n-3. Isotope enrichment was measured by GCC-IRMS. Three milligrams of ^{13}C-22:6-TG were given intragastrically to rats by intubation, while the control animals received 0.6 mL of cream-free milk which was used as the vehicle. At 3, 6, 12, and 18 h after ingestion, animals were sacrificed and approximately 10 mL of blood were drawn into ACD. Brains were quickly excised, then homogenized into the Folch mixture.

Fatty acid methyl esters were analyzed by GCC-IRMS to determine their ^{13}C/^{12}C ratios. Samples were injected into an HP 5890 gas chromatograph equipped with a Ross injector and a capillary column (SP-2380, 0.32 mm × 30 m). The temperature was held at 160°C for 1 min and raised at 20°C/min to 230°C where it was maintained for 5 min. At the end of the column, a heart-cut valve directed the helium flow either to the flame ionization detector or to a quartz tube filled with CuO and heated at 800°C to generate CO_2 from fatty acids by catalytic combustion. The effluent was driven into a water trap before ionization in the source of the isotope ratio mass spectrometer by electron impact (SIRA 12, VG Isogas, Middlewich, UK). Extraction and ion focusing were obtained by a series of electrostatic lenses. The different isotopic isomers of mass 44 (main ion: $^{12}C^{16}O^{16}O$), mass 45 ($^{13}C^{16}O^{16}O$, $^{12}C^{16}O^{17}O$), and mass 46 ($^{12}C^{17}O^{17}O$, $^{12}C^{16}O^{18}O$, $^{13}C^{16}O^{17}O$) were separated by a uniform magnetic field and collected onto three different collectors at m/z 44, 45 and 46 until the 44 signal returned to the baseline value. The heart-cut valve was controlled by a computer (HP 9816) which also calculated and reported the surface area of the main ion peak (m/z 44) and the ratios of 45/44 and 46/44 ions. A CO_2 reference of known enrichment, calibrated against the international standard (Pee Dee Belemnite: PDB) was automatically injected into the spectrometer before and after the CO_2 peaks generated from the fatty acids (16,17). The $^{13}C/^{12}C$ values of samples and the reference were used to calculate the δ per 1000 value ($\delta^{13}C$‰) with the following equation:

$$\delta^{13}C\text{‰} = \frac{^{13}C/^{12}C \text{ sample} - {}^{13}C/^{12}C \text{ reference}}{^{13}C/^{12}C \text{ reference}} \times 10^3$$

where the reference is the international PDB standard. $\delta^{13}C$‰ was transformed to ^{13}C At% with the formula:

$$At\% = \frac{100 \times R\,(0.001\delta^{13}C\text{‰}S + 1)}{1 + R\,(0.001\delta^{13}C\text{‰}S + 1)}$$

where R is the $^{13}C/^{12}C$ of the international PDB reference ($R = 0.0112372$) and S refers to the sample.

Results and Discussion

Precision of $\delta^{13}C$‰ values of natural and commercially available n-3 PUFA was tested by daily evaluation of the molecules. The analyses were performed on 100–300 ng of PUFA methyl esters which generated a sharp peak of CO_2 (Figure 24.1). $\delta^{13}C$‰ values for 22:6n-3 were -27.58 ± 0.27 (mean ± SD) which gave a relative standard deviation of 0.97%. Relative standard deviations obtained on other n-3 PUFA and biological samples were very close.

A standard curve was generated from dilutions of ^{13}C-22:6n-3 methyl esters with natural 22:6n-3. To increase the $^{13}C/^{12}C$ ratio from 1.0800 to 1.1300 At% 0.1–4.9% of ^{13}C-22:6n-3 enriched to 2.0420 At% was added to natural 22:6n-3 at 1.0800 At%. The line obtained by plotting measured At%, against calculated At% was linear with a correlation coefficient of 0.999. For three standard points analyzed in triplicate at minimum, maximum, and the middle of the curve, the standard deviation was less than 0.0002 At%. Isotopic enrichment as low as 0.0010 At% could be detected. The limit of detection reported here is in a good agreement with that obtained for glucose (17) and highly enriched stearic acid (18).

Fatty acid incorporation in plasma and cell lipids, as well as the metabolic conversion of PUFA have been studied with deuterated (19,20) and ^{13}C fatty acids (21)

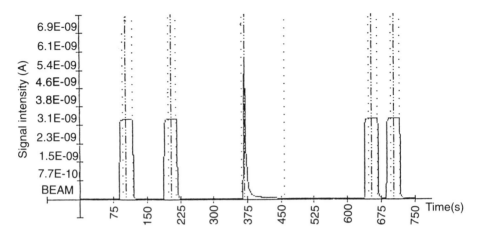

Figure 24.1. A typical GCC-IRMS plot of signal from m/z 44 detector for an injection of 22:6n-3 methyl ester. The four square-shaped peaks were calibrated CO_2, and the peak at 375 sec was issued from 22:6n-3 combustion.

using selective ion monitoring-GC-MS. Although these techniques have the advantage over GCC-IRMS to report on the specific IE of a particular atom, much higher detection thresholds (around 0.5 At%) were reported. Similarly, ^{13}C nuclear magnetic resonance spectroscopy, which has been used to study the uptake of [U-^{13}C]PUFA by rat tissues, can monitor the fate of defined carbons but has the disadvantage of low sensitivity (22).

To validate the method *in vivo*, a kinetic tracer experiment was performed in rats ingesting a single dose of 3 mg ^{13}C-22:6-TG, equivalent to around 1 mg of ^{13}C-22:6n-3. The fatty acid composition of ^{13}C-22:6-TG showed that 22:6n-3 represented 35 mol% of total fatty acids and was the sole n-3 PUFA detectable. The IE of ^{13}C-22:6n-3 according to the initial dilution of [1-^{13}C] glucose, was 2.0420 At%, which is roughly double the natural abundance. The 22:6n-3 labeling in triglycerides of the chylomicrons + VLDL fraction was highest, reflecting the intestinal hydrolysis of ^{13}C-22:6-TG and the newly synthesized triglyceride secretion (Figure 24.2). A fast labeling decrease in this fraction was concomitant with increased IE in HDL-TG, which is in good agreement with the known relationship between these two lipoprotein classes.

Phospholipid labeling was more linear with time for the period studied. The incorporation of ^{13}C-22:6n-3 in red blood cell phosphatidylcholine (PC) increased up to 18 h (Figure 24.3), and was already apparent 3 h after ingestion (not shown). In contrast, a weak incorporation of ^{13}C-22:6n-3 in phosphatidylethanolamine (PE) was observed only at 18 h and was not detectable at earlier times (not shown). Similarly, labeling in the brain was detected in PC, and not in PE, which may be due to higher dilution of the molecule in the endogenous pool of PE or to different mechanisms for 22:6n-3 incorporation or transfer in this phospholipid versus PC (Figure 24.4).

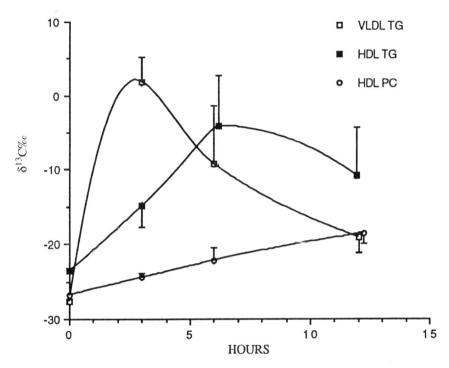

Figure 24.2. $^{13}C/^{12}C$ ratios ($\delta^{13}C$‰) in 22:6n-3 from rat lipoprotein lipids, as a function of time after ingestion of triglycerides containing 22:6n-3 enriched with ^{13}C. Values are means ± S.E.M. from five different animals. Abbreviations: TG, triglycerides; PC, phosphatidylcholine.

Finally, it has been possible to study the retroconversion of ^{13}C-22:6n-3. Although, no 20:5n-3 and 22:5n-3 were detectable in the triglyceride ingested, both ^{13}C fatty acids were detected in HDL-PC, 12 h after ingestion (Figure 24.5) and were measurable at earlier times. The retroconversion of 22:6n-3 has been described in rat liver using radiotracers (24), and in humans fed pure 22:6n-3 ethyl ester (25). In the latter case, 20:5n-3 was detected by GC in plasma phospholipids, after a 6-day supplementation. The sensitivity obtained with the ^{13}C-labeling allowed early detection of this event and would facilitate characterization of this pathway in humans.

To apply this technique to humans, the authors verified that $^{13}C/^{12}C$ ratios were measurable in all lipid fractions of lipoproteins and blood cells, even those more limited in terms of dose size of 22:6n-3, such as lysoPC-albumin, TG-HDL, or TG-LDL. The natural ^{13}C abundance is 1.10 At%, or a content of less than 2 g/kg. Thus, the ^{13}C enrichment generated in metabolic studies has always had a baseline. The mean values of 22:6n-3 $\delta^{13}C$‰ in all lipid components of lipoproteins and blood

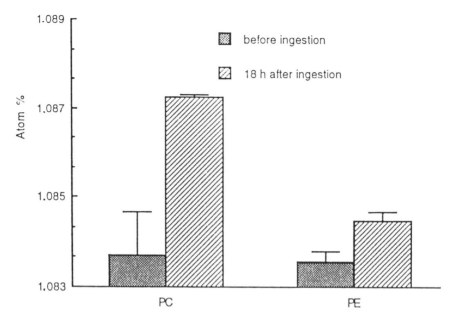

Figure 24.3. Incorporation of ^{13}C-22:6n-3 in rat red cell phosphatidylcholine (PC) and phosphatidylethanolamine (PE), 18 h after ingestion of triglycerides containing 22:6n-3 enriched with ^{13}C. Values (At%) represent the isotope enrichments and are means ± S.D. from three different animals.

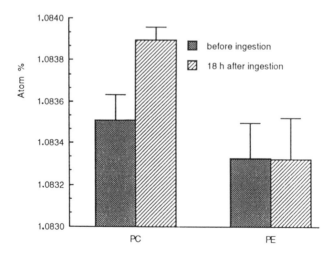

Figure 24.4. Incorporation of ^{13}C-22:6n-3 into rat brain phosphatidylcholine (PC) and phosphatidylethanolamine (PE), 18 h after ingestion of triglycerides containing 22:6n-3 enriched with ^{13}C. Values represent the isotope enrichments (At%) and are means ± S.D. from three different animals.

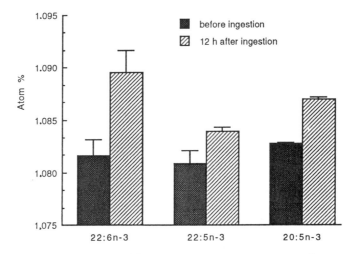

Figure 24.5. Retroconversion of ^{13}C-22:6n-3 into ^{13}C 22:5n-3 and ^{13}C-20:5n-3 in rat 12 h after ingestion of triglycerides containing 22:6n-3 enriched with ^{13}C. The ^{13}C/^{12}C ratios (At%) in 22:6n-3, 22:5n-3 and 20:5n-3 from rat HDL phosphatidylcholine were measured 12 h after ingestion of triglycerides containing 22:6n-3 enriched with ^{13}C. Values are means ± S.D. from three different animals.

cells of three different subjects was −27.03 ± 1.01 (1.0815 ± 0.0011 At%). This natural abundance could be measured in all components from 10 mL of blood. The standard deviation of 0.0011 At%, including dispersion between subjects, showed that precision was preserved throughout the treatment of biological samples.

Since the ingestion of a small amount of ^{13}C-22:6-TG with a ^{13}C abundance double the natural abundance enabled study of the lipid componental metabolism of 22:6n-3 in animals, this type of triglyceride appears suitable for human studies. Various fatty acids enriched with stable isotopes could be obtained by growing different strains of microalgae, a rapidly expanding technology (26). The much lower detection limits of IE obtained by GCC-IRMS compared to selecting ion monitoring-GCMS implies utilization of smaller quantities of tracers or of tracers having lower ^{13}C enrichment thus reducing the cost of experiments.

References

1. Salem, N., and Ward, G.R. (1993) in *World Rev. Nutr. Diet.,* Simopoulos, A.P., Karger, Basel, vol. 72, pp. 128–147.
2. Neuringer, M.G., Anderson, J., and Connor, W.E. (1988) *Ann. Rev. Nutr. 8:* 517–541.
3. Bazan, N.G. (1990) *Nut. and the Brain 8:* 1–24.
4. Carlson, S.E., and Salem, N. (1991) in *Health Effect of ω-3 Polyunsaturated Fatty Acids in Seafoods. World Rev. Nutr. Diet.,* Simopoulos, A.P., Kifer, R.R., Martin, R.E., and Barlow, S.M., Karger, S., Basel, vol. 66, pp. 74–86.
5. Neuringer, M., Connor, W.E., Lin, D.S., Barstad, L., and Luck, S. (1986) *Proc. Natl. Acad. Sci. USA 83:* 4021–4025.

6. Uauy, R., Birch, D., Birch, E., Hoffman, D., and Tyson, J. (1992) in *Essential Fatty Acids and Eicosanoids,* Sinclair, A., and Gibson, R., American Oil Chemists' Society, Champaign, Illinois, pp. 197–202.
7. Carlson, S.E., Werkman, S.H., Rhodes, P.G., and Tolley, E.A. (1993) *Am. J. Clin. Nutr. 58:* 35–42.
8. Lands, W.E.M. (1986) *Fish and Human Health,* Academic Press, New York.
9. Harris, W.S., Connor, W.E., and Goodnight, S.H. (1981) *Progr. Lipid Res. 20:* 75–79.
10. Li, J., Wetzel, M.G., and O'Brien, P.J. (1992) *J. Lipid Res. 33:* 539–548.
11. Wolfe, R.R. (1992) *Radioactive and Stable Isotope Tracers in Biomedicine: Principles and Practice of Kinetic Analysis,* Wiley-Liss, New York.
12. Chung, B., Segrest, J.P., Ray, M.J., Brunzell, J.D., Hokanson, J.E., Krauss, R.M., Beaudrie, K., and Cone, J.T. (1986) *Methods Enzymol. 128:* 181–209.
13. Bligh, E.G., and Dyer, W.J. (1959) *Can. J. Biochem. Physiol. 37:* 911–917.
14. Moore, S.A., Yoder, E., Murphy, S., Dutton, G.R., and Spector, A.A. (1991) *J. Neurochem. 56:* 518–524.
15. Croset, M., Bayon, Y., and Lagarde, M. (1992) *Biochem. J. 281:* 309–316.
16. Jumeau, J., Fredman, P.A., Hall, K., Guilly, R., Pachiaudi, C., and Riou, J.P., U.S. Patent 4,916,313 (1990).
17. Normand, S., Pachiaudi, C., Khalfallah, Y., Guilluy, R., Mornex, R., and Riou, J.P. (1992) *Am. J. Clin. Nutr. 55:* 430–435.
18. Goodman, K.J., and Brenna, J.T. (1992) *Anal. Chem. 64:* 1088–1095.
19. Emken, E.A., Adlof, R.O., Rakof, H., Rohwedder, W.K., and Gulley, R.M. (1990) *Biochem. Soc. Trans. 18:* 766–769.
20. Boustani, S., Causse, J.E., Descomps, B., Monnier, L., Mendy, F., and Crastes de Paulet, A. (1989) *Metabolism 38:* 315–321.
21. Emken, E.A., Adlof, R.O., Hachey, D.L., Garza, C., Thomas, M.R., and Brown-Booth, L. (1989) *J. Lipid Res. 30:* 395–402.
22. Bougneres, P.F., and Bier, D.M. (1982) *J. Lipid Res. 23:* 502–507.
23. Cunnane, S.C., McDonagh, R.J., Naratan, S., and Kyle, D.J. (1993) *Lipids 28:* 273–277.
24. Schlenk, H., Sand, D.M., and Gellerman, J.L. (1969) *Biochim. Biophys. Acta 187:* 201–207.
25. Von Schacky, C., and Weber, P.C. (1985) *J. Clin. Invest. 76:* 2446–2450.
26. Radwan, S.S. (1991) *Appl. Microbiol. Biotechnol. 35:* 421–430.

Chapter 25

A Quick Method for Sterols Titration in Complex Media

D. Pioch, P. Lozano, C. Frater, and J. Graille

Laboratoire de Lipotechnie, CIRAD-CP, BP 5035, 34032 Montpellier Cedex, France.

Introduction

Sterols are minor lipid components most frequently analyzed for corporate rule and dietary reasons in quality-control laboratories. It is well known that the sterol fraction composition is the fingerprint of an oil or fat, hence proof of its origin (1–4). The statutory field has taken advantage of this feature to control margarines and other dairy product substitutes.

In the medical field, both doctors and dieticians recommend limiting cholesterol ingestion. As a result of "cholesterol awareness," a large number of dairy-product substitutes were brought onto the market; these products are often made by replacing all or part of the milk fat by an aqueous phase and/or a vegetable fat. Consequently, quality-control analysis of these elaborated products that contain a small proportion of animal fat, hence a small amount of cholesterol, often involves analysis of the sterol fraction.

For over 50 years the role of sterols has involved a substantial research effort in the analytical field (5). A bibliographical study shows that the numerous titration methods for sterols cannot be used to institute a method for rapid qualitative or semiquantitative control of the sterol fraction of a wide range of products. For example, gravimetric methods, precipitation of complexes with digitonin, (5) cannot be used to determine the composition of the sterol fraction. However, the existence of phytosterols and cholesterol can be detected by examining sterol crystals under a microscope (5). Colorimetric titration of cholesterol after oxidation using various reagents has been described in many publications (7–9). An enzymatic method using cholesterol oxidase (10,11) has been marketed in the form of a reagent kit. Among the chromatographic methods, high-performance liquid chromatography (HPLC) is promising (13,14). However, the individual titration of each sterol by gas chromatography (GC) is most frequently used (15–18). There are also methods involving refractometry and fluorimetry. A fair number of the listed methods are used for specific applications and cannot be used for the general foodstuff analysis.

Processing the sample is complicated, due to the large number of steps: fat extraction, saponification, extraction of the unsaponifiable fraction, sterol separation by thin-layer chromatography (TLC), sterol desorption, derivatization into silylated esters and GC analysis. Moreover, the diversity of the products to be analyzed (milks, creams, margarines, powders, etc.) also complicates fat extraction (19–21). All the disadvantages led the authors to set up a simplified procedure based on dissolving the sterols in dimethylformamide (DMF) as described in previous papers (22–24).

Material and Methods

Reagents and solvents were of analytical grade, and the samples analyzed were purchased from shops. The DMF method is based on placing the sample, containing 0.05–0.5 mg of sterols, in solution or suspension in 100 mL DMF without prior fat extraction (22–24). The process includes saponification for 2 min at 65°C in a closed flask with 0.5 mL of the DMF solution, 0.5 mL of a 2N KOH ethanol 96% solution, and 0.5 mL of cholestanol in DMF as the GC standard; extraction of the unsaponifiable fraction with 2 mL of hexane and 1 mL of water; evaporation of the solvent under a nitrogen stream; derivatization of sterols into silyl esters (BSTFA/TMCS, 80/20) for 30 min at 65°C; and GC analysis (Figure 25.1) on a J-W DB 1701 column (30 m; i.d. 0.33 mm; film thickness 0.25 µm; oven temperature 270°C; glass needle injector 300°C; flame-ionization detector [FID] 350°C; helium as carrier gas at 2 mL/min).

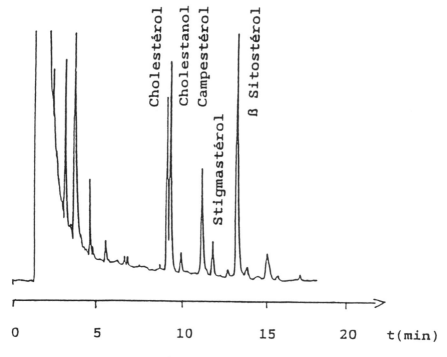

Figure 25.1. GC insapogram of a margarine sample.

For the reference method, the fat was extracted using the standardized ether-ammonia method (20,21), except for the dried milk and the β-cyclodextrin (bCD) processed products to which the Wolff and Castera method (25) and the ether-hydrochloric acid method (20,21) were applied respectively. The method of Mordret et al. (26) was used for extraction of the unsaponifiable fraction and sterol silylation before GC analysis. The dairy products had been processed with βCD as in Graille et al. (27–28).

Results and Discussion

Fat extraction must be carried out according to a protocol adapted to each type of product to be analyzed, for example, milk, powdered milk, sweetened condensed milk, butter (or margarine), cream cheese, lactoserum, and yogurt. The lack of a generalized fat extraction method considerably complicates the work of analysts, and stimulated the authors' research effort to simplify sample processing prior to actual titration.

The most direct analytical method is to replace the fat-extraction stage by dissolving the sample or by forming a suspension in an appropriate solvent. Among various solvents (dioxane, acetone, dichloromethane, DMF, methanol, isopropanol, n-butanol, t-butanol, and amyl alcohol), the selection is based on the appearance of the solution or suspension obtained from various dairy products, such as existence of two liquid phases, large amount of flocculent on the surface, high melting point (t-butanol), and incomplete wetting. DMF is a good compromise, but it is still not the ideal solvent. In the case of aqueous products, the sample weight must not exceed 10 g/100 mL of solution to obtain a micella or a suspension stable enough not to influence sampling or risk incomplete dissolution of the sterol fraction. If necessary, contact effectiveness between the sample and the solvent may be enhanced by using a high-speed plug-type blender. For high levels of protids and/or carbohydrates, the sample is taken after decanting the insoluble matter.

To process the DMF solution and sterol titration, a micromethod including GC analysis, originally developed by Mordret et al. (26) has been adapted. The reproducibility of the DMF method was tested. Three DMF solutions, corresponding to three samples taken from the same batch of milk cream, underwent duplicate saponification/derivatization reactions. Each of the six solutions obtained was injected four times into the gas chromatograph. The coefficient of variation was only 5%. Generally speaking, the coefficient of variation calculated for cholesterol titration from various dairy products, such as full-cream milk, skimmed milk, butter, lactoserum, cream cheese, and yogurt, ranged from 2–10% and averaged 5.3%.

Application of the DMF method to dairy products (22)

The process was applied to cholesterol titration for a range of commercially available dairy products. These products varied in consistency and composition; they were chosen to cover virtually all possibilities, water/oil and oil/water emulsions, dehydrated products, and products with a solid, paste, or liquid consistency. Comparison

TABLE 25.1
Titration of Cholesterol in Dairy Products[a] (22)

Sample	Main Components (wt%)			Method	
	Fat	Water	Other	DMF	Ref.
Liquid					
Full Cream Milk	3.6	88.5	7.9	125	115
Condensed Milk	7.5	76	16.5	260	240
Paste					
Milk Cream	31	63	6	920	890
Sweet Cond. Milk	9	69	22	310	310
Solid					
Butter	82	16	2	2060	2170
Half Skimmed Milk Powder	14	9	77	160	180

[a]mg of cholesterol/kg of product

of the data concerning the actual cholesterol titration (Table 25.1), obtained using the DMF method and the reference method, reveals that the results tally well for this range of dairy products. The same goes for other dairy products and by-products, such as skimmed milk, buttermilk, lactoserum, yogurt, and cream cheese.

βCD-Processed and/or Dehydrated Samples (23)

Using the HCl/Et$_2$O/petroleum ether standardized method (26) for a cholesterol titration of a cream containing 3 wt% βCD leads to a default value when compared to the actual cholesterol content determined for the cream sample (Table 25.2). Buttermilk analyzed under the same conditions produces similar results. The loss of cholesterol could be the result of partial inhibition by the βCD under the analytical conditions. The cholesterol loss was verified as not resulting from the medium heterogeneity, because there was no creaming or precipitation of a phase containing the βCD*cholesterol complex. Table 25.2 shows the results from the DMF method and the reference method (HCl/Et$_2$O/petroleum ether) for a milk cream. In the DMF method, the result is not modified by the presence of βCD. Therefore, the method was applied with no trouble to various dairy products, such as cream, butter, and buttermilk, either fresh or dehydrated (Table 25.2).

TABLE 25.2
Titration of Cholesterol by the DMF Method in βCD-Processed Dairy Product[a]

Sample	Fresh		Dehydrated
Milk Cream[c]	1.04	(1.07[b])	1.09
Cream + 3% βCD	1.05	(0.87[b])	1.03
Buttermilk	0.150	(0.095[b])	0.155
Butter	0.22		0.21
Butter[c]	2.20		2.06

[a]g/kg of fresh sample.
[b]Reference procedure.
[c]Not βCD-processed.

From the analytical results, it is clear that the DMF method is suitable for βCD-processed dairy products, such as low-cholesterol butter, since the traces of cholesterol are even more difficult to titrate because of the inhibitive action of the polyglucoside. This method also works with dehydrated samples and samples with low-fat and high-protein contents for which classical extraction procedures are often critical.

Application to Phytosterol Titration (24)

The satisfactory results for dairy products led to the method being tested on food products made from vegetable fats, which have a more complex sterol fraction composition. In addition to numerous phytosterols, these products may contain cholesterol in varying amounts, because certain so-called dietetic or low-fat preparations are made from a mixture of animal and vegetable fats. The range of products studied includes sunflower-oil-based margarine, a "mixed" spread (hydrogenated vegetable oils and milk fat) and two milk substitutes, coconut cream and soybean milk. A comparison of the results in Table 25.3 expressed as the composition of the sterol fraction and limited to cholesterol and the six main phytosterols, indicates a good correlation between the titrations carried out using the DMF method and the reference method. It should be noted that the DMF method enables checking of the origin of the main fat component and detection of any mixture, pollution, or even fraud. For example, a spread made with a fat mixture from vegetable and animal origin is indicated by the presence of 20% cholesterol. With regard to margarine, the small but still significant amount of cholesterol could come from the use of palm-oil and/or milk-fat-derived additives. The procedure was applied successfully to cosmetics, a liposome solution and a water/oil emulsion (24).

Conclusion

The original idea of direct product "dissolution" without prior fat extraction led to an acceptable method that was much quicker than the standardized procedures. It also produced substantial solvent savings and its simple implementation significantly reduced analyst exposure to harmful chemicals. The DMF protocol may be

TABLE 25.3
Composition of Sterol Fractions in the Dairy Product Substitutes Analyzed[a]

Sample	Spread		Margarine		Coconut Cream		Soybean Milk	
	DMF	Ref.	DMF	Ref.	DMF	Ref.	DMF	Ref.
Cholesterol	20	21	1.0	0.6	2.7	3.3	12	11
Campesterol	18	18	12	11	4.1	4.7	25	25
Stigmasterol	4.3	5.1	5.8	6.1	15	15	21	20
β-Sitosterol	47	46	58	58	25	25	44	44
δ-5Avenast.	2.3	2.2	3.2	3.0	36	35	tr	tr
δ-7Stigmast	7.7	9.0	18	19	18	18	tr	tr
δ-7Avenast.	—	—	2.3	2.1	—	—	—	—

[a]wt%

applied to a wide range of products irrespective of consistency, liquid, paste, or solid (block and powder), and the continuous phase of the emulsions analyzed. At this stage, the DMF method is suitable for checking industrial manufacturing and can be used as a routine procedure.

References

1. Mordret, F., Prevot, A., and Wolff, J.P. (1977) *Ann. Fals. Exp. Chim. 70:* 87–100.
2. Castang J., and Estienne, J. (1980) *Rev. Fse Corps Gras 27:* 437–441.
3. Blanchard, F., Castang, J., Derbesy, M., Estienne, J., Olle, M., Solere, M. (1979) *Ann. Fals. Exp Chim. 72:* 25–37.
4. Wolff, J.P. (1979) *Rev. Fse Corps Gras 26:* 23–28.
5. Sweeney, J.P., and Weilhrauch, J.L. (1977) *Critical Reviews in Food Science and Nutrition 8:* 131–159.
6. Association of Official Analytical Chemists (1980) *Official Methods of Analysis,* 13th edn., Arlington, VA, USA.
7. Malaspina, J.P., Fourche, J., Jensen, H., and Neuzil, E. (1980) *Ann. Biol. Clin. 38:* 207–213.
8. Burke, R.W., Diamondstone, B.I., Velapoldi, R.A., and Menis, O. (1974) *Clin. Chem. 20:* 794–802.
9. Zlatkis, A., and Zak, B. (1969) *Anal. Biochem. 29:* 143–148.
10. Naudet, M., and Hautfenne, A. (1985) *Pure and Appl. Chem. 57:* 899–904.
11. Chemin-Douaud, S., and Karleskind, A. (1979) *Rev. Fse Corps Gras 26:* 313–316.
12. Ritsch, J., and Entressangles, B. (1980) *Rev. Fse Corps Gras 27:* 185–188.
13. Perrin, J.L., and Raoux, R., (1980) *Rev. Fse Corps Gras 35:* 329–333.
14. Saaricsallany, A., Kindom, S.E., and Addis, P.B. (1989) *Lipids 24:* 645–651.
15. Mordret, F., Prevot A., Le Barbanchon, N., and Barbati, C. (1977) *Rev. Fse Corps Gras 24:* 467–475.
16. Kaneda, T., Nakajima, A., Fujimoto, K., Kobayashi, T., Kiriyama, S., Ebihara, K., Innami, T., Tsuji, K., Tsuji, E., Kinumaki, T., Shimma, H., and Yoneyama, S. (1980) *J. Nutr. Sci. Vitaminol 26:* 497–505.
17. French Standard NFT 60-249 (1989).
18. French Standard NFT 60-232 (1975).
19. Folch, J., Lees, M., and Sloane-Stanley, G.H. (1957) *J. Biol. Chem. 226:* 497–509.
20. French Standard NF V04-261 (1984).
21. French Standard NF V04-125 (1983).
22. Pioch, D., Lozano, P., Frater, C., and Graille, J. (1992) *Fat Sci. Technol. 94:* 268–272.
23. Pioch, D., Lozano, P., Frater, C., and Graille, J. (1992) *6th International Cyclodextrin Symposium,* Chicago.
24. Pioch, D., Lozano, P., Frater, C., and Graille, J. (1991) *Rev. Fse Corps Gras 38:* 381–386.
25. Wolff, R.L., Castera-Rossignol, A.F.M. (1987) *Rev. Fse Corps Gras 34:* 123–132.
26. Mordret, F., Coustille, J.L., and Taconne, L. (1984) *Rev. Fse Corps Gras 31:* 503–505.
27. Graille, J., Pioch, D., Serpelloni, M., and Mentink, L., European Patent EP 0406101A1 (1990).
28. Graille, J., Pioch, D., Serpelloni, M., and Mentink, L., European Patent EP 0408411A1, (1990).

Chapter 26

Some Improvements in Contaminants Analytical Methodology

F. Lacoste, A. Castera, J.L. Perrin, and J.L. Coustille

Institut Des Corps Gras, Rue Monge, Parc Industriel, F33600 Pessac, France.

Introduction

The presence of mineral and organic contaminants in oils and fats may have several origins, such as from the environment, a source of metals (Cd, As, and Pb) and polycyclic aromatic hydrocarbons from industrial wastes, grease-removal solvents, and agricultural fertilizers; production processes, which yield catalysts and extraction solvents; transport and storage of products in contaminated containers, a source of mineral oil and chemicals from previous cargoes; and packaging, which produces metal migration (stabilizers in plastics), monomer residues, and trace solvents of inks and glues. Although the potential toxicity of such pollutants is real, poisoning risks are rather limited due to efficient elimination during the oil-refining steps, careful conditioning, and top-quality packaging.

Nevertheless, under the present economical conditions, product quality control can be used as an excuse to enforce artificial importation barriers. Therefore, dispersal of highly sensitive analytical methods able to detect and measure potential contaminants in commercial products is essential to enhance export prospects. In this field, ITERG has developed methods to determine trace toxic metals, polycyclic aromatic hydrocarbons, and organic solvents in oils and fats.

Methodological Approach in the Determination of Trace Contaminants

Determination of trace contaminants generally requires a two-step procedure to extract and analyze the contaminants from the matrix. The extraction must be selective. It may become a problem with contaminants, such as polycyclic aromatic hydrocarbons (PAH), which have a strong affinity with lipids. It also must be quantitative and nonpolluting. For example, it is necessary to use special solvents with low metal contents to determine trace metal and in this case all the glassware has to be washed with diluted nitric acid. The analysis should be specific whenever possible. It has to be sufficiently sensitive to detect ppm (mg/kg) or ppb (µg/kg), and analysis must be without interference.

Toxic Metals: Cadmium, Lead, Arsenic, Tin, and Chromium

Among the different analytical techniques that can be used for metals, atomic absorption spectrometry with electrothermal atomization appears to be the method of choice for the determination of low levels of trace metals due to its high sensitivity and specificity. When dealing with oils and fats, direct determination of the oil sample dispersed in an organic solvent is possible. It ensures the lowest risk of contamination and drastically reduces analysis time.

In the case of elements such as cadmium, lead, and arsenic some difficulties do exist. These metals are rather volatile, and losses may occur if high temperatures are applied during the charring step (mineralization). Reducing the charring temperature will not eliminate enough of the matrix. The remaining matrix components may cause nonspecific interferences and contribute to a high background. Another problem comes from the low concentrations to be determined, ranging from as little as one ppb.

Some equipment and procedural improvements have brought solutions to these problems. Furnace pretreatment with refractory carbides, such as niobium or tantalum, increases the thermal stability of volatile elements. In the case of cadmium, atomic absorption is increased, while the background is lowered slightly (Figure 26.1). Moreover, the background due to matrix components can be reduced by using an oxygen-rich atmosphere during charring, while atomic absorption is maintained at the same level (Figure 26.2). Finally, atomization yield is improved by the use of a L'vov platform (Figure 26.3) and results in increased sensitivity. Operating conditions for the determination of cadmium, lead, tin, arsenic, and chromium are listed in Table 26.1.

In spite of the use of the improved methods and equipment, it is sometimes necessary to add a matrix modifier to improve the thermal stability of elements such as lead and arsenic. Performance of the methods developed in this laboratory have been studied (Table 26.2). Detection limits are between 0.25 and 10 ppb, depending on the element.

The cadmium area of linear response is rather narrow, compared to other metals. Generally, the cadmium concentration in oils and fats is under 20 ppm. In the case of higher contamination levels, it is possible to dilute the sample more. Relative standard deviations are between 2 and 10%, which is rather good.

PAH

Detecting PAH in vegetable oils requires sensitivity in the trace component range, since the individual contamination levels generally are about 1 ppb. In addition to the necessity for a sensitive detector, a selective extraction technique is required due to the affinity of PAH for lipids. Until recently, existing methods have proposed numerous liquid–liquid extractions followed by a silica cleanup procedure. The PAH measurement is then performed by reversed phase HPLC with fluorescence detection. Some drawbacks of these methods are the use of large quantities of

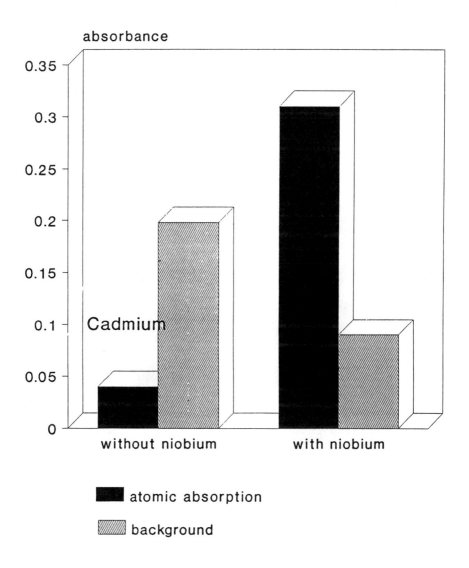

Figure 26.1. Some improvements: Furnace pretreatment with Niobium or Tantalum.

expensive and sometimes toxic solvents, variable recoveries (from 30 to 120%) resulting from degradation or loss of the PAH during the concentration steps, and a time-consuming and tedious sample preparation.

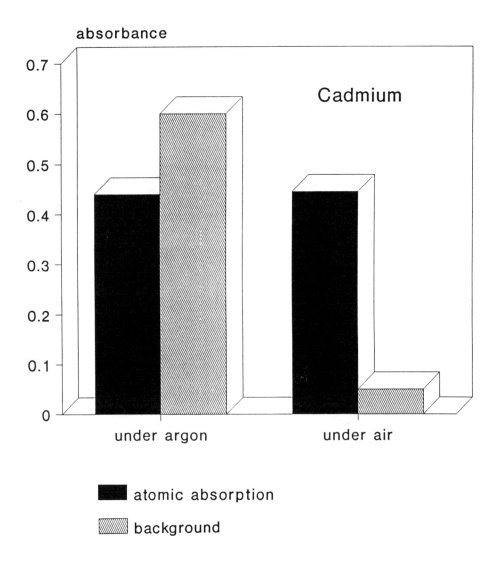

Figure 26.2. Some improvements: Charring under oxygen-rich atmosphere.

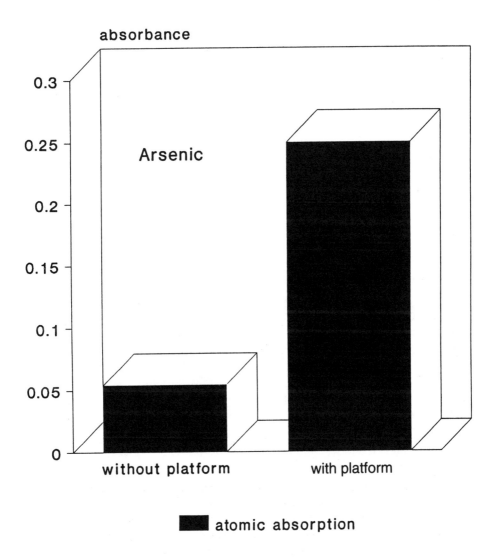

Figure 26.3. Some improvements: Use of L'vov platform.

TABLE 26.1
Determination of Toxic Metals Operating Conditions

Metal	Cd	Pb	Sn	As	Cr
Wavelength (nm)	228.8	283.3	224.6	193.7	357.9
Matrix modifier	—	Lecithins	—	Pd + diethyl-dithio-carbamic acid	—
Platform	+	+	+	+	+
Refractory carbide	Nb	Nb	Ta	—	—
Charring temperature (°C)	600/900	750	1200	600/1100	1200
Oxygen-rich conditions	+	—	—	—	—
Atomization (°C)	2400	1700	2600	2500	2600

TABLE 26.2
Determination of Toxic Metals Methods Performances

Metals	Cd	Pb	Sn	As	Cr
Detection limit	0.25 ppb	5 ppb	10 ppb	2 ppb	1 ppb
Characteristic mass (0.0044 A.S.)	1.2 pg	7 pg	25 pg	7 pg	4 pg
Linearity area	0.25–20 ppb	5–200 ppb	10–150 ppb	2–150 ppb	1–250 ppb
RSD for 10 ppb[a]	3%	2%	10%	6%	10%

[a]RSD = Relative Standard Deviation ($n = 10$).
Source: Lacoste, F., Castera, A., Lespagne, J., Rev. Fr. Corps Gras 40, (1–2), 19–31 (1993).

A rapid and comparatively simple PAH isolation procedure by donor-acceptor complex chromatography (DACC) has been developed. Direct injection of an oil sample dispersed in an organic solvent is performed on a tetrachlorophtalimido-propyl (TCPI) modified silica column which was provided by Felix (CNRS UA 35 Bordeaux I University). Selective elution of triglycerides and minor components, such as sterols and tocopherols, is obtained with a hexane-methyltertiobutyl ether mixture (75:25 v/v), while molecular complexes between PAH and the stationary phase are fixed on the column. Afterwards, the PAH fraction is recovered using a methylene chloride/hexane mobile phase (95:5 v/v) and analyzed after a concentration step by reversed phase HPLC with time-programmed fluorescence detection (Figure 26.4).

Figure 26.5 shows a chromatogram of an oil sample spiked with 1 ppb of various PAH, such as benzo(*a*)anthracene and benzo(*a*)pyrene. To test the efficiency of the DACC cleanup method, recoveries of the seven most toxic PAH were determined from an oil sample spiked with 1, 5, and 10 ppb of each PAH (Table 26.3).

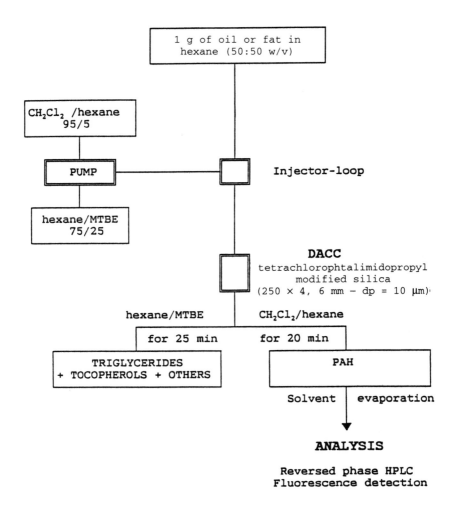

Figure 26.4. Some improvements: Flow chart of the isolation procedure for the determination of PAH.

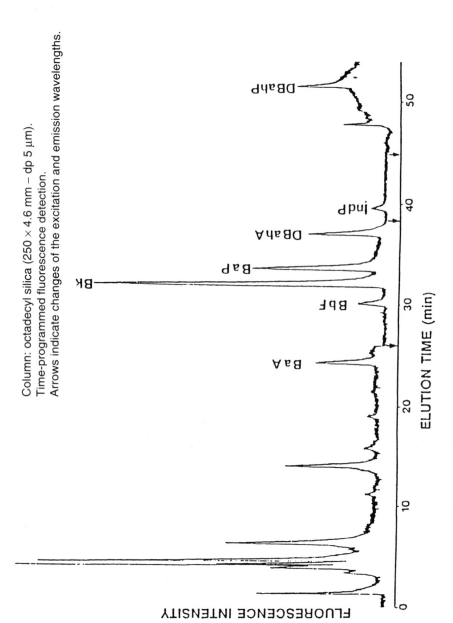

Figure 26.5. Determination of oil spiked with 1 ppb of each PAH (1).

TABLE 26.3
Determination of PAH, Recovery from Spiked Oil[a]

Compound	Level, ppb					
	1		5		10	
	Recovery (%)	RSD[b] (%)	Recovery (%)	RSD[b] (%)	Recovery (%)	RSD[b] (%)
Benzo(a)anthracene	78	20	95	11	106	4
Benzo(b)fluoranth.	82	17	104	11	107	9
Benzo(k)fluoranth.	99	14	106	4	113	6
Benzo(a)pyrene	94	13	103	4	111	5
Dibenzo(a,h)anthracene	100	16	103	10	114	6
Indeno(1,2,3-c,d)pyrene	82	28	117	14	121	7
Dibenzo(a,h)pyrene	80	25	112	23	117	27

[a]Each value is the average of four determinations.
[b]RSD: Relative Standard Deviation ($n = 4$).

Recoveries were between 78 and 121%. Relative standard deviations were between 13 and 28% for 1 ppb, and around 6% for 10 ppb except for dibenzo(a,h) pyrene. These results demonstrate the accuracy, precision, and repeatability of this procedure.

Organic Solvents

A preliminary step of contaminant extraction is required for organic solvent trace measurements. Then gas chromatography (GC) analysis can be performed.

The static headspace technique is the simplest and most frequently used method. Two standardized methods to determine pollutants in fats involves this type of desorption. However, depending on the contamination level of the samples, it is ineffective because of limited intrinsic sensitivity. Then, it is preferable to use a dynamic headspace system which preconcentrates volatile components on a porous polymer, improving the sensitivity. In both cases, there may be chromatographic interferences with flavor and autooxidation products desorbed from the matrix. In most cases, these problems can be solved through the use of a thick-film stationary phase capillary column and specific detectors, such as electron capture (ECD) or mass spectrometry (MS). Determination of chlorinated solvents can be performed by static headspace GC/ECD (Figure 26.6).

Tetrachloroethylene determination had been studied a few years ago in the authors' Institute. The problem now is to quantify all the chlorinated solvents in one analysis to satisfy European regulations. It is possible by using a very thick film of an apolar stationary phase (5 µm) and a specific detector. Detection limits depend on the chemical structure of the compounds. The more chlorine atoms and unsaturations the molecule has, the lower the detection limit is.

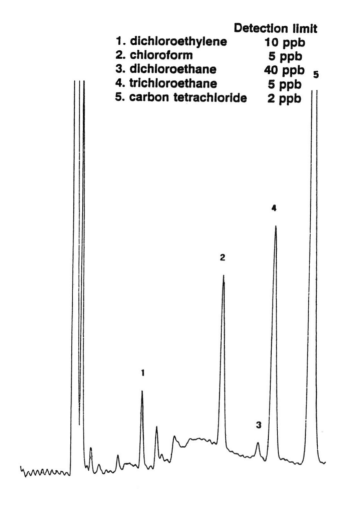

Oil spiked with: 1 (20 ppb) – 2 (25 ppb) – 3 (40 ppb) – 4 (20 ppb) – 5 (25 ppb)

Operating conditions

Desorption: 1 hour at 80°C
Injected headspace volume: 0.5 mL
Wide-bore methyl silicone column (5 µm)

Injector temp.: 150°C
ECD temp.: 250°C
Oven temp.: 50°C

Figure 26.6. Determination of chlorinated solvents by static headspace/GC/ECD.

TABLE 26.4
Determination of Hexane Residues

	IUPAC 2.607 Method	Improved Method
Desorption	Static headspace	Dynamic headspace (DCI)
	1 h at 80°C	15 min at 100°C
		Tenax Temp.: -20°C
GC capillary column	Methyl silicone	Methyl silicone
	Film thickness: 0.2 µm	Film thickness: 3 µm
Detector	FID	FID
Quantification limit	10 ppm	1 ppm
Field application	10–1500 ppm	1–20 ppm
RSD[a]	20% for 3 ppm	12% for 1 ppm

[a]Relative standard deviation.

In the case of hexane residue measurement in oils, IUPAC Standardized Method 2.607 is not suitable for hexane concentrations under 10 ppm, since European regulations now require concentrations to be under 5 ppm. It is therefore preferable to use a dynamic headspace system, in this case the Desorption-Concentration-Injection system (DCI) improving sensitivity about tenfold (Table 26.4).

Chromatograms presented in Figure 26.7 have been obtained with a refined oil spiked with 100 ppb and 1 ppm of technical hexane. The quantification limit has been fixed at 1 ppm in order to get a good estimation of each isomer of technical hexane. Determination of aromatic hydrocarbons in oils can be performed by DCI/GC/FID, but mass spectrometry should be used whenever possible since it offers a specific identification by an ion selection (Figure 26.8). Mass spectrometry detection also permits quantification of a compound even when interferences remain. This is the case of benzene which is coeluted with an autooxidation product (Figure 26.9). Ethylbenzene is a component of technical xylene and interferes with pentanoic acid. That is why GC/FID determination of these compounds is not suitable for concentrations under 100 ppb, with MS detection the quantification limit is about 10 ppb.

European Regulations

Table 26.5 lists some European regulations on mineral and organic contaminants. The European Economic Community regulations limit the concentration of hydrogenation catalysts, and of some solvents, such as hexane and chlorinated compounds. In this field, the Codex Alimentarius recommendations only concern two toxic metals. European regulations appear to be less restrictive, when compared to other countries, such as the United States.

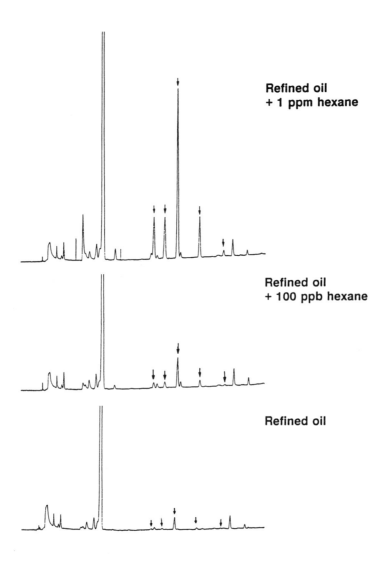

Figure 26.7. Determination of hexane residues by DCI/GC/FID.

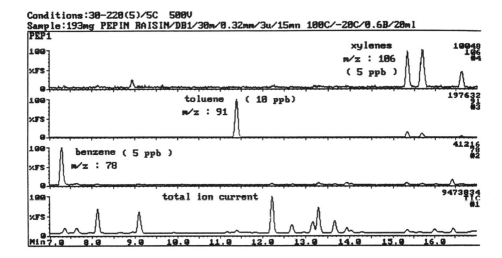

Operating conditions:
  DCI: 15 min at 100°C (Tenax = –20°C)
  GC: methylsilicone column (3 µm)
  MS: 30–120 uma (70 eV)

Figure 26.8. Determination of aromatic hydrocarbons by DCI/GC/MS. Quantification Limit: 10 ppb. Refined oil spiked with benzene, toluene, and xylene.

TABLE 26.5
Regulations for Contaminants in Oils and Fats

Contaminants	Maximum Concentration Limits	
	EEC	(Codex Alimentarius)
Metals		
Hydrogenation catalysts	200 ppb	
Chromium	50 ppb	
Lead		100 ppb
Arsenic		100 ppb
Solvents		
Hexane	5 ppm	
Chlorinated solvents	200 ppb (100 ppb each)	

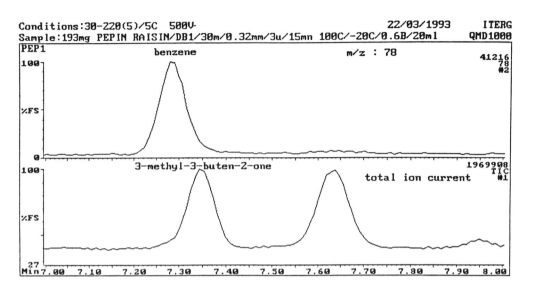

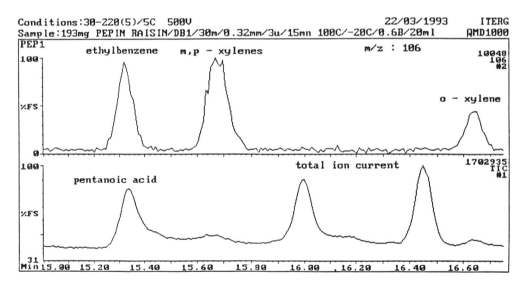

Figure 26.9. Some interferences of the determination of aromatic hydrocarbons by DCI/GC/MS.

References

1. Perrin, J.L., Poirot, N., Liska, P., Thienpont, A., and Felix, G. (1993) *Fat Sci. Technol.* 95: 46–51.
2. Lacoste, F., Castera, A., and Lespagne, J.(1993) *Rev. Fr. Corps Gras* 40: 19–31.

Chapter 27

Size Exclusion Chromatography Applied to the Analysis of Lipoproteins

Philip J. Barter

Department of Medicine, University of Adelaide, North Terrace, Adelaide SA, Australia.

Introduction

The evidence linking abnormalities of plasma lipoproteins with the development of premature coronary heart disease (CHD) is overwhelming, as is the evidence that treatment of such disorders translates into a reduction in coronary risk. This has stimulated a vast body of research into techniques designed to separate the different classes of plasma lipoproteins. These techniques have enabled quantification of the various lipoprotein classes in health and disease and investigation of their structure, function, regulation, and the mechanism of their relationship to CHD.

Plasma lipoproteins function to transport water-insoluble triglyceride and cholesterol between tissues through the aqueous blood plasma. Triglyceride is an important natural source of energy and is transported in plasma as a component of chylomicrons and very low density lipoproteins (VLDL). Cholesterol is present in all human tissues since it plays an essential role in the normal structure and function of cell membranes. Its concentration in these membranes is tightly regulated. Problems occur if there is either too little or too much cholesterol. The cholesterol levels in the membranes of extrahepatic tissues are controlled to a large extent by the rate at which cholesterol is transported to tissues by plasma low-density lipoproteins (LDL) and from tissues by high-density lipoproteins (HDL). Thus, plasma lipoproteins play fundamental roles in both the delivery of energy to tissues and the homeostasis of cell membranes in tissues throughout the body.

The different classes of plasma lipoproteins may be separated from each other on the basis of a number of physicochemical characteristics: by ultracentrifugation on the basis of differences in hydrated density, by agarose gel electrophoresis on the basis of differences in charge, by immunoaffinity chromatography on the basis of differing apolipoprotein composition, and by size-exclusion chromatography (SEC) or nondenaturing polyacrylamide gradient gel electrophoresis on the basis of differences in particle size. This report is concerned with the use of SEC to separate lipoprotein particles of differing size.

Plasma Lipoproteins

As outlined previously, the plasma lipoproteins function as transport vehicles for water-insoluble triglyceride and cholesterol. All plasma lipoproteins have the same general structure: a hydrophobic core of triglyceride and cholesteryl esters (the main form in which plasma cholesterol is transported) surrounded by a surface layer of apolipoproteins, phospholipids, and unesterified cholesterol (Figure 27.1). The surface layer is sufficiently hydrophilic to "solubilize" the entire particle. The various classes of lipoprotein differ size, density, apolipoproteins and whether the predominant core lipid is triglyceride or cholesteryl ester (Table 27.1).

Chylomicrons and Chylomicron Remnants

These particles are of intestinal origin. Their main function is to transport dietary triglyceride to tissues throughout the body and dietary cholesterol to the liver. They are the largest and least dense plasma lipoproteins. They have mainly triglyceride in the core. The main apolipoprotein of chylomicrons is apoB$_{48}$, a truncated form of the apoB$_{100}$ in VLDL and LDL. When a proportion of the chylomicron triglyceride has been hydrolyzed by lipoprotein lipase located at the surface of endothelial cells, the partially hydrolyzed catabolic product remaining is called a chylomicron remnant. Chylomicron remnants contain apoE in addition to apoB$_{48}$. The apoE is acquired from

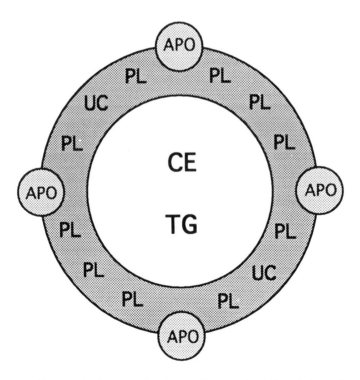

Figure 27.1. Schematic diagram of a plasma lipoprotein. Abbreviations: Apo, apolipoprotein; PL, phospholipid; UC, unesterified cholesterol; CE, cholesteryl ester; and TG, triglyceride.

TABLE 27.1
Plasma Lipoproteins

Class	Main Lipids	Main Apolipoprotein	Diameter (nm)
Chylomicrons	TG[a]	B_{48}, C	75–1000
Chylomicron remnants	TG, Chol	B_{48}, C, E	50–200
VLDL	TG	B_{100}, C	30–100
IDL	TG, Chol	B_{100}, E	25–30
LDL	Chol	B_{100}	20–25
HDL	Chol	AI, AII	7.4–12

[a]TG, Triglyceride; Chol, Cholesterol.

HDL and is necessary for the hepatic uptake of chylomicron remnants. Relative to chylomicrons, chylomicron remnants contain less triglyceride and more cholesterol.

VLDL

Very low density lipoproteins are of hepatic origin. Their main function is to transport triglyceride synthesized in the liver to extrahepatic tissues for use as a source of energy. Like chylomicrons, VLDL have a core rich in triglyceride. However, they differ from chylomicrons in that they are smaller and denser, and they contain $apoB_{100}$ rather than $apoB_{48}$ as their main apolipoprotein. Each VLDL particle contains a single molecule of $apoB_{100}$ which remains an integral component of the particle during its catabolic conversion to IDL and LDL. Very low density lipoprotein particles also acquire apoE as they are catabolized; about 80% of VLDL particles contain apoE in addition to $apoB_{100}$.

Intermediate Density Lipoproteins

Intermediate density lipoproteins (IDL) are formed from VLDL particles which have lost a proportion of their triglyceride and gained some additional cholesteryl esters which are transferred from HDL. They are smaller and denser than VLDL and have a core rich in both triglyceride and cholesteryl esters. They contain $apoB_{100}$ and apoE.

LDL

Low-density lipoproteins are the major transporters of cholesterol in human plasma. They are taken up by tissues in a process dependent on the tissue expression of a specific LDL, receptor. Low-density lipoproteins are the end-product of the catabolism of VLDL/IDL particles which have lost most of their triglyceride and all of their apoE. They are smaller and denser than IDL and contain a core which is predominantly cholesteryl ester. Each LDL particle contains a single molecule of $apoB_{100}$ as its sole protein component. The LDL fraction comprises a spectrum of particles of varying size and density. Low-density lipoprotein particles of differing size may be more or less atherogenic. Low-density lipoprotein particles are subject to a range of modifications during their transit in the plasma. Some of these modifications, including oxidative changes, may increase the atherogenicity of the particle. The LDL fraction also includes lipoproteins known as lipoprotein (a) (Lp[a]) in which a protein, apo(a), is linked to apoB$_{100}$. A possible relationship between Lp(a) and CHD has stimulated enormous interest in this lipoprotein.

HDL

The HDL are the smallest and densest of the plasma lipoproteins. Their main function is to transport cholesterol from extrahepatic tissues to the liver in a process known as reverse cholesterol transport. They include a number of discrete subpopulations of particles in which the predominant core lipid is cholesteryl ester. In terms of their apolipoprotein content, there are two main classes of HDL: those which contain apoA-I without apoA-II (LpA-I) and those which contain both apoA-I and apoA-II (LpA-I, A-II). Both classes exist in several discrete subpopulations that differ in size, density, and in their content of minor apolipoproteins. High-density lipoproteins are often classified solely on the basis of their size and density into the smaller and more dense HDL$_3$ and the larger and less dense HDL$_2$. High-density lipoproteins also exist transiently in a lipid-poor form in particles which contain surface components but no core lipids. These particles appear in the electron microscope as discs rather than spheres and are probably the form in which HDL are first secreted from the liver and intestine. Once in the plasma they are rapidly converted into spheroidal particles by the enzyme lecithin:cholesterol acyltransferase (LCAT) which esterifies their cholesterol and provides them with a core of cholesteryl esters.

Atherogenicity of Plasma Lipoproteins

Chylomicrons and Chylomicron Remnants

There is little indication that chylomicrons are atherogenic as evidenced by the lack of accelerated atherosclerosis in subjects with lipoprotein lipase deficiency and massive elevations of chylomicrons in plasma. Chylomicron remnants, by contrast, may be atherogenic. When chylomicron remnants are incubated with human arterial smooth muscle cells, there is an increase in cholesterol esterification and an accumulation of cholesteryl esters in the cells.

VLDL

It is currently uncertain whether VLDL are atherogenic. While subjects with elevated levels of VLDL tend to have an increased risk of developing CHD, it has been argued that the relationship may be indirect. For example, elevations of VLDL are often associated with reduced concentrations of HDL cholesterol and with small, dense LDL which may be responsible for the CHD. Furthermore, subjects with elevated VLDL levels may also have increased concentrations of potentially atherogenic IDL particles.

IDL

By definition, IDL are intermediate in density and size between VLDL and LDL. There is no clear division between IDL and VLDL. There are several lines of evidence linking IDL to CHD.

1. They have been shown in vitro to promote an accumulation of cholesterol in macrophages,
2. They appear to be responsible for the development of atherosclerosis in a number of animal species fed diets rich in cholesterol,
3. They are present in increased concentrations in human subjects with dys-β-lipoproteinaemia and familial combined hyperlipidaemia, conditions with a very high frequency of CHD,
4. It has been reported in human subjects that the presence, the severity, and the progression of CHD are all predicted by the concentration of IDL.

LDL

The fact that elevated levels of LDL predispose individuals to atherosclerosis has been established beyond reasonable doubt by the observations of patients with familial hypercholesterolaemia. These subjects have well-defined genetic defects of the LDL receptor which result in marked increases in the plasma concentration of LDL. They also have an extremely high incidence of premature CHD.

The mechanism by which LDL may cause atherosclerosis is uncertain. For example, when studied in vitro, unmodified LDL do not promote the accumulation of cholesteryl esters in macrophages and do not lead to the formation of the foam cells which are believed to initiate the development of atherosclerosis. However, if the LDL are first modified by oxidation, self-aggregation, or derivative formation, they are then able to promote the formation of foam cells. Evidence is accumulating that such modifications of LDL may take place within the artery wall, and that they may provide the link between LDL and atherosclerosis.

There is also evidence that different subpopulations of LDL may vary in their atherogenic potential. For example, subjects with CHD are frequently found to have an increased concentration of small, dense LDL. Such LDL are often associated with hypertriglyceridaemia, especially in subjects with noninsulin dependent diabetes mellitus or truncal obesity, and it is uncertain whether the small, dense LDL or some other factor is the true cause of CHD. It has been shown, however, that small, dense LDL are more readily oxidized than larger and less dense LDL are, suggesting that they may be directly related to the development of atherosclerosis.

HDL

High-density lipoprotein particles are clearly not atherogenic. The epidemiological observations that the risk of CHD correlates inversely with the concentration of HDL cholesterol has led to a view that HDL may protect against CHD. Such a suggestion has gained recent support from studies of transgenic mice in which overexpression of human apoA-I led to an increase in the concentration of HDL and an obvious protection against diet-induced atherosclerosis. However, it has been suggested that the relationship between HDL levels and CHD in human subjects may be indirect and may reflect an association of low-HDL levels with other factors (e.g., smoking, obesity, noninsulin dependent diabetes mellitus, and increased concentrations of β-VLDL) which may be the true cause of the CHD. Clearly, there is a need for further research into the relationship between HDL and CHD.

Separation of Plasma Lipoproteins by SEC

Size-exclusion chromatography on columns of agarose gel has been used successfully in the separation of plasma lipoproteins for many years (1–4). The technique effectively separates VLDL from LDL and LDL from HDL but does not separate VLDL from chylomicrons or effectively isolate IDL. As described originally, the method required a long elution time (more than 16 h) and a large elution volume to separate the major lipoprotein classes from plasma. More recently, these problems have been overcome by the use of an improved agarose gel matrix, which greatly reduces the separation time and improves the analysis of plasma lipoproteins (5). The highly cross-linked nature of the individual agarose beads in the improved gel ensures its overall rigidity, enabling elution at high pressure with a marked increase in eluent flow rate. Furthermore, the fractions collected after chromatography with the improved gel are sufficiently concentrated to be assayed directly for lipoprotein constituents without any requirement for prior concentration of the sample.

Unless stated otherwise, the chromatographic separations described in this report were conducted on columns of Superose 6B or Superose 6 HR 10 (Pharmacia, Uppsala, Sweden) connected to a high-pressure P-500 pump (Pharmacia). Lipoproteins were eluted with 0.05 M Tris HCl (pH 7.4) containing 0.15 M NaCl at flow rates of 30–45 mL/hr. In general, fractions of 0.6 mL were collected for analysis. Recoveries of lipids and apolipoproteins ranged from 80–90%. With this technique, lipoprotein separation is completed in 2–3 h. In some experiments, chromatography was conducted on two HR 10/30 columns of preparative grade Superose 6 and Superose 12 connected in series to a high-pressure pump.

When the human plasma fraction containing all classes of lipoproteins (the fraction isolated as the supernatant after ultracentrifugation at a density of 1.25 g/mL) is subjected to chromatography on a column of Superose 6 HR 10, there is a clear separation into peaks of VLDL, LDL, and HDL (Figure 27.2). When human whole plasma is applied to such a column, the absorbance 280 profile is dominated

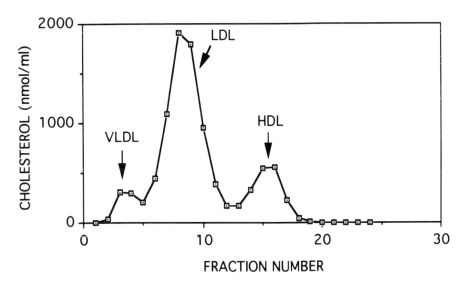

Figure 27.2. Separation of plasma lipoproteins by SEC. The plasma fraction of density < 1.25 g/mL was isolated by ultracentrifugation and then separated by SEC on a column of Superose 6HR 10/30 attached to a high-pressure pump. Lipoproteins were eluted with 0.05 M Tris-HCl (pH 7.4), containing 0.15 M NaCl at a flow rate of 30 mL/h and fractions of 0.6 mL each were collected for assay of cholesterol. The three peaks represent the VLDL, LDL, and HDL fractions.

by nonlipoprotein proteins. However, the elution of cholesterol under such conditions is identical to that after chromatography of the plasma fraction of d < 1.25 g/mL (5). Comparable profiles are also obtained for the distribution of other lipids and apolipoproteins in whole plasma (6).

The completeness of the separation of LDL and HDL by SEC is illustrated in Figure 27.3. The human plasma fraction of density 1.006–1.25 g/mL (containing the LDL plus HDL fractions) was applied to a column of Superose 6 HR 10. Fractions were assayed for apolipoprotein (apo) B as a marker for the elution of LDL and for apoA-I as a marker for the elution of HDL. It is apparent that there is virtually no overlap of the LDL and HDL fractions (Figure 27.3).

It is also possible by this technique to recover subfractions of HDL. While HDL elutes from the Superose 6 HR 10 column as a single peak, when individual fractions within the HDL peak are recovered and subjected to nondenaturing gradient gel electrophoresis, it becomes apparent that the chromatography has isolated several HDL subpopulations ranging from the largest of the HDL_2 to the smallest of the HDL_3 (7).

High-pressure SEC is undoubtedly an efficient and effective preparative technique for isolating discrete plasma lipoprotein fractions for quantification and for characterization. It is also a powerful tool in metabolic studies of plasma lipoproteins.

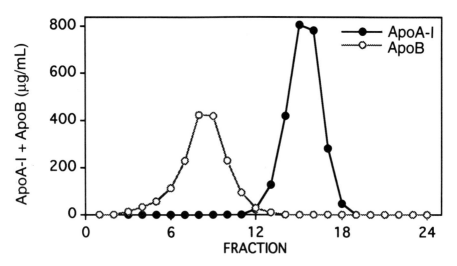

Figure 27.3. Separation of LDL and HDL by SEC. The plasma fraction of density 1.006–1.25 g/mL was isolated by ultracentrifugation and then subjected to SEC on a column of Superose 6 HR 10/30 as described in the legend to Figure 27.2. Fractions were assayed for apoB (open circles) as a marker of LDL and for apoA-I (closed circles) as a marker of HDL.

Studies of Cholesteryl Ester Transfer Protein

The cholesteryl ester transfer protein (CETP) promotes transfers and exchanges of cholesteryl esters and triglycerides between plasma lipoprotein fractions (8). Such transfers are readily demonstrated using SEC to separate plasma lipoprotein fractions. This technique has been used to obtain insights into the function of CETP in studies of rats, a species which is naturally deficient in activity of CETP (9). Rats were injected with a preparation of CETP isolated from human plasma. The effects of CETP on the distribution of lipid and apolipoprotein constituents among rat plasma lipoprotein classes were determined by analysis of fractions separated by SEC on columns of Superose 6B (10,11). In control animals, most of the plasma cholesterol was recovered in the HDL peak. After CETP injection, there was a reduced concentration of cholesterol in the HDL peak and an increase in concentration of cholesterol the VLDL and LDL fractions. Furthermore, a delay in the elution of HDL following injection of CETP indicated a reduction in the HDL particle size in the treated animals.

Studies of Lipoprotein Lipase and Hepatic Lipase

Activities of the heparin-releasable endothelial lipases, lipoprotein lipase and hepatic lipase, are known to play important roles in plasma lipoprotein metabolism. Both enzymes hydrolyze triglyceride and phospholipids in all lipoprotein fractions.

There is evidence from kinetic studies that lipoprotein lipase has a preference for lipids in VLDL and chylomicrons, while hepatic lipase has a preference for HDL lipids. However, most studies of substrate specificity have been conducted with artificial substrates or with isolated lipoprotein fractions. Size exclusion chromatography on columns of Superose 6HR 10 has been used to compare the effects of lipoprotein lipase and hepatic lipase when present in incubations containing mixtures of human lipoproteins (12). It was apparent that even when all lipoprotein fractions are present, lipoprotein lipase has a preference for triglyceride in VLDL, while hepatic lipase has a preference for the triglyceride in HDL.

Dissociation of ApoA-I from HDL

High-density lipoprotein particles are subject to extensive remodeling by a number of plasma factors. Newnham and Barter showed several years ago that CETP, hepatic lipase, and VLDL interact synergistically with HDL in a process which leads to a marked reduction in HDL particle size (13). The mechanism of this process is as follows. Cholesteryl ester transfer protein promotes an exchange of HDL cholesteryl esters for VLDL triglyceride. This leads to the formation of HDL which are depleted in cholesteryl esters and enriched in triglyceride. Hepatic lipase subsequently hydrolyzes a proportion of the HDL triglyceride to produce HDL particles which are depleted of core lipids and reduced in particle size. They also showed that there is a loss of a substantial proportion of the apoA-I from the HDL that coincides with a reduction in HDL size (14,15). However, this result was first observed after prolonged ultracentrifugation, a procedure which has been reported to strip apolipoproteins from HDL. Size exclusion chromatography, a much gentler technique was used to demonstrate a dissociation of apoA-I from HDL. Using columns of Superose 6 HR and Superose 12 HR connected in series, it was shown following incubation of HDL in the presence of CETP, VLDL, and hepatic lipase that a proportion of the apoA-I eluted after both the main HDL peak and albumin in fractions which were essentially free of lipid (14).

References

1. Van't Hooft, F., and Havel, R.J. (1981) *J. Biol. Chem. 256:* 3963–3968.
2. Fainaru, M., Havel, R.J., and Imaizumi, K. (1977) *Biochem. Med. 17:* 347–353.
3. Bowden, J.A., and Fried, M. (1970) *Comp. Biochem. Physiol. 32:* 391–400.
4. Rudel, L.L., Lee, J.A., Morris, M.D., and Felts, J.M. (1974) *Biochem. J. 139:* 89–95.
5. Ha, Y.C., and Barter, P.J. (1985) *J. Chromatog. 341:* 154–159.
6. Barter, P.J., Chang, L.B.F., and Rajaram, O.V. (1990) *Atherosclerosis 84:* 13–24.
7. Clifton, P.M., MacKinnon, A.M., and Barter, P.J. (1987) *J. Chromatog. 414:* 25–34.
8. Barter, P.J., Hopkins, G.J., and Calvert, G.D. (1982) *Biochem. J. 208:* 1–7.
9. Ha, Y.C., and Barter, P.J. (1982) *Comp. Biochem. Physiol. 71B:* 265–269.
10. Ha, Y.C., Chang, L.B.F., and Barter, P. (1985) *J. Biochim. Biophys. Acta. 833:* 203–210.
11. Ha, Y.C., and Barter, P.J. (1986) *Comp. Biochem. Physiol. 83B:* 463–466.

12. Newnham, H.H., Hopkins, G.J., Devlin, S., and Barter, P.J. (1990) *Atherosclerosis 82:* 167–176.
13. Newnham, H.H., and Barter, P.J. (1990) *Biochim. Biophys. Acta. 1044:* 57–64.
14. Clay, M.A., Newnham, H.H., and Barter, P.J. (1991) *Arterio. and Thromb. 11:* 415–422.
15. Clay, M.A., Newnham, H.H., Forte, T.M., and Barter, P.J. (1992) *Biochim. Biophys. Acta. 1124:* 52–58.

Chapter 28

Determination of Lipoprotein-Size Distribution by Polyacrylamide Gradient Gel Electrophoresis

Laurent Lagrost

Laboratoire de Biochimie des Lipoprotéines, INSERM CJF 93-10, Faculté de Médecine, Dijon, France.

Introduction

In fasting human plasma, three major lipoprotein classes have been identified: very low-density lipoproteins (VLDL), low-density lipoproteins (LDL), and high-density lipoproteins (HDL). Different lipoprotein fractions can be analyzed according to various criteria by using different processes. They can be separated on the basis of their density by using ultracentrifugation, their charge by using agarose electrophoresis, their apolipoprotein content by using immunological techniques, and their size by using gradient gel electrophoresis, gel filtration chromatography, or ultrafiltration. In particular, the recent use of polyacrylamide gradient gel electrophoresis revealed that each plasma lipoprotein class is composed of a large number of discrete lipoprotein subclasses. The heterogeneity of plasma lipoproteins, in particular LDL and HDL, could have some important implications since variations in the relative distribution of distinct subclasses of either LDL or HDL plasma fractions are associated with variations in the risk for coronary artery disease. The present paper will describe the electrophoretic procedure currently used to separate plasma lipoproteins on polyacrylamide gradient gels. In addition, the effect of a specific factor, the cholesteryl ester transfer protein (CETP), on the gradient gel patterns of both HDL and LDL fractions will be addressed.

General Method for Separating Lipoproteins Using Nondenaturing Polyacrylamide Gradient Gel Electrophoresis

Specific polyacrylamide gradients are used to study the size distribution of either LDL or HDL particles in human plasma. Low-density lipoprotein analysis is conducted on gradient gels usually ranging from 2–18% polyacrylamide (1,2) while HDL analysis can be carried out on more concentrated gels ranging from 4–30% polyacrylamide (3).

Recently, a practical method has been described to prepare homemade gels ranging from 3–31% polyacrylamide (4). This type of gel is particularly interesting since it allows simultaneous determination of the size distribution of both LDL and HDL plasma fractions. On gradient gels, separated lipoprotein subfractions can be localized by using two staining procedures. Lipoprotein particles can be stained for lipids by using Suddan Black B (1,5) or stained for proteins by using Brilliant Blue G Commassie (3). Finally, lipoprotein distribution profiles can be analyzed by laser densitometric scanning of the gels on a laser densitometer attached to an integrator. This latter step can provide information on the number, the mean apparent diameter, and the relative abundance of distinct lipoprotein subpopulations (3,6). The apparent diameter of the separated lipoprotein subfractions can be determined by comparison with protein and carboxylated latex bead standards subjected to electrophoresis with the samples. In addition, the relative proportions of various lipoprotein subpopulations can be obtained by determining the relative areas under the scan curves.

Size Distribution of Plasma Lipoprotein Fractions

HDL

In early ultracentrifugation studies, HDL were described as a population of particles with densities ranging from 1.063–1.21 g/mL. Subsequently, they were subdivided into two major subclasses, HDL_2 and HDL_3. More recently, an even greater heterogeneity was shown after separation of HDL subspecies on the basis of particle size by using polyacrylamide gradient gel electrophoresis. Indeed, Blanche and co-workers (3) revealed that plasma HDL can be subdivided into at least five subfractions with distinct apparent diameters: HDL_{2b} (9.7–12.9 nm), HDL_{2a} (8.8–9.7 nm), HDL_{3a} (8.2–8.8 nm), HDL_{3b} (7.8–8.2 nm), and HDL_{3c} (7.2–7.8 nm). Recent data indicated that the apolipoproteins may be important in determining the HDL particle size. A good correspondence between the distribution of HDL subpopulations separated either by gradient gel electrophoresis (3) or by immunoaffinity chromatography (7) has been observed. Figure 28.1 shows typical gradient gel profiles of plasma HDL obtained after separation of total plasma lipoproteins on 4–30% polyacrylamide gradient gels. Five distinct portions, corresponding to the five HDL subpopulations previously described (3), can be identified. The heterogeneity of HDL could have some important physiopathological implications, and recent observations suggest that HDL particles of different size could have different metabolic properties. For instance, strong inverse correlations were found between the plasma levels of HDL_{2b} and both the severity and the rate of progression of coronary lesions (8). Conversely, the abundance of the smallest HDL subpopulations, HDL_{3b} and HDL_{3c}, was shown to be significantly increased in patients with coronary artery disease as compared with healthy controls (9).

LDL

Plasma LDL have also been shown to consist of several subfractions of distinct size, density, and composition. The LDL heterogeneity has been demonstrated by ultracentrifugation and nondenaturing gradient gel electrophoresis (6) and up to seven

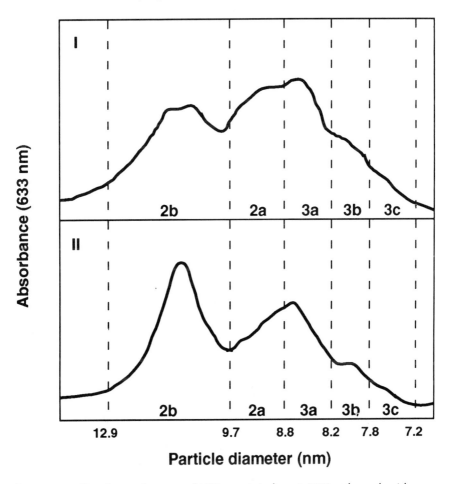

Figure 28.1. Densitometric scans of HDL separated on 4–30% polyacrylamide gradient gels. High-density lipoproteins were ultracentrifugally isolated from two normolipidemic plasmas (profile I: female, 39 years old, total cholesterol 200 mg/dL, triglycerides 28 mg/dL, HDL-cholesterol 72 mg/dL; Profile II: female, 35 years old, total cholesterol 157 mg/dL, triglycerides 47 mg/dL, HDL-cholesterol 59 mg/dL).

distinct LDL subpopulations were identified in human plasma (5). Classifications were proposed on the basis of distribution profiles of LDL subpopulations as obtained by polyacrylamide gradient gel electrophoresis of the plasma LDL fraction. Three main LDL types, varying in the degree of heterogeneity of LDL subpopulations, were described in normolipidemic sera: type 1, characterized by the presence of only one major band; type 2, characterized by the presence of two close major bands; and type 3, characterized by the presence of at least three major bands (1). In fact, according to this classification, it appears that the plasma LDL fraction is distributed either as a monodisperse (type 1) or a polydisperse (type 2 and type 3) population of particles. In addition, two main typical LDL patterns, pattern A and pattern B, have also been described on the basis of gradient gel distribution profiles. In that latter case, the dichotomous classification of gradient gel LDL patterns was based not only on their general shapes but also on the mean apparent diameter of the major plasma LDL subfraction. Pattern A contains mainly large particles with a diameter greater than 25.5 nm and pattern B contains mainly small particles with a diameter lower than 25.5 nm (10) (see Figure 28.2). As with HDL, LDL heterogeneity has been

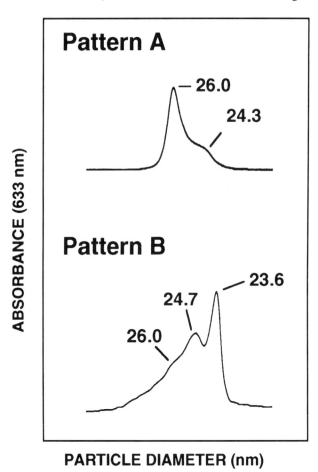

Figure 28.2. Densitometric scans of LDL separated on 2–16% polyacrylamide gradient gels showing typical LDL patterns A and B. Low-density lipoproteins were ultracentrifugally isolated from total plasmas (LDL pattern A: female, 41 years old, total cholesterol 206 mg/dL, triglycerides 50 mg/dL, HDL-cholesterol 63 mg/dL; LDL pattern B: male, 22 years old, total cholesterol 179 mg/dL, triglycerides 134 mg/dL, HDL-cholesterol 39 mg/dL).

shown to present important physiopathological implications. Indeed, LDL pattern B, primarily composed of small-sized particles, is significantly associated with an increased risk for coronary artery disease (11–13). In addition, plasma LDL subpopulations of distinct size may present different metabolic properties and, for instance, various LDL subspecies can vary in their ability to interact with the cellular LDL receptor (14). Recent studies also demonstrated that small sized LDL subspecies are more susceptible to oxidative modifications than larger particles (15,16).

Effect of CETP on the HDL and LDL Distribution Profiles

Recent observations in humans raised considerable interest in identifying mechanisms which induce alterations in distribution profiles of plasma lipoproteins, and in vitro studies have implicated several factors in changing the size of the particles. In fact, the distribution profiles of either LDL or HDL appear to result from complex processes depending on several factors, among them lecithin-cholesterol acyl transferase (LCAT), lipoprotein lipase (LPL), hepatic lipase (HL), phospholipid transfer protein (PTP), and CETP. In particular, several lines of evidence, such as infusion of human CETP in CETP-deficient animal models, CETP-transgenic animals, and inherited CETP-deficiency in humans, suggest that CETP may constitute a very important factor in determining the size distribution of lipoprotein fractions. The implication of CETP in lipoprotein remodeling appears to be of particular interest since this protein is able to modify the atherogenic potential in human plasma by transferring cholesteryl esters from the antiatherogenic HDL fraction toward the lower density proatherogenic fractions, VLDL and LDL (17).

CETP and Size Distribution of HDL

Recent in vitro studies demonstrated that purified human CETP can induce size redistribution of HDL particles (18,19). This CETP-mediated size redistribution of HDL has been shown to occur in the absence of other lipoprotein substrates and in the absence of other enzyme activities known to affect lipoprotein structure. In addition, the direct involvement of CETP in the HDL size redistribution process has been confirmed by using specific anti-CETP monoclonal antibodies which previously have been shown to inhibit the cholesteryl ester transfer reaction (18).

As shown in Figure 28.3, incubation of HDL for 24 h at 37°C in the presence of purified CETP induces dramatic changes in the densitometric profiles of these particles. Whereas the ultracentrifugally isolated HDL_3 fraction initially appeared as a relatively homogeneous population of HDL_{3a} particles (mean apparent diameter, 8.5 nm), increasing concentrations of CETP progressively induced the redistribution of HDL_{3a} toward particles of both larger (HDL_{2a}; mean apparent diameter, 9.4 nm) and smaller (HDL_{3b}; mean apparent diameter, 7.8 nm) size. At the highest CETP concentration studied, a discrete subpopulation of even smaller size (HDL_{3c}, mean apparent diameter, 7.4 nm) also was distinct. The physiological significance

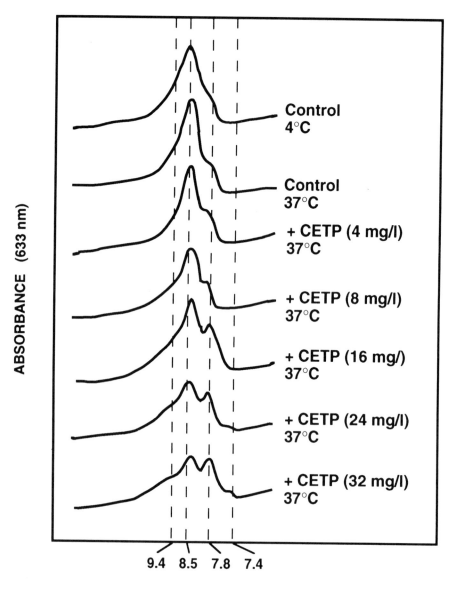

Figure 28.3. Effect of CETP concentration on gradient gel profiles of isolated HDL_3. High-density lipoprotein 3 was either kept at 4°C (control 4°C) or incubated for 24 h at 37°C in the absence (control 37°C) or in the presence of various concentrations of purified CETP (final concentration, 4, 8, 16, 24, or 32 mg/L). After incubation, HDL_3 distribution profiles were obtained by laser densitometric scanning of 4–30% polyacrylamide gradient gels.

of these observations was investigated by comparing the gradient gel distribution profile of HDL with CETP activity in total plasma from 27 normolipidemic subjects (20). The relative abundance of HDL_{3b} particles correlated positively with plasma CETP activity ($r = 0.542$; $P < 0.01$) while it did not relate to the activity of other enzymes, such as LCAT (20). Therefore, these data suggest that a specific increase in the small-sized HDL_{3b} fraction could constitute a marker of increased activity of the potentially atherogenic CETP protein.

The ability of CETP to induce the size redistribution of HDL particles in vitro was shown to vary considerably from one CETP preparation to another, suggesting that additional factors co-purified with CETP could influence the HDL redistribution process. In fact, additional studies revealed that small amounts of nonesterified fatty acids can alter the CETP-mediated size redistribution of HDL (19). These observations extended previous data which indicated that nonesterified fatty acids can also affect the CETP-mediated cholesteryl ester transfer reaction (21). Figure 28.4 shows the CETP-mediated size redistribution of isolated HDL_{3a} particles when incubated for 24 h at 37°C in the presence of CETP with or without myristic acid supplementation. Whereas myristic acid alone did not modify the gradient gel profile of HDL_{3a}, it markedly enhanced the CETP-mediated redistribution process. In particular, at the highest myristic acid concentration studied (75 µmol/L), an abundant HDL_{3c} subpopulation (7.4 nm diameter), which did not appear in the presence of CETP alone, was generated (Figure 28.4). Subsequent studies indicated that the effect of nonesterified fatty acids on the HDL redistribution process, and in particular on the formation of very small-sized HDL_{3c} particles, was markedly dependent on the concentration of nonesterified fatty acids as well as on the length and the degree of unsaturation of their acyl carbon chain (22).

CETP and Size Distribution of LDL

Evidence for the direct involvement of CETP in determining the size distribution of the plasma LDL subpopulations was also recently described (1,20). Indeed, in vitro incubations revealed that purified CETP is able to promote the shift of isolated LDL towards LDL subpopulations of larger size (1). Subsequent studies conducted by incubating total normolipidemic plasma at 37°C showed that CETP activity can induce the progressive disappearance of minor small-sized LDL subpopulations which are shifted towards LDL particles of larger size (20). After a 24-h incubation, LDL with an initial polydispersed LDL pattern exhibited a monodispersed profile as observed by using polyacrylamide gradient gel electrophoresis (Figure 28.5). The implication of CETP in determining the LDL pattern in human plasma was confirmed by comparing the shape of the gradient gel LDL patterns with plasma CETP activity among a population of normolipidemic subjects. Indeed, subjects with a monodispersed LDL pattern presented a significantly higher plasma CETP activity than subjects with a polydispersed LDL pattern (301 ± 85 vs. 216 ± 47%/h/mL, respectively; $P < 0.02$ [20]).

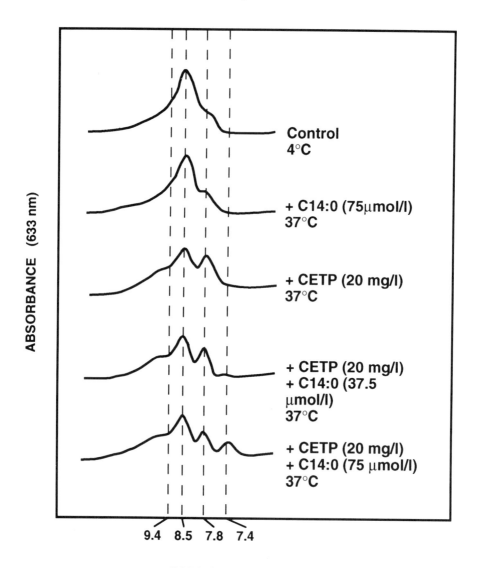

Figure 28.4. Effect of nonesterified fatty acids on the CETP-mediated redistribution of HDL_3. High-density lipoprotein 3 was incubated for 24 h at 37°C with or without CETP (final concentration, 20 mg/L) and in the presence or in the absence of myristic acid ($C_{14:0}$; final concentration, 37.5 or 75.0 µmol/L). After incubation, HDL_3 distribution profiles were obtained by laser densitometric scanning of 4–30% polyacrylamide gradient gels.

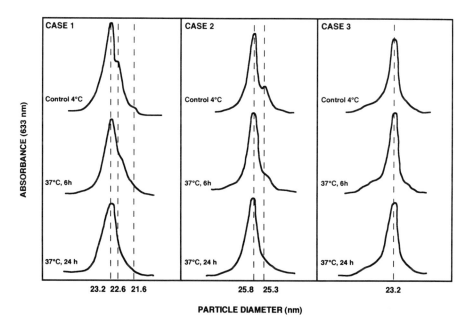

Figure 28.5. Densitometric scans showing effect of the incubation of total plasma on the LDL distribution pattern. Total plasma from normolipidemic subjects was incubated for 6 or 24 h at 37°C. Controls were maintained at 4°C. At the end of incubation, lipoproteins were subjected to electrophoresis on 20–160 g/L polyacrylamide gradient gels. Case 1: female, 28 years old, total cholesterol 183 mg/dL, triglycerides 62 mg/dL, HDL-cholesterol 58 mg/dL. Case 2: female, 35 years old, total cholesterol 157 mg/dL, triglycerides 47 mg/dL, HDL-cholesterol 59 mg/dL. Case 3: male, 32 years old, total cholesterol 231 mg/dL, triglycerides 67 mg/dL, HDL-cholesterol 50 mg/dL.

While it is now obvious that CETP can alter the size distribution of LDL particles, in vitro experiments conducted in the presence of CETP activity alone did not reveal the formation of small-sized LDL particles which have been shown to be associated with an increased risk for coronary artery disease (11–13). This suggests that a supplementary event is required to account for the size reduction of LDL particles. Several years ago, a general model for the size reduction of isolated LDL particles involving triglyceride hydrolysis in addition to CETP activity was proposed on the basis of results obtained by using analytical ultracentrifugation (23). As a first step, LDL particles would be progressively enriched with triglycerides through the CETP-mediated neutral lipid exchange reaction. In a second step, triglyceride-enriched LDL would be lipolyzed by either lipoprotein lipase or hepatic lipase, producing the reduction of LDL diameter observed by using electron microscopy (23). Recent studies demonstrated that the combined effects of neutral lipid transfers and triglyceride hydrolysis can favor the formation of the small-sized LDL pattern B (24), which is associated with an increased risk for coronary artery disease.

Conclusion

During the last decade, polyacrylamide gradient gel electrophoresis arose as one of the most convenient tools to study the size distribution of lipoprotein particles in human plasma. In addition, this method has been used to identify and characterize the factors which may affect lipoprotein structure. Since recent studies indicated that the size distribution of lipoprotein particles fluctuate markedly under various physiological and pathological conditions, gradient gel lipoprotein patterns may allow evaluation of the global risk for atherosclerosis in human plasma.

References

1. Gambert, P., Bouzerand-Gambert, C., Athias, A., Farnier, M., and Lallemant, C. (1990) *J. Lipid Res. 31:* 1199–1210.
2. Krauss, R.M., and Burke, D.J. (1982) *J. Lipid Res. 23:* 97–104.
3. Blanche, P.J., Gong, E.L., Forte, T.M., and Nichols, A.V. (1981) *Biochim. Biophys. Acta 665:* 408–419.
4. Rainwater, D.L., Andres, D.W., Ford, A.L., Lowe, W.F., Blanche, P.J., and Krauss, R.M. (1992) *J. Lipid Res. 33:* 1876–1881.
5. McNamara, J.R., Campos, H., Ordovas, J.M., Peterson, J., Wilson, P.W.F., and Schaefer, E.J. (1987) *Arteriosclerosis 7:* 483–490.
6. Krauss, R.M., and Blanche, P.J. (1992) *Curr. Opin. Lipidol. 3:* 377–383.
7. Cheung, M.C., and Albers, J.J. (1984) *J. Biol. Chem. 259:* 12201–12209.
8. Johansson, J., Carlson, L.A., Landou, C., and Hamsten, A. (1991) *Arterioscler. Thromb. 11:* 174–182.
9. Cheung, M.C., Brown, B.G., Wolf, A.C., and Albers, J.J. (1991) *J. Lipid Res. 32:* 383–394.
10. Austin, M.A., and Krauss, R.M. (1986) *Lancet 2:* 592–595.
11. Austin, M.A., Breslow, J.L., Hennekens, C.H., Buring, J.E., Willet, W.C., and Krauss, R.M. (1988) *JAMA 260:* 1917–1921.
12. Tornvall, P., Karpe, F., Carlson, L.A., and Hamsten, A. (1991) *Atherosclerosis 90:* 67–80.
13. Campos, H., Genest, J.J., Blijlevens, E., McNamara, J.R., Jenner, J.L., Ordovas, J.M., Wilson, P.W.F., and Schaefer, E.J. (1992) *Arterioscler. Thromb. 12:* 187–195.
14. Nigon, F., Lesnik, P., Rouis, M., and Chapman, M.J. (1991) *J. Lipid Res. 32:* 1741–1753.
15. Chait, A., Brazg, R.L., Tribble, D.L., and Krauss, R.M. (1993) *Am. J. Med. 94:* 350–356.
16. DeGraaf, J., Hendriks, J.C.M., Demacker, P.N.M., and Stalenhoef, A.F.H. (1993) *Arterioscler. Thromb. 13:* 712–719.
17. Tall, A.R. (1993) *J. Lipid Res. 34:* 1255–1274.
18. Lagrost, L., Gambert, P., Dangremont, V., Athkials, A., and Lallemant, C. (1990) *J. Lipid Res. 31:* 1569–1575.
19. Barter, P.J., Chang, L.B.F., Newnham, H.H., Rye, K.A., and Rajaram, O.V. (1990) *Biochim. Biophys. Acta 1045:* 81–89.
20. Lagrost, L., Gandjini, H., Athias, A., Guyard-Dangremont, V., Lallemant, C, and Gambert, P. (1993) *Arterioscler. Thromb. 13:* 815–825.
21. Sammett, D., and Tall, A.R. (1985) *J. Biol. Chem. 260:* 6687–6697.
22. Lagrost, L., and Barter, P.J. (1991) *Biochim. Biophys. Acta 1082:* 204–210.
23. Deckelbaum, R.J., Eisenberg, S., Oschry, Y., Butbul, E., Sharon, I., and Olivecrona, T. (1982) *J. Biol. Chem. 257:* 6509–6517.
24. Lagrost, L., Gambert, P., and Lallemant, C. (1994) *Arterioscler. Thromb. 14:* 1327–1336.

Chapter 29
Preparative Ultracentrifugation of Plasma Lipoproteins: A Critical Overview

P. Michel Laplaud

INSERM Unité 321 "Lipoprotéines et athérogénèse", Hôpital de la Pitié, 83 boulevard de l'Hôpital, 75651 Paris Cedex 13, France.

Introduction

In plasma, lipids (essentially esterified and nonesterified cholesterol, glycerides, and phospholipids) are transported in the form of lipoprotein particles from their organs of synthesis (e.g., the liver), or absorption (intestine) to their sites of utilization. Lipoproteins, in which lipids and specific amphipathic proteins (called apolipoproteins) are linked together by noncovalent bonds, thus ensure solubilization of fat in aqueous plasma. Taking into account the difference in density between pure lipids (typically 0.90 g/mL) and pure proteins (approximately 1.35 g/mL), it is conceivable that the respective proportions of lipids and proteins within a given lipoprotein determine the hydrated density of the resulting particle. A universally adopted classification of lipoproteins has been established, dividing the human plasma lipoprotein spectrum into several density classes, namely chylomicrons ($d < 0.94$ g/mL), very low density lipoproteins (VLDL) ($0.94 < d < 1.006$ g/mL), low-density lipoproteins (LDL) ($1.006 < d < 1.063$ g/mL), and high-density lipoproteins (HDL) ($1.063 < d < 1.21$ g/mL). Subdivisions and additions to this classification have also been introduced, such as those categorizing LDL of lower density as intermediate density lipoproteins (IDL) ($1.006 < d < 1.019$ g/mL), subdividing HDL into HDL_2 ($1.063 < d < 1.125$ g/mL) and HDL_3 ($1.125 < d < 1.21$ g/mL), or defining very high density lipoproteins (VHDL) as particles where $d = 1.21-1.25$ g/mL. Whatever the lipoprotein considered, each class of particles has its own, quantitatively and qualitatively unique content of molecular species of lipids and apolipoproteins, as well as its specific role in lipid transport in plasma (Table 29.1).

In humans, each lipoprotein density class defined is implicated in the pathogenesis of various dyslipidemias. For example, an increase in the plasma concentration of cholesteryl ester-rich LDL is associated with the formation of atherosclerotic lesions in the arterial wall. In the coronary arteries, such lesions ultimately lead to the occurrence of myocardial infarction, the major cause of morbidity and mortality in most developed countries.

For more than 40 years, such a situation has created a need for the development of techniques capable of purifying the different plasma lipoprotein species. While chemical precipitation, chromatographic and electrophoretic techniques are per-

TABLE 29.1
Plasma Lipoprotein Classification and Characteristics

Particle Name	Synonyms (based on electrophoretic mobility)	Hydrated Density Range (g/mL at 20°C)	Flotation Rate	Composition Lipid		Composition Protein		Major Site(s) of Synthesis	Major Site(s) of Final Catabolism	Major Function
				Major Lipid(s)	%	%	Major Apolipoprotein			
Chylomicrons	—	<0.94	$S_f^a > 400$	Triglyceride	98	2	Apo B-48	Intestine	Liver	Transport exogenous lipids
Very low density Lipoproteins (VLDL)	Pre-β-lipoproteins	0.94–1.006	S_f 20–400	Triglyceride	90	10	Apo B-100	Liver	(see LDL)	Transport endogenous lipids
Low-density lipoproteins (LDL)	β-lipoproteins	1.006–1.063	S_f 0–20	Cholesteryl ester	75–80	20–25	Apo B-100	Intravascular from VLDL	Liver and other tissues	Supply peripheral cells with cholesterol
High-density lipoproteins (HDL)	α_1-lipoproteins	1.063–1.210	F^b0–9	Cholesteryl ester, phospholipid	60–40	40–60	Apo A-I	Intravascular, from chylomicrons and VLDL. Liver and intestine	Liver	Reverse cholesterol transport

^{a,b}Flotation coefficients, expressed in Svedberg units and measured in the analytical ultracentrifuge at 26°C and a solvent density of 1.063 g/mL (a), or 1.21 g/mL (b).

formed in many laboratories, nowadays, preparative ultracentrifugation is undoubtedly the most widely-used methodology.

All ultracentrifugal techniques applied to lipoprotein isolation take advantage of the difference in density between the various lipoprotein classes and other macromolecular components of plasma. The two most widely used techniques are sequential and density-gradient preparative ultracentrifugation. Irrespective of the technique used, the general principle is to adjust the density of the centrifugation medium to obtain flotation at the top of the ultracentrifuge tube of the desired lipoprotein fraction, with a density less than that of the solvent, while plasma proteins and/or other lipoprotein classes sediment at the bottom (sequential ultracentrifugation in a fixed angle rotor); or isopycnic equilibrium of lipoproteins at their respective hydrated densities in a density gradient (density gradient ultracentrifugation in a swinging bucket rotor).

In this context, defining suitable densities for ultracentrifugation media, as well as achieving accurate and reproducible density measurements, is obviously of paramount importance. Adjustment of density is usually obtained by mixing the lipoprotein sample with a saline solution of known molality, prepared from anhydrous NaCl and/or NaBr, or with solid salts (e.g., KBr). In both cases, well-established mathematical formulas exist in the literature, allowing calculations of the required amount of salt, with various degrees of accuracy. Indeed, with regard to density adjustment achieved using saline solutions, the simplest and most commonly used equation is

$$d_f v_f = d_1 v_1 + d_2 v_2$$

where d_f, d_1, and d_2 are the densities of the final mixture, the plasma (or already isolated lipoprotein fraction), and the diluting solution, respectively; v_f, v_1, and v_2 are the corresponding volumes. Once having decided on d_f, d_2 can be found, or calculated from computer-calculated tables giving density as a function of molality and molarity for NaCl or NaBr. Such tables can be found in the *Manual of Laboratory Operations of the Lipid Research Clinics Programs of the NIH (1)* or in the book by Mills et al. (2). Although such calculations are adequate for most bioclinical applications, they do not take into account the nonlinearity of the relation existing between the concentration and density of saline solutions. Therefore, if highly precise density determinations are contemplated, such as for physicochemical studies of lipoprotein subfractions, the use of iterative procedures, such as that of Lindgren (3), is necessary. In any case, actual density values of the saline solutions used must be verified either by controlled temperature, precision refractometry (for monosalt solutions only) or, more conveniently, by use of an electronic density meter.

Since the final density actually obtained will be more difficult to check, the use of solid salts is recommended only when dilution of the sample must be avoided. If this method is chosen, calculations will be made according to the formula of Radding and Steinberg (4):

$$M = \frac{v(d_f - d_i)}{1 - \bar{v}d_f}$$

where v is the volume of the solution, d_f and d_i the final and initial densities, respectively, and $\bar{v}$ the partial specific volume of the salt, measured at the relevant temperature and concentration. Again, $\bar{v}$ values for NaCl, NaBr, and KBr can be found in Mills et al. (2).

During density adjustment several important points are frequently overlooked, resulting in imprecision of the real nature of the lipoprotein fraction finally isolated. First, when dealing with total plasma, one must remember that approximately 6% of its volume is occupied by proteins. Therefore, it is the partial solvent volume (i.e., 0.94 times the actual volume) that should be used in whatever formula is used when calculating the amount of salt to add to raise the density.

Another essential point is that of the differential thermal expansivity of lipoproteins and their solvents. For example, as discussed at length by Mills et al. (2), the ratio between the density decrease of LDL and the corresponding value for a NaCl solution for 5–25°C is roughly two- to fourfold (-7×10^{-4} g/[mL/°C] and -1.65×10^{-4} to -3.93×10^{-4} g/[mL/°C], respectively). Therefore, care must be exercised to perform all density adjustments at the same temperature as the ultracentrifugal run. Finally, not even ions are insensitive to extremely high centrifugal fields generated in an ultracentrifuge, and redistribution of salts will occur during the run, leading to a decrease in solvent density at the top of the tube. The magnitude of the error in density resulting from neglect of these factors will depend on the precise experimental conditions, but may be as high as 1×10^{-2} g/mL. Such an error will be especially detrimental to experiments aimed at characterizing discrete, narrow density range lipoprotein subfractions, such as those currently favored for LDL studies.

The other main concern when ultracentrifugally purifying lipoproteins involves the preservation of the physicochemical integrity of the particles during the whole preparative process. Indeed, lipoprotein degradation can occur from different causes, such as microbial degradation, oxidation, and the exposure to high centrifugal fields. Attempts to prevent microbial degradation usually include addition of sodium azide (0.01%); sodium merthiolate (0.001%); and antibiotics, such as chloramphenicol or preferably, gentamicin sulfate (60–100 mg/L each). Proteolytic enzymes, such as serine proteases, originating from contaminating microbial organisms or co-purifying with lipoproteins may operate during ultracentrifugation. Their action is usually prevented by adding phenylmethylsulfonyl fluoride (PMSF) or aprotinin (final concentrations 1 mM or 10 units/mL, respectively). Use of sodium azide has been criticized on the basis that it may accelerate oxidation of LDL, and lead to increased degradation of apo B (5). However, inclusion of EDTA (final concentration 3 mM), which acts as an antioxidant by chelating heavy metal ions such as Cu^{++}, may counteract such detrimental effects (6). With current emphasis placed on the pathophysiological role of oxidized lipoproteins (especially LDL subfractions), the need to avoid any oxidative modification of lipoproteins during isolation has led to more frequent use of supplementary antioxidants, such as glutathione, ε-amino caproic acid, and butylated hydroxytoluene (BHT), at final concentrations of 50, 5, and 130 µg/mL, respectively.

It is well known that lipoprotein particles may lose some of their lipid or protein components when subjected to repeated, long-lasting and high-speed centrifugations. Such fragility results from the fact that their constitutive molecular species are only bound together by noncovalent forces. In this context, most studies have dealt with the fate of apo A-I, and several investigators have demonstrated losses of this protein after ultracentrifugation; for example Kunitake and Kane (7) who also provided evidence that the high ionic strengths commonly used in ultracentrifugal preparation of lipoproteins may offer some protection to these particles against the detrimental effects of the centrifugal held. Apo A-I, when leaving the HDL density range, is an artifact recovered in the "bottom" (d >1.21 g/mL) fraction. Recently, Asztalos et al. (8) have shown that this phenomenon does not occur randomly among HDL particles, but rather that specific subpopulations of apo A-I-containing lipoproteins are affected. This may particularly concern those with pre-β electrophoretic mobility and whose role in reverse cholesterol transport is now thought to be crucial (9). Other investigators (10) have demonstrated that ultracentrifugation is actually responsible for the previously reported presence of apo A-IV in the d >1.21 g/mL fraction, while this protein is physiologically associated with HDL, and with chylomicrons in post-prandial plasma. Reported detrimental consequences of ultracentrifugation also include losses of apo E from different lipoproteins (11), modifications in both the chemical composition and particle distribution of VLDL (12), and redistribution of the enzyme lecithin-cholesterol-acyl-transferase (LCAT) between HDL subfractions (13).

Sequential Ultracentrifugation

This is undoubtedly the most widely used ultracentrifugal method in lipoprotein research, probably because it is comparatively easy to perform and can result in the preparation of relatively large volumes of lipoprotein solutions. It primarily has been applied to successive isolation of VLDL, LDL, and HDL, and typical schemes for such purifications can be found in a number of references, for example in Hatch and Lees (14) or the laboratory manual of Lindgren (1). Suitable adjustment of the density of the medium theoretically allows isolation of whatever fraction is desired, such as IDL (d = 1.006–1.019 g/mL), HDL_2 (1.063–1.125 g/mL), or HDL_3 (1.125–1.210 g/mL).

Sequential ultracentrifugation is usually performed in fixed angle rotors, allowing preparation of various amounts of lipoproteins in a single run. Usual conditions are rotor speeds of 40,000–60,000 rpm, and centrifugation times between 16 h (for VLDL) and 36 h (for HDL_3). In addition, it is often suggested that the purity of the floating lipoprotein fraction be increased by subsequent "washing," that is recentrifugation of the fraction at the same density, after collecting it from the initial ultracentrifuge tube. This may help eliminate contaminating lipoproteins of higher density and/or adsorbed plasma proteins, but it also increases the risk of degradation of the desired lipoprotein particles considerably.

It is noteworthy that many types of open or sealed ultracentrifugation tubes exist, differing in material, size, and preferred application. In any case, use of clear tubes is essential as the resulting floating lipoprotein fraction must be seen (either by eye only or with the help of a light beam) to be quantitatively recovered by aspiration using either a fine bore Pasteur pipette or a micropipette. Performing such a recovery satisfactorily is not an easy exercise, and it is suggested that the reader reads detailed descriptions, such as that made by Mills et al. (2), before attempting to begin work in this area.

Density-Gradient Ultracentrifugation

The superior resolution power of this technique has prompted the development of a number of corresponding methodologies in the field of plasma lipoproteins. Such experiments may be broadly divided into two main classes, those attempting to separate large plasma lipoprotein classes from each other in a single run, and those with a goal of subfractionating a lipoprotein density class previously isolated from plasma. In both cases, density-gradient ultracentrifugation is usually performed in a high-performance swinging bucket rotor made from titanium. Advantages of such devices include possible use of the long tubes necessary to accommodate high-resolution gradients, consecutive generation of very high intensity centrifugal fields (about 200,000 g), and decreasing the so-called "wall effect" (i.e., convective flow produced by collision of particles with the tube wall), which is detrimental to the final resolution of the technique.

Gradients are usually of the discontinuous, or "step," type, made from different saline solutions of decreasing density (15–18). Buffered sucrose solutions are not favored for such applications, due to their viscosity and to the ability of sucrose to bind irreversibly to the protein moiety of lipoproteins (19).

In addition to the precautions already listed for sequential ultracentrifugation, an essential point for this technique is the correct evaluation of the minimum number of g's delivered to the sample. Indeed, this parameter must be calculated to attain isopycnic equilibrium, otherwise cross-contamination and contamination of lipoprotein fractions by plasma proteins will occur. Unfortunately, some of the techniques reported in the literature do not fulfill this criterion. Once the correct combination of radial centrifugal field (RCF) and centrifugation time has been determined, experimental conditions must be adhered to rigorously if reproducibility of the separation is to be expected. These conditions include acceleration, full-speed, and deceleration times, as the gradient may still be in a dynamic state at the end of the run, that is continuing to approach equilibrium while lipoprotein particles have already reached their equilibrium positions in the gradient.

When correctly performed, density-gradient ultracentrifugation is a powerful tool in lipoprotein research, although it is applicable only to small volume samples. For example, the author's laboratory has reported two such methodologies. The first method (18), used a five-step gradient, prepared by layering 2 mL NaCl-KBr solution ($d = 1.240$ g/mL), 3 mL serum with the nonprotein solvent density increased to 1.210 g/mL by addition of solid KBr, 2 mL NaCl-KBr solution ($d = 1.063$ g/mL), 2.5 mL NaCl-KBr solution ($d = 1.019$ g/mL), and 3 mL NaCl solution ($d = 1.006$ g/mL) successively in the rotor tube. The serum sample was placed on a "cushion" of higher density because simultaneous sedimentation of plasma proteins and isopycnic banding of lipoproteins was expected to occur, expanding the final distance between the two classes of macromolecules and limiting the possible contamination of the various lipoprotein species, especially by albumin. Then the gradients were spun for 48 h at 40,000 rpm in a swinging bucket rotor (56.7×10^7 g average minimum) and resulted in their reorientation into smooth profile gradients of which two-thirds were linear (Figure 29.1). Plasma lipoproteins were separated by a single centrifuge run into four main bands, corresponding to VLDL, LDL, HDL_2, and HDL_3, respectively (Figure 29.2), as evidenced by a number of methodologies, including chemical, physical, and immunological analysis (18). An additional advantage of the method was that it resulted in a multi-fold concentration of native plasma lipoproteins, the precise value being a function of the ratio between the original plasma volume used to construct the gradient and the volume in which each lipoprotein fraction was finally recovered.

Research work dealing with the implications of plasma lipoproteins on the atherogenic process has made use of many different animal species, including nonhuman primates, the rat, rabbit, pig, dog, bovine, pigeon, and others (20). In most of these species, the density spectrum of plasma lipoproteins may differ considerably from that in man, regarding the respective density limits of the various classes of particles. Therefore, density gradient ultracentrifugation is admirably suited for animal lipoprotein analysis.

The author has made extensive use of the technique in studies dealing with the hormonal regulation of lipoprotein metabolism in species exhibiting spontaneous seasonal endocrine variations, notably the European badger (21) and the European hedgehog (22). Unfortunately in these latter species, as well as in others, the respective density distributions of LDL and/or HDL particles (defined as lipoproteins containing apo B and apo A-I, respectively, as their main apolipoprotein component), significantly overlap each other. Separation of these two lipoprotein classes by any technique based on density manipulation is not feasible, and additional methodologies, such as affinity or immuno-affinity chromatography, should be used.

Similar difficulties arise with regard to isolation of lipoprotein (a) (Lp(a)), a particle of considerable interest due to its implication in both atherogenesis and thrombosis (23). Indeed, the density range for cholesteryl ester-rich Lp(a) particle is approximately 1.050–1.100 g/mL, encompassing the respective domains of denser LDL and HDL_2. Again, additional preparative methodologies. such as chro-

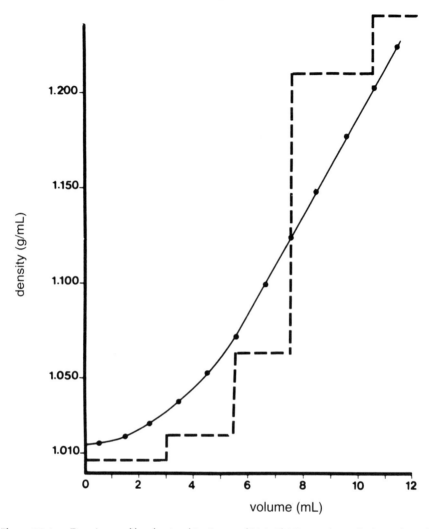

Figure 29.1. Density profile obtained in "control" NaCl-KBr gradients before (dotted line) and after (solid line) ultracentrifugation for 48 h. Density (g/mL) is plotted on the ordinate against volume (from meniscus downwards) on the abscissa. Densities were determined at 15°C on successive 1-mL fractions with a digital precision density meter. Points represent the means of determinations on two gradients from each of four ultracentrifugal runs: measurements were made in duplicate or triplicate on individual fractions. Adapted from the *Journal of Lipid Research* (18) and reproduced with permission.

matofocusing, gel filtration or affinity chromatography on lysine-Sepharose, are required for complete purification of Lp(a).

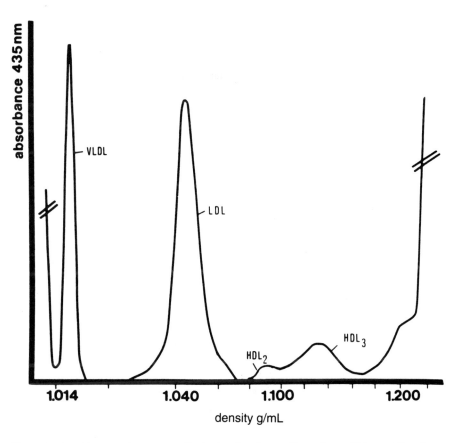

Figure 29.2. Lipoprotein mass profile in the density-gradient after a 48 h centrifugal separation of a normolipidemic serum from a 33-yr old healthy male (serum cholesterol and triglyceride 160 and 53 mg/dL, respectively). After photography on "Kodachrome" or "Ektachrome 50T," the positive was scanned at 435 nm. Absorbance is plotted against density (g/mL), the latter being derived from Figure 29.1 as a function of tube length. Each absorption maximum is labeled according to its subsequent identification by chemical, physical, and immunological analyses. Reproduced with permission from the *Journal of Lipid Research* (18).

Use of density-gradient ultracentrifugation for subfractionation of lipoprotein density classes was prompted by the demonstration of their structural and metabolic heterogeneity. Most importantly, LDL have been shown to consist of a number of particle species, each of which has its own physicochemical characteristics, as well as its own atherogenicity, for example through its affinity for the LDL receptor or its susceptibility to oxidation. In this context, a number of studies have been published, dealing with the ultracentrifugal resolution of LDL heterogeneity (for a review see Chapman et al. [24]).

The author's laboratory has presented a density-gradient technique allowing the fractionation of the LDL distribution in normolipidemic males into 15 subfractions. Among them, the six quantitatively major fractions, representative of the central part of the distribution, could be analyzed in detail and shown to be physicochemically different (Table 29.2 [24]). For this purpose, LDL (d = 1.024–1.050 g/mL) were first isolated from plasma by sequential ultracentrifugation, followed by dialysis against a saline solution (d = 1.006 g/mL). The nonprotein solvent density of the sample was then adjusted to 1.040 g/mL using solid KBr. Discontinuous density gradients were then constructed by successively pumping into the ultracentrifuge tubes 4.5 mL NaCl-KBr solution, (d = 1.054 g/mL), 3.5 mL sample, 2 mL NaCl solution (d = 1.024 g/mL), and finally 2 mL NaCl solution (d = 1.019 g/mL). After a 44-h run at 40,000 rpm, the gradient fractions were recovered for further analysis.

Such density-gradient methodologies are of course applicable to the abnormal lipoprotein distributions observed in various dyslipidemic states (for a review see Dejagen et al. [25]). For example, the author's laboratory has recently reported (25) the use of one of the techniques (18) in patients affected by combined hyperlipidemia (CHL). By analysis of the LDL spectrum from subjects presenting this metabolic disorder, a specific increase in the relative proportions of two dense LDL subfractions (with d = 1.039–1.050 g/mL and 1.050–1.063 g/mL, respectively) was demonstrated. The latter was remarkable by its low resistance to oxidative modifications as indicated by the short lag time preceding the propagation phase of conjugated diene formation. Such results are indicative of the capability of density-gradient ultracentrifugal techniques to remain essential to modern forms of lipoprotein analysis.

Rate-zonal ultracentrifugation has been proposed to overcome the limitation of conventional density-gradient techniques to small volume samples. This technique has been used successfully for fast, large scale preparation of VLDL (26), or for the subfractionation of total HDL into either HDL_2 and HDL_3 or three and five subfractions, of each of these two major density subclasses, respectively, (27). However, mostly because it requires a heavy capital investment in rotor, loading equipment, and apparatus to measure and record the lipoprotein distribution as the rotor is unloaded, this method has not become very popular in lipoprotein research laboratories.

Some words of caution must be said about the special problems arising when attempting either to separate chylomicrons from VLDL, or to subfractionate one of these lipoprotein classes. The operational limit between chylomicrons and VLDL is often reported in the literature as being Sf 400, where Sf is a flotation coefficient defined for analytical ultracentrifugation purposes at 26°C in a medium of d = 1.063 g/mL. This limit has to be adapted to the actual parameters used in most method, that is 20°C and 1.006 g/mL, otherwise unacceptable error will take place. Mills et al. (2) suggest that applying a 2.25×10^6 g minimum is suitable for removing chylomicrons (Sf > 400) from native plasma under these conditions. Unfortunately, this results in the flotation of chylomicrons heavily contaminated by serum albumin. Removal of this protein necessitates repeated "washings," with the risk that some of the lipoprotein components (especially low molecular weight apolipoproteins)

TABLE 29.2
Average Molecular Weights, Molecular Compositions, and Molecular Diameters of Human LDL Particle Subspecies Isolated by Density Gradient Ultracentrifugation[a]

Subfraction No.	Density g/mL	Molecular Weight $\times 10^{-6}$	Average Number of Molecules of Each Component per Particle of Each LDL Subspecies					Molecular Diameter Å
			Cholesteryl Ester	Free Cholesterol	Triglyceride	Phospholipid	Protein	
5	1.0260	2.96	1872	872	188	791	1.2	209.6
6	1.0286	2.86	1795	776	161	779	1.2	207.2
7	1.0314	2.75	1790	824	120	717	1.1	204.1
8	1.0343	2.62	1725	745	108	686	1.1	200.7
9	1.0372	2.48	1621	686	108	608	1.1	196.9
10	1.0409	2.33	1505	572	104	583	1.1	192.7

[a]Reproduced with permission from the *Journal of Lipid Research* (24).

may separate from the native particles. In addition, it is well known that, according to nutritional factors, lipoprotein particles originating from the liver may be found in the $d < 0.94$ g/mL range attributed to chylomicrons. while some of the latter, intestinally derived particles, may be present in the VLDL density domain ($d = 0.94–1.006$ g/mL). Clearly, a density-based separation technique applicable at any time to any patient and resulting in consistent separation of intestinal from hepatic triglyceride-rich lipoproteins is not conceivable.

Subfractionation of chylomicrons or VLDL can be performed by sequential centrifugation in a fixed angle rotor (2,28). Such experiments can result in the recovery of two chylomicron (Sf > 5000 and Sf 400–5000, respectively), and three VLDL subfractions (Sf 100–400, 50–100, and 20–50, respectively). An alternative here is the use of density gradients in nonequilibrium conditions (29), although the sophistication of these latter techniques with superior resolution makes them suitable only for highly specialized laboratories. Whatever the method chosen, it is absolutely essential to apply the lowest possible lipoprotein load in the centrifuge tubes.

During recent years, developments in the field of lipoprotein centrifugation have included the appearance of vertical rotors and of highspeed bench-top ultracentrifuges. Vertical-tube rotors have been developed following the observation that isopycnic gradients centrifuged in a fixed angle rotor with a shallow angle gave better resolution than gradients centrifuged in a similar centrifugal field in swing-out rotors. Also, in vertical rotors, the maximum path length of the particles is no more than the inside diameter of the tube, and the design allows the centrifugal field to be of high intensity (by increasing the radial distance between the tubes and the axis of rotation) and identical at any point along the height of the tubes. This results in much shorter run times (45–150 min) according to the rotor used, for separating VLDL, LDL, and (partly) HDL_2 and HDL_3 from native plasma (30). Essential practical considerations when using vertical rotors include paying particular attention to tube sealing and using a centrifuge capable of very low acceleration and deceleration since it is by too rapid reorientation of the gradient during these phases of the run that loss of resolution occurs. The major disadvantage of the use of vertical rotors lies in the wall adherence of VLDL and albumin, with consecutive loss of the former and contamination of the different lipoprotein fractions by the latter. Fixed angle rotors with very small tube angle, "Near Vertical Rotors" (NVT), have been developed in an attempt to overcome these problems, but the usefulness of such rotors for lipoprotein research has yet to be confirmed. In any case, reviews like those by Rickwood (31) or Chung et al. (32) provide comprehensive understanding of the principles and applications of vertical rotors.

The increasing need for ultrafast, small volume, multisample centrifugation for clinical laboratory and metabolic experiments (e.g., analysis of lipoproteins secreted by cells in culture) has prompted the development of high-speed, air- or electrically driven bench-top centrifuges capable of speeds up to 100,000 rpm. In turn, rotors of the different types have been developed for these instruments. Use of this technology is claimed to allow purification of VLDL from 175 µL plasma aliquots in 3 hr, or sequential purification of VLDL, LDL, Lp(a), and HDL from O.5 mL serum samples in about 10 h (33).

Thus, preparative ultracentrifugation of lipoproteins presently consists of a large variety of techniques, each being adapted to a particular purpose. It must be kept in mind that preparative ultracentrifugation has unavoidable drawbacks, such as equipment cost, need for skilled and meticulous laboratory staff, and possible deleterious effects of extremely high centrifugal fields on lipoprotein particles. However, its unique capabilities for purifying the different lipoprotein classes and subclasses are responsible for its continuing success in most lipoprotein laboratories worldwide.

References

1. Lindgren, F.T. (1974) Appendix I, in *Manual of Laboratory Operations, Lipid Research Clinics Program,* vol. 1, National Heart and Lung Institute, National Institutes of Health, Bethesda, MD.
2. Mills, G.L., Lane, P.A., and Weech, P.K. (1984) *A Guidebook to Lipoprotein Technique. Laboratory Techniques in Biochemistry and Molecular Biology,* Burdon, R.H., and Van Knippenberg, P.H., Elsevier, Amsterdam, vol. 14, pp. 30–31.
3. Lindgren, F.T. (1975) in *Analysis of Lipids and Lipoproteins,* Perkins, E.G., American Oil Chemists' Society, Champaign, IL, pp. 204–234.
4. Radding, C.M., and Steinberg, D. (1960) *J. Clin. Invest. 39:* 1560–1569.
5. Schuh, J., Fairclough, G.F., Jr., and Haschemeyer, R.H. (1978) *Proc. Natl. Acad. Sci. USA 75:* 3173–3177.
6. Edelstein, C., and Scanu, A.M. (1986) in *Methods in Enzymology,* Segrest, J.P., and Albers, J.J., Academic Press, Orlando, FL, vol. 128, pp. 151–155.
7. Kunitake, S.T., and Kane, J.P. (1982) *J. Lipid Res. 23:* 936–940.
8. Asztalos, B.F., Sloop, C.H., Wong, L., and Roheim, P.S. (1993) *Biochim. Biophys. Acta 1169:* 291–300.
9. Castro, G.R., and Fielding, C.J. (1988) *Biochemistry 27:* 25–29.
10. Lagrost, L., Gambert, P., Boquillon, M., and Lallemant, C. (1989) *J. Lipid Res. 30:* 1525–1534.
11. Mahley, R.W., and Holcombe, K.S. (1977) *J. Lipid Res. 18:* 314–324.
12. Herbert, P.N., Forte, T.M., Shulman, R.S., La Piana, M.J., Gong, E.L., Levy, R.I., Fredrickson, D.S., and Nichols, A.V. (1975) *Prep. Biochem. 5:* 93–129.
13. Chen, C.H., and Albers, J.J. (1982) *Biochem. Biophys. Res. Commun. 107:* 1091–1096.
14. Hatch, F.T., and Lees, R.S. (1968) *Adv. Lipid Res. 6:* 1–68.
15. Lindgren, F.T., Jensen, L.C., and Hatch, F.T. (1972) in *Blood Lipids and Lipoproteins,* Nelson, G.J., Wiley-Interscience, New York, pp. 221–248.
16. Redgrave, T.G., Roberts, D.C.K., and West, C.E. (1975) *Anal. Biochem. 65:* 42–49.
17. Foreman, J.R., Karlin, J.B., Edelstein, C., Juhn, D.J., Rubenstein, A.H., and Scanu, A.M. (1977) *J. Lipid Res. 18:* 759–767.
18. Chapman, M.J., Goldstein, S., Lagrange, D., and Laplaud, P.M. (1981) *J. Lipid Res. 22:* 339–358.
19. Edelstein, C., Pfaffinger, D., and Scanu, A.M. (1984) *J. Lipid Res. 25:* 630–637.

20. Chapman, M.J. (1986) in *Methods in Enzymology,* Segrest, J.P., and Albers, J.J., Academic Press, Orlando, FL, vol. 128, pp. 70–143.
21. Laplaud, P.M., Beaubatie, L., and Maurel, D. (1982) *Biochim. Biophys. Acta 711:* 213–223.
22. Laplaud, P. M., Saboureau, M., Beaubatie, L., and El-Omari, B. (1989) *Biochim. Biophys. Acta 1005:* 143–156.
23. Lawn, R.M. (1992) *Scientific American,* June, 54–60.
24. Chapman, M.J., Laplaud, P.M., Luc, G., Forgez, P., Bruckert, E., Goulinet, S., and Lagrange, D. (1988) *J. Lipid Res. 29:* 442–458.
25. Dejager, S., Bruckert, E., and Chapman, M.J. (1993) *J. Lipid Res. 34:* 295–308.
26. Danielsson, B., Ekman, R., and Johansson, B.G. (1978) *Prep. Biochem. 8:* 295–319.
27. Patsch, W., Schonfeld, G., Gotto, A.M., Jr., and Patsch, J.R. (1980) *J. Biol. Chem. 255:* 3178–3185.
28. Gustafson, A., Alaupovic, P., and Furman, R.H. (1965) *Biochemistry 4:* 596–605.
29. Lindgren, F.T., Jensen, L.C., and Hatch, F.T. (1972) in *Blood Lipids and Lipoproteins: Quantitation, Composition, and Metabolism,* Nelson, G.J., Wiley, New York, pp. 181–274.
30. Chung, B.H., Wilkinson, T., Geer, J.C., and Segrest, J.P. (1980) *J. Lipid Res. 21:* 284–291.
31. Rickwood, D. (1982) *Anal. Biochem. 122:* 33–40.
32. Chung, B.H., Segrest, J.P., Ray, M.J., Brunzell, J.D., Hokanson, J.E., Krauss, R.M., Beaudrie, K., and Cone, J.T. (1986) in *Methods in Enzymology,* Segrest, J.P., and Albers, J.J., Academic Press, Orlando, FL, vol. 128, pp. 181–209.
33. David, J.A., Paksi, J., and Naito, H.K. (1986) *Clin. Chem. 32:* 1094.

Index

Alkylmagnesium bromide. See Grignard reagents
Animal fats, fatty acid content, 1, 103–104
AOCS Cg 2-83 sensory assessment method, 265
AOCS method for gas chromatography (GC) of fatty acids, 185
AOCS methods for *trans* fatty acid content, 182–183, 185, 187–188, 232
Apolipoproteins
    lipoprotein class, 337–340, 342–345, 347, 348, 357–358
    ultracentrifugation, 360, 361, 363, 366, 367
Argentation. See Silver-ion chromatography
Aromatic hydrocarbons. See also Polycyclic aromatic hydrocarbons (PAH)
    contaminants in oils and fats, 333, 335, 336
ASTM STP 758, standard method for sensory assessment of fats and oils, 265
Atherosclerosis. See Cardiovascular disease
Atom percent excess (APE) measurement, 300, 302
Atomic absorption spectrometry of metals, 324–328

Biosynthesis. See Metabolism
Butylmagnesium bromide. See Grignard reagents

Cardiovascular disease
    cholesteryl ester transfer protein (CETP), 351, 355
    fish oils, 87, 309
    lipid metabolism, 303, 304
    lipoproteins, 340–342, 347, 348, 350–351, 355, 356, 357
    oxysterols, 290–291
    trans fatty acids, 181
Chevreul, Michael E., 2–3
Chiral chromatography. See Stereospecific analysis
Cholesterol and cholesterol esters. See also Sterols
    analytical methods, 317
    autoxidation, 33–34
    biosynthesis, 303
    chromatography, 25–29, 46, 47, 48, 53–54, 56, 317
    dimethylformamide (DMF) method of analysis, 318–322
    lipoproteins, 337–340, 367
    mass spectrometry (MS), 64
    solid-phase extraction (SPE), 69
    structure, 292
    trans fatty acids effect on levels, 181
    vegetable fats, 321
Cholesterol oxides. See Oxysterols
Cholesteryl ester transfer protein (CETP), 344, 345, 347, 351–355
Chromarods, 22, 24–36
Chromatography. See also individual types of chromatography
    detection methods, 17–18
    history of, 2, 6–8, 17
Chylomicrons
    cardiovascular disease, 340
    function and structure, 337–339, 357, 358
    isotope studies, 305, 310, 312
    lipase action on, 338, 340, 345
    ultracentrifugation, 366–368
Codex Alimentarius recommendations on contaminants in fats and oils, 333, 335
Contaminants in oils and fats, 323–336
Coronary heart disease (CHD). See Cardiovascular disease
Cosmetics, sterol analysis, 321
Cottonseed oil
    chromatography, 43, 85
    stereospecific analysis, 111–116
Cyclic fatty acid monomers (CFAM)
    chemical techniques of analysis, 285–287
    gas chromatography-mass spectrometry (GC-MS), 192–193, 194
    mass-analyzed ion kinetic energy (MIKE) spectrometry, 199–203
Cyclic fatty acids
    gas chromatography-Fourier transform infrared spectrometry (GC–FTIR), 234–236
    gas chromatography-mass spectrometry (GC-MS), 195–197
    nuclear magnetic resonance (^{13}C NMR) spectroscopy, 258–260
β-cyclodextrin, dairy product processing, 319, 320–321

Dairy products. See also Milk
    β-cyclodextrin processed, 319, 320–321
    fatty acid content, 181
    lipid analysis, 12, 14
    sterol analysis, 317–322
    substitutes for, 317, 321
Degree of isomerization (DI), 164, 169, 178
Deodorization, of oils, 156, 158–160, 162–172, 177–178
Desorption-Concentration-Injection system (DCI), 333, 334–336
Diabetes mellitus, lipid metabolism, 306, 307, 341
Diacylglycerol kinase, 95–96
Diacylglycerols
    chromatography, 26–28, 46, 48
    diastereomeric derivatives, 97–100
    mass spectrometry (MS), 229
    nuclear magnetic resonance (^{13}C NMR) spectroscopy, 260–262

Diglycerides. *See* Diacylglycerols
Dimethylformamide (DMF) method of sterol titration, 318–322
3,5-Dinitrophenylurethane (DNPU) derivatives for triacylglycerol analysis, 100–102, 113–116, 118, 127–129
Docosahexaenoic acid, 309–316
Donor-acceptor complex chromatography (DACC), 328–331

Electrophoresis of lipoproteins, 343, 347–356
Equivalent chain length (ECL)
 fatty acids, 191
 linolenic acid isomers, 79–80, 160, 161, 168
 petroselinic and oleic acids, 148, 150
 pinolenic acid isomers, 173, 176–178
 polyunsaturated fatty acids (PUFA), 153–155
Essential fatty acids, 151, 153–155
Ethylmagnesium bromide. *See* Grignard reagents
European regulations on contaminants in fats and oils, 333, 335
Evaporative light-scattering detectors (ELSD)
 high-performance liquid chromatography (HPLC), 46–48, 50–51, 53–56, 206, 219–220
 Mars detector, 43
 silver-ion chromatography, 61, 69
Exclusion chromatography. *See* High-performance size-exclusion chromatography (HPSEC)
Extraction
 of lipids, 10–16, 319
 of phospholipids, 49
 of polycyclic aromatic hydrocarbons (PAH), 324–325

Factorial Correspondence Analysis (FCA), sensory measurement, 265, 268, 269
Fats
 definitions, 10
 diets deficient in, 151, 153–155
 extraction of, 10–16, 319
 high-performance size-exclusion chromatography (HPSEC), 81–92
 sensory assessment, 265–276
 trans fatty acid content, 181–190
Fatty acid esters, 65–68, 147–180
Fatty acid isopropyl esters (FAIPE), 148–150, 156–157, 160, 161, 162, 170–172
Fatty acid methyl esters (FAME), 5, 232–240
 borage oil, 170–172
 chemical techniques, 278, 279, 280
 Fourier transform infrared (FTIR) spectroscopy, 184
 gas liquid chromatography (GLC), 7, 76, 134–138, 145, 148–150, 151, 153, 156–168, 185, 243–244

high-performance size-exclusion chromatography (HPSEC), 86
localization of double bonds, 193, 195
mercuric adduct, 281, 283, 285, 286
polyunsaturated fatty acids (PUFA), 310, 311
reduction and ozonolysis of, 280
silver-ion chromatography, 59, 60–61, 64–66, 68, 70–71, 75–77
solid-phase extraction (SPE), 69
supercritical fluid chromatography (SFC), 69
Fatty acid phenacyl esters, 65–66, 67–68, 70–71, 75–78, 148
Fatty acid picolinyl esters, 195–196, 280
Fatty acids. *See also Trans* fatty acids
 adipose tissue, 103–104
 biological studies, 167–169
 branched and cyclic, 258–260
 cholesteryl ester transfer protein (CETP) reactions, 353, 354
 classical chemical techniques, 277–289
 derivatives, 196–197, 198, 244–248, 280
 essential, 309
 extraction, 10
 fats and oils content, 102–104, 112, 118–120
 gas chromatography-Fourier transform infrared spectrometry (GC-FTIR), 232–241
 high-performance liquid chromatography (HPLC), 46, 47, 48, 56
 history of analysis, 1–9
 mass-analyzed ion kinetic energy (MIKE) spectra, 197–201, 202
 metabolism, 305
 nuclear magnetic resonance (^{13}C NMR) spectroscopy, 250–264
 oxidation of, 301–302
 polymerized, 86
 short chain, 242–249
 structural analysis of unusual, 285–287
 structural modifications, 191–197
 thin-layer chromatography-flame ionization detection (TLC–FID), 26–29, 35
 unsaturated, 87, 191–204, 197, 200, 278–280, 281–285, 285–286
Fermentation products, gas–liquid chromatography, 137–138, 140
Fish oils
 cardiovascular disease, 87, 309
 chromatography, 63–64, 87, 277
 fatty acid content, 1
 hydrogenated, 181, 188
 nuclear magnetic resonance (^{13}C NMR) spectroscopy, 252, 254, 255
 stereospecific analysis, 101
Flame ionization detection (FID)
 gas chromatography (GC), 76, 146, 223–225, 244–245

high-performance liquid chromatography (HPLC), 41, 43
thin-layer chromatography (TLC), 19, 22, 24–37
Fluorescence detection
  high-performance liquid chromatography (HPLC), 42, 43
  reversed phase high-performance liquid chromatography (HPLC), 55, 324, 328–331
  thin-layer chromatography (TLC), 19–20, 21, 22–23
Folch extraction method, 13, 14, 292
Food emulsifiers, 260–263
Fourier transform infrared (FTIR) spectroscopy
  gas chromatography (GC), 182, 186–187, 188, 232–241
  high-performance liquid chromatography (HPLC), 43
  trans fatty acid content, 184
Fractional chain length (FCL) of fatty acids, 153–155, 161, 162, 176

Gas chromatography combustion-isotope ratio mass spectrometry (GCC-IRMS), 310–315
Gas chromatography (GC). *See* Gas liquid chromatography (GLC)
Gas chromatography-Fourier transform infrared spectrometry (GC–FTIR), 232–241
Gas chromatography-infrared (GC-IR), 183–184, 186–187, 188
Gas chromatography-isotope ratio-mass spectrometry (GC-IRMS), 301, 302
Gas chromatography-mass spectrometry (GC-MS), 146, 151, 291, 303–307
Gas chromatography-mass spectrometry/mass spectrometry (GC–MS/MS), 200–201, 202
Gas chromatography/matrix isolation/Fourier transform infrared spectrometry (GC/MI/FTIR), 236–237
Gas liquid chromatography (GLC). *See also individual classes of compounds*
  capillary columns, 134–139, 146–180, 223, 331
  column temperature, 140–142
  concept proposed, 6–7
  derivatives for, 144, 146
  detectors, 7, 146
  electron capture detection (ECD), 146, 331–332
  flame ionization detector (FID), 223–225
  injection systems, 144, 145
  operating conditions, 133–146
  retention, 140–141
  sample pretreatment, 144, 146
  stationary phase, 141
Glycerol, 2–3

Glycerol esters, 260–262
Glycerol triesters. *See* Triacylglycerols
Glycol esters, 260–263
Glycolipids, 10, 18, 20, 55
Golay equation for gas liquid chromatography (GLC), 134, 141–142
Grignard reagents
  short-chain fatty acid analysis, 242–249
  triacylglycerol analysis, 96, 101–102, 111–115, 118, 127–128
  waxes, 242

Headspace techniques for organic solvent trace analysis, 331–336
Heart disease. *See* Cardiovascular disease
Height of an effective theoretical plate (HETP), 141–142
Hepatic lipase, 344–345, 351, 355
Hexane analysis, 137, 139, 333
High-density lipoproteins (HDL)
  analysis methods, 347
  cardiovascular disease, 342, 348
  cholesteryl ester transfer protein (CETP), 345, 351–354
  function and structure, 337, 340, 357, 358
  metabolism, 305, 310, 311–313, 315
  polyacrylamide gradient gel electrophoresis, 347–348, 349
  remodeling by plasma factors, 345
  size exclusion chromatography (SEC), 342–344
  ultracentrifugation, 366, 368, 369
High-performance liquid chromatography (HPLC), 8, 38–44. *See also individual classes of compounds;* Silver-ion high-performance liquid chromatography (HPLC)
  comparison to thin-layer chromatography (TLC), 17–18, 23
  detectors, 41–44
  lipid class separations, 45–58
High-performance liquid chromatography-mass spectrometry (HPLC–MS), 43, 131, 225–229
High-performance size-exclusion chromatography (HPSEC), 81–92
  See also Size exclusion chromatography (SEC)
High-performance thin-layer chromatography (HPTLC), 19
High-speed liquid chromatography (HSLC), 39
Hilditch, T.P., 2, 3–6
History
  of lipid analysis, 1–9
  of thin-layer chromatography (TLC), 17
Hydrazine reduction of polyunsaturated fatty acids, 281–285, 287
Hydrogenation of lipids, 4, 181, 191–194, 321

Iatroscan instruments, thin-layer chromatography-flame ionization (TLC–FID), 25
Infrared (IR) spectroscopy
  chromatography detector, 42, 43, 45–46, 83
  trans fatty acid content, 182–184, 186–188
Intensity tests, sensory measurement, 267–268
Intermediate density lipoproteins (IDL), 339, 341, 357
Iodine value, 3, 4–5
IOOC method for quality control of olive oils, standard methods, 269–270, 272–273
ISO 8586-1 and 2 methods for sensory assessment of fats and oils, 265
ISO 6586 difference tests in sensory measurement, 266
ISO/Dis 11035 method of descriptor identification, 268
Isotope Ratio-Mass Spectrometry (IR-MS) of ^{13}C-labeled compounds, 299, 300, 301–308, 308
Isotopes in metabolic studies, 299–308, 309–316
IUPAC Commission on Oils, Fats, and Derivatives, polymerized triglycerides (TG) analysis, 84–85
IUPAC method of total *trans* fatty acid content, 182–185
IUPAC Standardized Method 2.607 of hexane measurement, 333
IUPAC-IUB Commission on nomenclature of glycerolipids, 93

Kapok seed oil, 259–260

Lecithin-cholesterol acyltransferase (LCAT), 340, 351, 353, 361
Light scattering detectors (LSD). *See also* Evaporative light-scattering detectors (ELSD)
  silver-ion chromatography, 75–77
Light-pipe interface for gas chromatography-Fourier transform infrared spectrometry (GC-FTIR), 232–236
Linoleic acid
  gas–liquid chromatography (GLC), 136, 139
  identification of, 3
  silver-ion chromatography, 61–62, 75
  thin-layer chromatography-flame ionization detection (TLC–FID), 27
Linolenic acid
  gas–liquid chromatography (GLC), 136, 139
  identification of, 3
Linolenic acid geometrical isomers (LAGI)
  gas–liquid chromatography (GLC), 156–172
  isomerization mechanism, 165, 178
  silver-ion high-performance liquid chromatography (HPLC), 75–80
Linseed oils
  early studies, 5
  epoxidized, 257–258
  heat treatment, 156, 164, 166, 172, 192–193, 199–200, 234–236, 286
  silver-ion high-performance liquid chromatography (HPLC), 63, 65
Lipases
  cardiovascular disease, 340
  lipoprotein metabolism, 338, 344–345, 351, 355
  regiospecific analysis of triacylglycerols, 94–95
  stereospecific analysis of triacylglycerols, 95–96, 110–113
Lipoprotein lipase (LPL)
  cardiovascular disease, 340
  lipoprotein metabolism, 338, 344–345, 351, 355
Lipoproteins
  cardiovascular disease, 337, 340–342
  cholesteryl ester transfer protein (CETP), 351–355
  classification of, 337–340, 357, 358
  extraction of, 10
  metabolism, 181, 303–304, 310, 311–313, 315, 363
  polyacrylamide gradient gel electrophoresis, 347–356
  size exclusion chromatography (SEC), 337–346
  ultracentrifugation, 357–370
Low-caloric fats, 89–91
Low-density lipoproteins (LDL)
  cardiovascular disease, 341, 350–351
  cholesteryl ester transfer protein (CETP), 353, 355
  function and structure, 337, 339–340, 357, 358
  metabolism, 310, 313
  polyacrylamide gradient gel electrophoresis, 347–351
  size exclusion chromatography (SEC), 342–344
  ultracentrifugation, 365–366, 367, 369
L'vov platform, metal analysis, 324, 327
Lysophosphatidylcholine, high-performance liquid chromatography (HPLC), 49, 50, 53

Margarines
  chromatography, 61, 318
  fatty acid content, 185
  sterol analysis, 321
Mars detector. *See* Evaporative light-scattering detectors (ELSD)
Mass detector. *See* Evaporative light-scattering detectors (ELSD)
Mass spectrometry (MS). *See also* Gas chromatography-mass spectrometry (GC-MS); High-performance liquid chromatography-

mass spectrometry (HPLC–MS); Thin-layer chromatography-mass spectrometry (TLC–MS)
  fatty acids, 194–201
  of triacylglycerols, 225–227
Mass-analyzed ion kinetic energy (MIKE) spectra, 197–203
Mercuric adducts
  cyclic fatty acid monomers (CFAM), 285–286
  polyunsaturated fatty acids, 281, 283–285, 287
Metabolism
  fatty acids, 181, 309–316, 341
  lipids in humans, 299–308
  lipoproteins, 338–340, 344–345
Metal contaminants in oils and fats, 323–328
Methylmagnesium bromide. *See* Grignard reagents
Milk. *See also* Dairy products
  fatty acids, 103–104, 242–244, 247–248, 303, 305
  optical activity of triacylglycerols, 94
  phospholipids, 55
  silver-ion chromatography of triacylglycerols, 61, 63
Mobile phase. *See* Solvent systems
Monoacylglycerols, 46, 48, 260–262
Multiple ion detection (MID), 299, 301

Near infrared reflectance (NIR), fat content determination, 14–15
Neutral lipids
  defined, 10
  thin-layer chromatography (TLC), 18, 35
Newman-Keuls test, intensity tests in sensory measurement, 267–268
NF ISO 8587 ranking tests in sensory measurement, 267
Nonaqueous reversed phase (NARP) chromatography, triacylglycerols, 211–212
Nonaqueous reversed phase (NARP) chromatography of triacylglycerols, 223, 228–229
Nuclear magnetic resonance (^{13}C NMR) spectroscopy
  fatty acids and lipids, 250–264, 307
  polyunsaturated fatty acids (PUFA), 312
Nuclear magnetic resonance (^{1}H NMR) spectroscopy of triacylglycerols, 94
Nuclear magnetic resonance (^{31}P NMR) spectroscopy of phospholipids, 263

Obesity, lipoprotein levels, 341
Octadecylsilyl (ODS) chromatography support, 39, 59–60, 209. *See also* Reverse phase high-performance liquid chromatography (HPLC)
Oil flavors, 268–269

Oils and fats, sensory assessment, 265–276
Oleic acid
  identification of, 3
  separation from petroselinic acid, 148–150
  thin-layer chromatography-flame ionization (TLC–FID), 27–29
Olive oil
  fatty acid distribution, 103
  high-performance size-exclusion chromatography (HPSEC), 89
  organoleptic assessment of, 269–273
  triacylglycerols, 141, 143
Organic solvent trace analysis in oils and fats, 331–333
Ovarian follicle fat, 104
Oxidation
  of fats, 84–89
  of fatty acid methyl esters (FAME), 193, 195
  of fatty acids, 4
Oxysterols
  analysis, 290–298
  extraction, 292–296
  gas chromatography-mass spectrometry (GC-MS), 295–297
  medical significance of, 290–291
  thin-layer chromatography-flame ionization detection (TLC–FID), 33–35
Ozonolysis
  cyclic fatty acid monomers (CFAM), 286–287
  fatty acids, 148, 151
  unsaturated fatty acids, 278–280, 287

Palm oil
  fatty acid content, 246
  nuclear magnetic resonance (^{13}C NMR) spectroscopy of epoxidized, 257–258
  silver-ion chromatography, 62
Pancreatic lipase, 94–96
Panel, sensory measurement, 265
Panel rooms. *See* Test rooms
Partially hydrogenated oils
  nuclear magnetic resonance (^{13}C NMR) spectroscopy, 254
  ozonolysis, 278–279
  structural analysis by chemical techniques, 280
Partially hydrogenated vegetable oils (PHVO)
  fatty acid isomer content, 181, 183–184
  gas chromatography (GC), 185
  gas–liquid chromatography-infrared (GLC–IR), 186–188
  nuclear magnetic resonance (^{13}C NMR) spectroscopy, 260
Peanut oil
  gas chromatography, 220, 221
  high-performance liquid chromatography (HPLC), 220, 222

oxidation of, 87–88
sensory measurement, 269
structural analysis of, 106–132
Periodate oxidation. *See* Iodine value
Petroselinic acid, separation from oleic acid, 148–150
Phospholipases, 95–96, 110–113
Phospholipid transfer protein (PTP), 351
Phospholipids
    extraction of, 10
    fatty acid methyl esters (FAME) prepared from, 151, 153–155, 167–168
    high-performance liquid chromatography (HPLC), 33, 45, 47, 49–57
    hydrolysis by lipases, 344–345
    lipoproteins, 338, 367
    metabolism, 306, 307, 312–315
    nuclear magnetic resonance (^{31}P NMR) spectroscopy, 263
    phosphorus assay, 51–52
    silver-ion chromatography, 59
    thin-layer chromatography (TLC), 18, 20, 26–27, 29, 33, 35
Phytosterols, 317, 321
Pinolenic acid geometrical isomers (PAGI), 173–178
Plant lipids, 55
Polar lipids, 10
Polyacrylamide gradient gel electrophoresis of lipoproteins, 347–356
Polycyclic aromatic hydrocarbons (PAH), 323–324, 328–331
Polyunsaturated fatty acids (PUFA)
    analysis, 234, 281–285, 287
    esters of, 71–72
    gas–liquid chromatography, 150–155
    high-performance size-exclusion chromatography (HPSEC), 87
    metabolism, 181, 309–316
    silver-ion chromatography, 75–80
Principal Component Analysis (PCA), sensory measurement, 265, 268, 269, 270
Proton magnetic resonance spectroscopy of triacylglycerols, 94

Ranking test (NF ISO 8587), sensory measurement, 267
Rapeseed oil, 156, 158, 164
    fatty acid methyl esters (FAME), 134–135
    gas–liquid chromatography (GLC), 163
    high-performance liquid chromatography (HPLC), 219
    nuclear magnetic resonance (^{13}C NMR) spectroscopy, 260
    room odor, 273–275
    sensory profile, 269, 271
Refined oil flavors, 268–269

Refractive index detector
    high-performance liquid chromatography (HPLC), 41–43, 45–47
    high-performance size-exclusion chromatography (HPSEC), 83
Regiospecific analysis of triacylglycerols, 109–110
Reverse phase high-performance liquid chromatography (HPLC), 38–39
    fatty acid methyl esters (FAME), 310
    fatty acid phenacyl esters, 148
    polycyclic aromatic hydrocarbons (PAH), 324, 328–331
    silver-ion chromatography, 59–61, 70, 72
    sucrose polyester (SPE) fat substitute, 90–91
    triacylglycerol analysis, 114, 115, 118, 127–128
    triacylglycerols, 63, 108, 206–216, 229
Reverse phase liquid chromatography (RPLC). *See* Reverse phase high-performance liquid chromatography (HPLC)
Reverse phase thin-layer chromatography (RPTLC), 18
Robotics, sample-handling, 40
Room odor tests, sensory measurements, 273–274, 275

Saponification
    of lipids, 4–5
    of oxysterols, 294, 296
Seed oils. *See also individual oils*
    fatty acids, 1, 102
    gas–liquid chromatography (GLC), 148
    high-performance liquid chromatography (HPLC), 46
    nuclear magnetic resonance (^{13}C NMR) spectroscopy, 255–256, 257, 258
    silver-ion chromatography, 63–64
    stereospecific analysis, 102
Sensory assessment of fats and oils, 265–276
Silica gel chromatography. *See* High-performance liquid chromatography (HPLC); Thin-layer chromatography (TLC)
Silver-ion chromatography, 8, 207
Silver-ion high-performance liquid chromatography (HPLC), 59–74
    fatty acids, 75–80, 182
    triacylglycerols, 107–108
Silver-ion thin layer chromatography (TLC), 34–36, 59, 70
    fatty acid methyl esters (FAME), 151, 154, 158, 159
    fatty acid phenacyl esters, 148
    fatty acids, 168, 173–175, 177, 178, 182, 185, 187, 188, 281, 287
    triacylglycerols, 106–108, 110–111, 114, 127
Simple lipids, 10, 46–53

Sitosterol oxides, 190, 291, 295, 296
Size exclusion chromatography (SEC). *See also*
    High-performance size-exclusion chromatography (HPSEC)
  cholesteryl ester transfer proteins (CETP), 344
  edible fats, 81–92
  lipases, 345
  lipoproteins, 342–344
  polymerized fatty acids, 86
Solid-phase extraction (SPE), 69, 86, 98
Solvent systems
  composition restricted by detector, 45, 47, 50–51
  high-performance liquid chromatography (HPLC), 41, 47–55
  high-performance size-exclusion chromatography (HPSEC), 82–83
  reverse phase liquid chromatography (RPLC), 211–216
  silver-ion chromatography, 59
  thin-layer chromatography (TLC), 18, 21–22
  thin-layer chromatography-flame ionization detection (TLC–FID), 25, 26, 31–33
Soxhlet extraction of fats and lipids, 11, 12
Soybean oil, 156, 158, 164
  fatty acid methyl esters (FAME), 134, 137, 138
  gas chromatography, 223, 225
  human metabolism, 168–169
  nuclear magnetic resonance (^{13}C NMR) spectroscopy, 257–258
  silver-ion chromatography, 65–66
  trans fatty acid content, 183–184
  triacylglycerol analysis, 217, 218
Specific ion monitoring (SIM), mass spectrometry, 301
Sphingomyelin
  high-performance liquid chromatography (HPLC), 47, 49–51, 53, 54
  thin-layer chromatography-flame ionization detection (TLC–FID), 35
Standard methods
  determination of fat pollutants, 331–333
  fat extractions, 319
  fatty acid content, 182–185
  gas–liquid chromatography (GLC), 146
  hexane measurements, 333
  IOOC method for quality control of olive oils, 269–270, 272–273
  for lipid extraction, 10, 11, 14
  sensory assessment, 265–270, 272–273
  sensory assessment of fats and oils, 265
  sensory measurement, 274
  statistical tests for sensory measurement, 266–268
  of total *trans* fatty acid content, 182–185
  trans fatty acid determinations, 182–183, 185, 187–188, 232

Static headspace technique, for organic solvent trace analysis, 331–333
Stationary phases
  adsorption chromatography, 45
  gas chromatography, 7, 134, 136–137
  high-performance size-exclusion chromatography (HPSEC), 82
Statistical tests for sensory measurement, 266–268, 269, 270
Stereospecific analysis
  acyl migration, 128, 130–131
  high-performance liquid chromatography (HPLC) of chiral phases, 100–102
  triacylglycerols, 93–105, 110–131
Sterol esters
  high-performance liquid chromatography (HPLC), 56, 295
  silver-ion chromatography, 59
Sterol oxides. *See* Oxysterols
Sterols
  gas chromatography (GC), 136, 317–319
  high-performance liquid chromatography (HPLC), 46, 56, 295
  titration in complex media, 317–322
Sucrose polyester (SPE) fat substitute, 89–91
Sunflower oil, 200–201, 234
  oxidation of, 87–88
  sensory measurement, 269
  as standard in odor tests, 273–274
Supercritical carbon dioxide (SC-CO_2) for lipid extraction, 15
Supercritical fluid chromatography, 68–69, 238, 240–241

Tandem mass spectrometry, fatty acid analysis, 191–204
Technicon Infralyser instrument, fat content determination, 14–15
Test rooms for sensory measurement, 265–266
Tetrachloroethylene determinations in fats and oils, 331
Thin-layer chromatography (TLC). *See also individual classes of compounds;* Silver-ion thin layer chromatography (TLC)
  densitometers, 22–23
  detection methods, 19–21
  flame ionization detection (FID), 24–37
  history of, 8, 17
  retention value (R_f), 19
  techniques, 18–19
  typical applications, 20–22
Thin-layer chromatography-mass spectro-metry (TLC–MS), 23
α-tocopherol, metabolism, 306–307
Tracer instrument, direct deposition, 236
Tracor model 945 detector, high-performance liquid chromatography (HPLC) detectors, 43

Tracor transport-flame ionization detection, chromatography of lipids, 56
Trans fatty acids, 65–67
  determination of total content, 182–184
  gas chromatography, 185
  human health effects, 181
  nuclear magnetic resonance (^{13}C NMR) spectroscopy, 254, 260
  ozonolysis of, 279
  separation of methyl esters by silver-ion chromatography, 75–80
  structural analysis by chemical techniques, 287
Transport flame ionization detector, 41, 43, 53, 60
Triacylglycerols
  cholesteryl ester transfer protein (CETP), 344
  classes of, 205–206
  defined, 10
  detection of, 41, 43
  Grignard reagent, 242–248
  high-performance liquid chromatography (HPLC), 46–48, 53–54, 56, 205–231, 206, 217
  high-performance size-exclusion chromatography (HPSEC), 90
  high-speed liquid chromatography (HSLC), 39
  lipase action, 344–345, 355
  lipoproteins, 337–339, 367
  mass spectrometry (MS), 205–231, 225–229
  metabolism, 93, 303, 306, 310, 312, 313, 315
  nuclear magnetic resonance (^{13}C NMR) spectroscopy, 251–260, 263
  number of possible, 1, 106, 109, 206
  oxidized, 84, 87–89
  in peanut oil, 106–132
  regiospecific analysis, 109–110
  silver-ion chromatography, 59, 61–65
  solid-phase extraction (SPE), 69
  stereospecific analysis, 93–105, 110–131
  supercritical fluid chromatography (SFC), 69
  thin-layer chromatography-flame ionization detection (TLC–FID), 25–29, 35
Triglyceride polymers
  (TGP), high-performance size-exclusion chromatography (HPSEC), 84–88
Triglycerides. *See* Triacylglycerols
Trimethylsilyl ether derivatives, of fatty acid methyl ester (FAME) fragments, 193, 195
Trimethylsilyloxy derivatives, of fatty acid methyl esters (FAME), 150

Ultracentrifugation
  density-gradient, 359, 362–369
  high-speed bench-top models, 368, 369
  lecithin-cholesterolacyltransferase (LCAT), 361
  lipoproteins, 357–370
  near vertical rotors (NVT), 368
  principles of, 359–360
  rate-zonal, 366
  sequential, 359, 361–362
  vertical rotors, 368
Ultraviolet (UV) detectors
  chromatography of phospholipids, 51, 53
  high-performance liquid chromatography (HPLC), 41, 42, 45–46, 47, 50, 220
  high-performance size-exclusion chromatography (HPSEC), 83
  silver-ion chromatography, 61, 75–77
Ultraviolet (UV) spectroscopy, 5, 6
United States regulations on contaminants in fats and oils, 333
Universal detectors, high-performance liquid chromatography (HPLC), 41, 43, 46
Unsaturation
  average in fatty acids, 4–5
  measurement of average, 3
U.S. Food and Drug Administration, pending approval of sucrose polyesters (SPE), 89

Van Deemter equation. *See* Golay equation
Vegetable fats, sterol analysis, 321
Vegetable oils, 242–243
  nuclear magnetic resonance (^{13}C NMR) spectroscopy of epoxidized, 257–258, 259
  oxysterols, 291, 292
  silver-ion chromatography, 61–62, 75–80
Very low density lipoproteins (VLDL)
  analysis methods, 347
  apolipoproteins, 303–305
  cardiovascular disease, 340–341
  cholesteryl ester transfer protein (CETP), 351
  function and structure, 337, 339, 357, 358
  lipases, 345
  metabolism, 310–312
  size exclusion chromatography (SEC), 342–343
  ultracentrifugation, 361, 366, 368, 369
Vitamin E, 47, 306–307

Wall-coated open tubular (WCOT) capillary columns, for gas–liquid chromatography, 134–139
Waxes
  fatty acids, 258–259
  Grignard reagent, 242
  thin-layer chromatography (TLC), 18